AF574435

Belt Conveyors for Bulk Materials

Prepared by the Engineering Conference of the

Conveyor Equipment Manufacturers Association

Belt Conveyors for Bulk Materials

THIRD EDITION

Published by the Conveyor Equipment Manufacturers Association

Printed in the United States of America

Printing (last digit): 9 8 7 6 5 4 3

Library of Congress Cataloging in Publication Data

Conveyor Equipment Manufacturers Association.
Engineering Conference.
Belt conveyors for bulk materials.

Includes index.
1. Belt conveyors. 2. Bulk solids handling.
I. Title.
TJ1390.C65 1979 621.8'675 78-31987

ISBN 0-8436-1008-5

Contents

Preface

After more than two decades of widespread use throughout the world, *Belt Conveyors for Bulk Materials* has established its place as the premier manual in its field.

Based on the popularity of the First and Second Editions, the Board of Directors of the Conveyor Equipment Manufacturers Association (CEMA) authorized preparation of this Third Edition.

Leading belt conveyor executives and engineers of CEMA member companies assisted in the development of new and updated information for the new edition through the Association's Bulk Handling Systems and Equipment Section, Bulk Handling Components Section, and Engineering Conference.

Throughout the history of *Belt Conveyors for Bulk Materials,* the primary objective has remained unchanged: To provide basic engineering data for the design of belt conveyors in a book written for technically qualified people who may not be totally familiar with belt conveyor design.

The Third Edition retains all of the most desirable features of book design, information presentation, and illustration incorporated in the previous editions.

The table of contents enables the reader to review quickly the entire book and readily find the location of every subject covered.

Special efforts have been made to consolidate into one location all of the pertinent information on each subject. For instance, all information on belt tension, power, and drive engineering has been gathered together in Chapter 6.

Since reader experience with the first two editions favored numbering all illustrative material—photographs, diagrams, tables—this system has been retained. Each figure and table is identified first by the chapter in which it appears, then by its sequence within that chapter. In the case of figures, chapter and sequence numbers are set off by a decimal point; with tables, by a hyphen. Thus, Figure 3.2 is the second figure in Chapter 3, and Table 6-2 is the second table in Chapter 6.

The use of belt conveyors as an important and broadly accepted means of long-distance transport of bulk materials prompted the development of Chapter 1, where examples and financial analysis of various applications are presented.

Chapter 12 contains information on the use of programmable controllers, computers, and multiplexing in modern, automated belt conveyor systems. Chapter 13 presents the important subjects of operation, maintenance, and safety.

Acknowledgment

The Conveyor Equipment Manufacturers Association is indebted to the members of its Bulk Handling Components and Systems Section, Engineering Conference, Bulk Conveyor Engineering Committee, and the many other individuals who contributed their time and effort to the study and compilation of the data in this book; to the individual member companies who made available the time and talents of their engineers, draftsmen, and publication specialists; to the Rubber Manufacturers Association, Inc., the Mechanical Power Transmission Association, and other technical associations and professional societies for their assistance in making this book a reality.

Introduction

Over the years since the first edition of this book was published, belt conveying as a means of handling bulk materials has been greatly advanced and has assumed dominance in industry. The increased dependence of industry on belt conveyors is due primarily to the fact that they have had "everything going for them." Demands of the mining industries as well as the other high-volume basic materials industries have been calling for ever-higher capacities and for more extensive belt conveyor systems having far greater complexity and employing higher degrees of automated control.

Concurrent with the continuing improvements in application and performance standards, there have been more stringent industry requirements for economic justifications of capital investment and relative costs of operation and maintenance. Radical changes in the relative costs of equipment, labor, and energy have further enhanced the favoring of belt conveyors for many relatively long-distance, high-volume haulage applications. Moreover, emphasis on our national problems of energy-saving, environmental control, and safety has drawn a sharp focus on the many other advantageous features of belt conveyors.

Technological advances have paralleled and supported these increased application, performance, and economic demands. There has been a quiet revolution in the development of the available widths, normal operating speeds, and tension capabilities of conveyor belts. The mechanical components, too, such as idlers, drive machinery, pulleys, and many accessories have undergone developments to enhance performance and durability. Most recently, developments have been in the applications of automated, and even computerized, electrical controls employing solid state components, multiplex wiring, and many other modern techniques of electrical engineering and equipment.

In the first edition of this book, the Conveyor Equipment Manufacturers Association (CEMA) stated its aims as being to make available in a book the experience and technical knowledge of its members as a contribution toward the design and construction of conveyors of superior performance and to provide the basic data and fundamentals of design for application to the ordinary belt conveyor problems toward the achievement of successful performance. Adhering to these same aims and policy, CEMA is pleased to present this revised Third Edition of *Belt Conveyors for Bulk Materials*. It offers information on numerous examples of modern high-performance and complex systems and includes suggestions and guidelines for economic evaluations and conclusions. In this Third Edition, the presentation of belt conveyor design technology and engineering data has been updated both in its technology and in its presentation.

Chapter 1

Belt Conveyor General Applications and Economics

Contents

The subject of belt conveyors is of primary interest to all engineers, managers, and others who are responsible for selecting equipment for handling bulk materials. This book is primarily a design manual, but Chapter 1 is included to acquaint the reader with the many uses of belt conveyors and their advantages under widely varying conditions of operation.

Belt conveyors have attained a dominant position in transporting bulk materials due to such inherent advantages as their economy and safety of operation, reliability, versatility, and practically unlimited range of capacities. In addition, they are suitable for performing numerous processing functions in connection with their normal purpose of providing a continuous flow of material between operations. Recently, their conformity to environmental requirements has provided a further incentive for selection of belt conveyors over other means of transportation.

Low labor and low energy requirements are fundamental with belt conveyors as compared with other means of transportation. The dramatic increase in these operating costs has placed conveyors in an extremely favorable position for applications that were not considered a few years ago.

Belt conveyor manufacturers have consistently anticipated the needs of industry with improvements in designs and with components that have exceeded all known requirements. Reliability and safety are outstanding now that stronger and more durable belts are available, as well as greatly improved mechanical parts and highly sophisticated electrical controls and safety devices.

Illustrated and described in this chapter are some of the advantages of belt conveyors, which are performing a wide variety of intraplant functions better and/or in a more innovative manner than is possible with other means of transporting bulk materials. Also included are examples of relatively long-distance belt conveyor systems which are being used extensively because they combine such important benefits as reliability, safety, and low cost per ton of material transported.

Conveying of a Variety of Materials

The size of materials that can be conveyed is limited only by the width of the belt. Materials can range from very fine, dusty chemicals to large, lumpy ore, stone, coal or pulpwood logs. See Figure 1.1. Closely sized or friable materials are carried with minimum degradation. Because rubber belts are highly resistant to corrosion and abrasion, maintenance costs are comparatively low when handling highly corrosive materials or those that are extremely abrasive, such as alumina and sinter.

Materials that might cause sticking or packing if transported by other means are often handled successfully on belt conveyors. Even such hot materials as foundry shakeout sand, coke, sinter, and iron ore pellets are conveyed successfully.

FIGURE 1.1. *54-inch conveyor carries large lumps of abrasive ore on incline.*

Wide Range of Capacities

Currently available belt conveyors are capable of handling hourly capacities in excess of any practical requirement. See Figure 1.2. Yet they are also used

FIGURE 1.2. *96-inch conveyor at high-capacity coal-loading facility.*

economically in plants for transporting materials between process units at a wide range of rates—sometimes as little as a mere dribble.

Belt conveyors operate continuously—around the clock and around the calendar when required—without loss of time for loading and unloading or empty return trips. Scheduling and dispatching are unnecessary as the material is loaded to and unloaded from the belt conveyor automatically. Operating labor costs differ little, regardless of capacity ratings. Overall costs per ton decrease dramatically, however, as annual tonnage handled increases. Such economic considerations are illustrated later in this chapter.

For these reasons, belt conveyors are capable of handling tonnages of bulk materials that would be more costly and often impractical to transport by other means.

Adaptability to Path of Travel

Belt conveyor systems provide the means of transporting materials via the shortest distance between the required loading and unloading points. They can follow existing terrain on grades of 30 to 35%, compared with the 6 to 8% effective limits for truck haulage. See Figure 1.3. They can be provided with structures which prevent the escape of dust to the surrounding atmosphere and are weather-protected. Such structures are economical and are adaptable to special requirements. See Figures 1.4 and 1.5. Belt conveyors provide a continuous flow of material while avoiding the confusion, delays, and safety hazards of rail and motor traffic in plants and other congested areas.

FIGURE 1.3. *Regenerative conveyor lowers coal across existing terrain in direct path from mine to preparation plant.*

FIGURE 1.4. *Corrugated metal cover over the belt provides weather and environmental protection.*

FIGURE 1.5. *Cable-suspended bridge provides support for conveyor across river.*

Paths of travel can be quite flexible, and the length of the routes can be extended repeatedly, as required. In some open-pit mining operations, conveyors thousands of feet long are shifted laterally on the bench to follow the progress of excavation at the face.

Loading, Discharging, and Stockpiling Capabilities

Belt conveyors are very flexible in their capabilities for receiving material from one or more locations and for delivering it to points or areas, as required by plant flow sheets. They can provide the main transportation artery while being loaded at several points (Figure 1.6) or anywhere along their length by equipment which provides a uniform feed to the belt (Figure 1.7).

FIGURE 1.6. *Multiple loading stations feed ore to slope conveyor in open-pit mine.*

FIGURE 1.7. *Rail-mounted hopper with feeder provides loading along full length of conveyor.*

They are particularly useful in tunnels beneath stockpiles, from which they can reclaim and, where required, blend materials from various piles (Figure 1.8). Material can simply be discharged over the head end of each conveyor (Figure 1.9) or anywhere along its length by means of plows or traveling trippers (Figure 1.10).

FIGURE 1.8. *Multiple feeders in tunnel beneath stockpile provide efficient reclaiming and blending.*

FIGURE 1.9. *Material discharging over conveyor head pulley.*

FIGURE 1.10. *Power-driven tripper with dust seal discharge distributing coal to bunkers.*

Belt conveyors, with their stackers and reclaimers, have become the only practical means for large-scale stockpiling and reclaiming of such bulk materials as coal, ore, and taconite pellets. See Figures 1.11 and 1.12. The combination stacker-reclaimer in Figure 1.12 illustrates the trend in modern

FIGURE 1.11. *Double-winged stacker discharging into high-capacity stockpiles on either side of the feeding conveyor.*

FIGURE 1.12. *Combination stacker-reclaimer provides continuous, high-capacity feed to shiploading conveyor system.*

rail to ship terminals. Even the shiploaders are equipped with conveyors for filling and trimming the holds of vessels at controlled rates. See Figure 1.13.

Self-unloading ships and lake vessels (Figure 1.14) equipped with belt conveyors can be unloaded in all ports, even those which do not have dockside unloading equipment (Figure 1.15). Unloading capacities of such systems are usually greater than those of several grab bucket unloaders, requiring less turnaround time and lower labor and other operating costs.

In contrast with the above-mentioned high-capacity unloading systems, certain materials, such as foundry sand, can be plowed from the belts (Figure 1.16) at specific locations in quantities controlled by the requirements of the application.

FIGURE 1.13. *Shuttle belt conveyors load and trim taconite pellets onto ore vessel on Great Lakes.*

FIGURE 1.14. *Self-unloading ship with 78-inch discharge conveyor unloads iron ore pellets at 10,000 tph.*

FIGURE 1.15. *Rail-mounted ship unloaders feed 60-inch conveyor system at steel company.*

FIGURE 1.16. *V-type plow diverts foundry sand from flat belt conveyor.*

Process Functions

Although belt conveyors are generally used to transport and distribute materials, they are also used with auxiliary equipment for performing numerous functions during various stages of processing. A high degree of blending is accomplished as materials are bedded into and reclaimed from stockpiles. See Figure 1.17. Several dissimilar materials can be proportioned continuously onto a common collecting belt.

Figure 1.17. *Bridge-mounted bucket wheel reclaimer provides a feed of blended iron ores at steel plant.*

Accurate samples of the material conveyed can be obtained by devices which cut through the stream of material as it flows from one conveyor to the next. Magnetic objects can be removed from the material. While being transported on the conveyor, materials can also be weighed accurately and continuously or they can be sorted, picked, or sprayed. In many cases, such operations are not only performed more effectively in connection with belt conveyors but are the only practical means.

Reliability and Availability

The reliability of belt conveyors has been proved over decades and in practically every industry. They are operating with the utmost reliability, many serving vital process units whose very success depends on continuous operation, such as handling coal in power plants, and transporting raw bulk materials in steel plants, in cement plants, and to and from ships in ports, where downtime is very costly.

Belt conveyors are operated at the touch of a button (Figure 1.18), at any time of the day or week. When required, they can and often do operate continuously, shift after shift. They can be housed so that both they and the material being transported are protected from elements that would impede the movement of trucks and certain other means of transportation.

FIGURE 1.18. *Operator controls entire conveyor system from control center with graphic display panels and push-button console.*

Environmental Advantages

Belt conveyors are environmentally more acceptable than other means of transporting bulk materials; they neither pollute the air nor deafen the ears. They operate quietly, often in their own enclosures which, when desirable, can be located above the confusion and safety hazards of surface traffic or in small tunnels—out of sight and hearing. See Figure 1.19. Furthermore, they do not contaminate the air with dust or hydrocarbons. At transfers, dust can be contained within transfer chutes or collected with suitable equipment, if necessary. Finally, overland belt conveyor systems can be designed to blend into the landscape, resulting in an unscarred, quiet, and pollution-free operation. See Figure 1.20.

FIGURE 1.19. *Conveyor in completely enclosed gallery carries its load safely overhead, avoiding any interference from highway or rail traffic.*

FIGURE 1.20. *Overland conveyor system utilizing a concrete support structure provides a pleasing appearance blending with the landscape.*

Safety

Belt conveyors operate with an extremely high degree of safety. Few personnel are required for operation and they are exposed to fewer hazards than with other means of transportation. Material is contained on the belt and personnel are not endangered by falling lumps or the malfunction of huge, unwieldy transport vehicles. Such vehicles also involve public liability, whether they operate over highways or in other areas accessible to the public. Also, conveyors offer less hazards to careless personnel than is inherent in other means of transporting bulk materials.The conveyor equipment itself can be protected from overload and malfunction by built-in mechanical and electrical safety devices.

Low Labor Costs

The labor hours per ton required to operate belt conveyor systems are usually the fewest of any method of transporting bulk materials. Like other low labor-intensive, highly automated operations, belt conveyors have low operating costs and provide a higher return on investment than competitive methods. Most functions of the system can be monitored from a central control panel or controlled by computer, allowing a minimum number of operating personnel to inspect the equipment and report conditions that may require attention by the maintenance department.

The time required for maintenance personnel is also minimal. As noted below in regard to maintenance costs, repairs and replacements of the relatively small parts can be made quickly at the site. Most belts can even be replaced in one shift—and some belts have conveyed well over 100,000,000 tons before wearing out. Later in this chapter several examples of long-distance belt conveyor systems illustrate the effect of low labor costs.

Low Power Costs

The increasing cost of energy emphasizes the importance of power and its relation to the cost per ton for transporting bulk materials. Because belt conveyors are operated by electric power, they are less affected by the prices, shortages, and other limitations of liquid fuel. They consume power only when they are being used. There is no need for empty return trips or idling in line for the next load. On long systems the declined portion often assists in propelling an inclined or horizontal portion. Some conveyor systems are completely regenerative. See Figure 1.3. The cost of power for belt conveyor systems has always contributed to their extremely low operating costs, and this advantage has increased substantially with the rise in the cost of liquid fuels.

Low Maintenance Costs

Maintenance costs for belt conveyors are extremely low compared with most other means of transporting bulk materials. Extensive support systems, such as those commonly associated with truck haulage, are not required. Component parts are usually housed and have very long life compared with that of motor vehicles. Usually, they need only scheduled inspection and lubrication. Any repairs or replacements can be anticipated and unscheduled downtime avoided. Parts are small and accessible so replacements can be made on the site quickly and with minimal service equipment. Also, adequate inventories of spare parts can be maintained at a low cost and require relatively little storage space.

Long-Distance Transportation

The economic benefits of low operating costs for labor and energy, as well as some of the other advantages outlined above, have led to a widespread adoption of belt conveyor systems as a means of transporting bulk materials over increasingly long distances. Not only were these systems the best investments at the time they were installed, but the recent dramatic increases in the costs of both labor and liquid fuel have greatly enhanced their present value. A few of these systems are described below.

Seattle, Washington

The system shown in Figure 1.21 established a landmark in the use of belt conveyors for long-distance transportation of bulk materials. In the late

FIGURE 1.21. *Denny Hill conveyor system safely moves excavated material through commercial section of city to scows in harbor.*

1920s a contractor pioneered the use of belt conveyors for relatively long-distance haulage by transporting 5,000,000 yards of excavated material from Denny Hill in downtown Seattle to scows waiting in the harbor. It was so highly profitable and dependable that such systems were later adopted by other contractors. Now they are commonplace in the construction industry in cases where large tonnages must be transported economically.

Lost Creek Dam[1]

A recent example of belt conveyor haulage on a construction project is the 1974 installation at Lost Creek Dam on the Rogue River near Bedford, Oregon. (See Figure 1.22.) Army Corps of Engineers plans had indicated a 10,000-foot, 8% haul road down a steep mountainside for handling 7,000,000 tons of shot rock from the quarry to the dam site. However, the contractor used one 54-inch wide by 3,000 feet centers belt conveyor down a 17° decline to underbid his competitors by several million dollars. The following benefits were revealed by his economic study:

1. An initial saving of $1,800,000 investment in trucks was realized, and an estimated 1,375,000 gallons of diesel fuel over the life of the project was not required.
2. The cost of constructing the haul road was avoided, as well as the much greater cost of restoring the terrain as required by the Corps of Engineers specifications.
3. The conveyor was regenerative and its motor generators supplied enough electrical energy for the project and returned the excess to the local public utility.
4. A substantial saving in manpower enabled the contractor to bid a price $3,200,000 below his next competitor.

FIGURE 1.22. *54-inch conveyor, 3,000 feet long, lowers 2,000 tph of rock down a 17° decline while supplying regenerative electricity.*

5. The project was completed ahead of time because the conveyor handled 2000 tph, whereas the schedule was based on 1200 tph.
6. The safety hazards and maintenance cost for trucks operating down an 8% winding road were eliminated.

One man monitored all critical points of the system on closed-circuit TV from his station on the loadout hopper. The system was protected by numerous safety devices, including magnetic brakes which stopped the loaded conveyor within 23 feet from 450 fpm.

Oklahoma Cement Plant[2]

Another example of the economy and adaptability to terrain of the belt conveyor is its use in this 5½-mile system from quarry to mill at an Oklahoma cement plant (Figure 1.23). The estimated total direct and indirect costs per ton for railroad, truck, and conveyor haulage are as shown below:

Estimated Cost Per Ton

	Belt conveyors	*Truck*	*Railroad*
Half production	34.5¢		44.4¢
Full production	21.6	32.1	35.1

Trucks were ruled out not only because of their higher costs, but because of hilly terrain and the necessity for crossing two railroads and two highways. The disadvantage of rail transportation was the need for extensive trackage, switchers, loading and unloading, and storage. Also, scheduling and dispatching would pose problems for full production.

FIGURE 1.23. *5½ mile overland conveyor system follows natural terrain as it moves limestone from quarry to cement mill.*

After the first year of operation it was found that the cost per ton for conveyors for full production was only 13.3 cents, compared with an estimated 21.6 cents, or 2.4 cents per ton-mile! For half production it was 23.6 cents per ton. This was due to an extended depreciation period and the fact that the entire system was operated and maintained by only two men and a supervisor.

Texas Iron Ore Mine to Steel Mill[3]

A Texas steel plant was faced with mounting costs for transporting sticky, abrasive iron ore from the nearby mine to an ore-washing plant adjacent to the mill. When the operation began in 1946 the mine was about a mile from the plant, but by 1964 the haul had increased to 5.5 miles. At that distance the haul alone was costing 6 cents per ton-mile which represented 55% of the total cost per ton for mining and delivering the ore to the washer. It was then that an exhaustive feasibility study was made of all operations, from excavating and loading in the pit through transportation to the washer.

The elimination of trucks as a shuttle between loading machines and main-line haulage was not considered, due to the rapid progress of the mining machines in the shallow ore bodies and the need to mix ores from several locations. The problem was to determine the best method of transporting the ore from a truck dumping station at the mine to the washer. A railroad was ruled out because of the large capital investment in equipment and terminal facilities, as well as anticipated difficulties in handling the sticky, lumpy ore.

A pipeline for the main transport appeared promising according to preliminary estimates, but further consideration was abandoned because of:

1. The difficulties of crushing the sticky ore to minus 2 or 3 inches at the mine.
2. Unavailability of water and reservoir sites.
3. The problem of designing a pump suitable for such coarse, abrasive ore, as well as wear and maintenance costs of pipe and pumps.
4. Undesirability of creating excess fines and the need for their subsequent agglomeration.
5. Dewatering and storage of ore at the mill.
6. Clearing the pipeline in case of mill stoppages.

An all-truck haulage operation, with an investment in new, larger trucks, was estimated to be much more costly than belt conveyor haulage. Also, the difference would increase as the shallow ore bodies were mined out and the length of haul increased (Figure 1.24). Increases in the costs of labor and liquid fuels could not be anticipated at the time this study was made, or the estimated economic advantage of belt conveyors would have been much greater.

Jamaican Bauxite Mine to Seaport[4]

The adaptability and economy of belt conveyors for transporting bulk materials over rugged terrain despite adverse weather conditions is illustrated

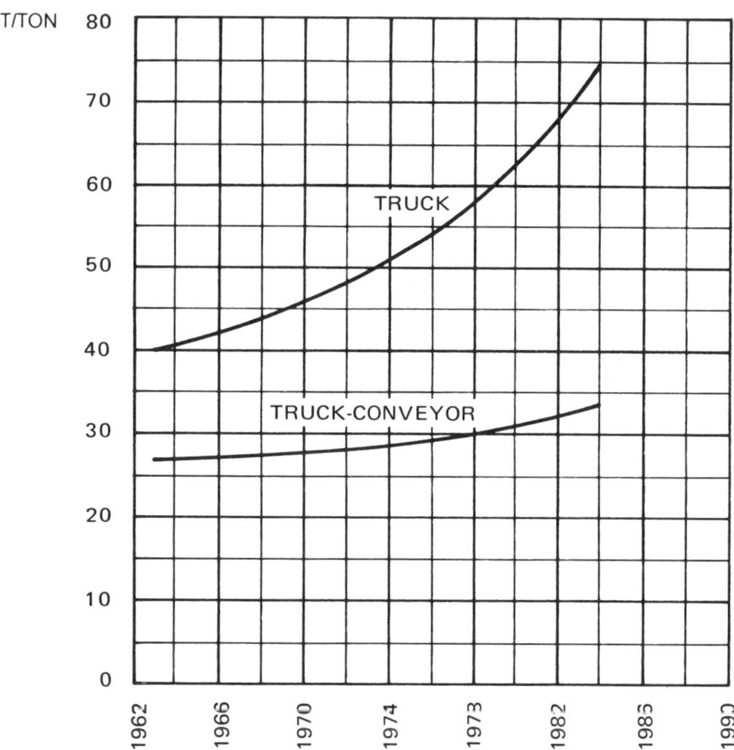

FIGURE 1.24. *Economic advantage of the truck-conveyor system over the all-truck system as the length of haul increases in the future.*

by this system designed to deliver 1300 tph of bauxite from ore dryers at the mine to a port facility in Jamaica. See Figure 1.25.

The investment decision for this system was based on an intensive economic study when it was decided to increase production at the mine. The

FIGURE 1.25. *6½ mile conveyor system transports 1,300 tph of bauxite ore over rugged terrain from mine to port facility in Jamaica.*

existing tramline was inadequate for the increase so the economics of truck, railroad, tramline, pipeline, and conveyor haulage were analyzed and such other factors as availability, reliability, and the effects of weather were considered. The results of the study are summarized below.

*Economic Comparison**

	Conveyor	*Railroad*	*Tramline*	*Truck*
Distance from mine to port (miles)	6.5	14.0	6.5	10.8
Relative transportation cost per ton-mile	1.00	0.58	2.29	1.30
Relative transportation cost per ton	1.00	1.26	2.29	2.16
Relative capital cost	1.00	1.30	0.81	0.97

The above truck, conveyor, and railroad systems are readily expandable to five million TPY. The cost of the tramline would increase substantially at this rate.
* Based on an annual production rate of three million net tons.

The pipeline method was rejected because of the serious problem of dewatering the slurried bauxite. Truck costs per ton were not only higher than conveyor costs but operating problems were anticipated due to tropical weather (e.g., rainfall of six inches per month and frequent dense fogs). A railroad would also entail high operating costs because of the circuitous route down long grades and over deep gullies subject to flooding. The profile of the conveyor route (Figure 1.26) suggests the considerably longer distances that would have been required for both trucks and railroad because of their inability to negotiate steep grades. The resulting higher costs are shown above. Actual power consumption by the belt conveyors is low since the fully loaded system is virtually in balance. The power generated by the overall descent of the material is about equivalent to the power demands to propel the belts. Spare parts are minimal because all belts are identical and there are only two sizes of speed reducers and three sizes or types of motors. The belts are covered to protect the dried bauxite from the weather and the return belts are of the "turnover" type to prevent buildup of ore on the return idlers.

Ore and Waste Haulage from an Arizona Copper Mine[5]

Characteristic of the open-pit copper mines in Arizona and other parts of the Southwest is their great depth. Often it is necessary to remove several hundred feet of waste rock and alluvial gravel to reach the uppermost ore in the pit. Early operations employed railroads to haul the ore and waste over a circuitous route to the surface. Then trucks became more economical for the haul from loading shovels, up 8% grades to the crest of the pit, and on to the primary crushers.

In 1968 a new mine was opened in Arizona where the ore was overlaid with 700 feet of waste rock and alluvial gravel. Based on exhaustive economic feasibility studies, belt conveyors were installed to transport minus-10-inch waste and ore from primary crushers in the pit to the surface and then to waste dump and mill. See Figures 1.27 and 1.6. The system successfully justifies preinvestment calculations. A fringe benefit of the belt

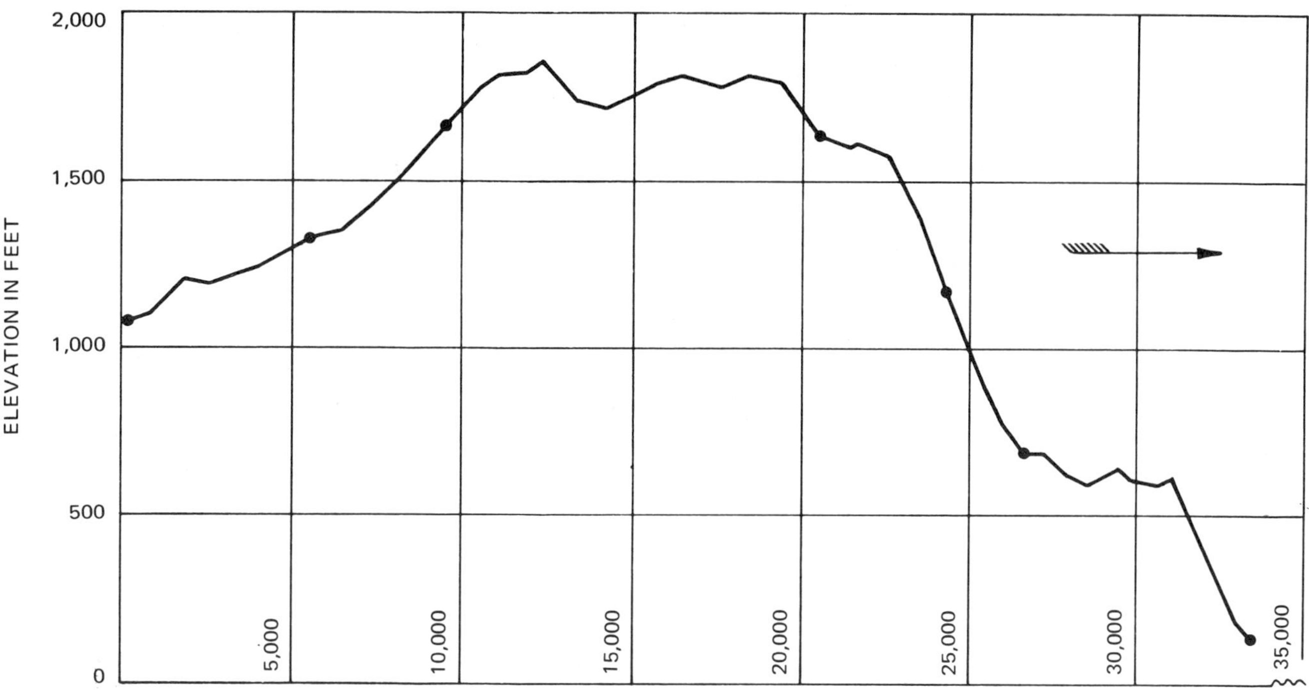

FIGURE 1.26. *Conveyor route profile.*

conveyor system is its environmental desirability—there is practically no dust discharged to the air and noise pollution is practically nonexistent. This is particularly important because of a nearby retirement community.

FIGURE 1.27. *High-capacity conveyor system lifts 8,000 tph of waste or copper ore from primary crushers in open pit and transports material to waste dump or mill.*

In light of the experience at this mine, a study was conducted in 1973 to compare conveyor and truck haulage for a similar but hypothetical operation. The study was based on recovering in 20 years 200 million tons of ore overlaid with 400 feet of waste rock and gravel, where the waste rock/ore ratio was 2:1. Individual studies were made for ore and waste, as well as for conveyor and truck systems. The combined results are summarized in the table below. The figures for each system include a 400-foot haul from pit crest to waste dump and a 1000-foot haul for ore to mill. The adverse grade to pit crest for trucks was figured at 8% and for conveyors 25%.

Comparison of truck versus conveyor haulage pit with annual tonnage of 30 million (20 million waste & 10 million ore tons)

	Trucking Costs		*Conveying Costs*		*Advantage (Disadvantage) Conveyor to Trucking*	
Lift In Pit	*Annual Cost*	*Cost Per Ton Handled*	*Annual Cost*	*Cost Per Ton*	*Annual*	*Per Ton*
100′	$1,889,000	$.0630	$2,252,000	$.0751	($ 363,000)	($.0121)
200′	$2,477,000	$.0826	$2,489,000	$.0830	($ 12,000)	($.0004)
300′	$3,128,000	$.1043	$2,726,000	$.0909	$ 402,000	$.0134
400′	$3,846,000	$.1282	$2,963,000	$.0988	$ 883,000	$.0294
500′	$4,619,000	$.1540	$3,203,000	$.1068	$1,416,000	$.0472
600′	$5,458,000	$.1819	$3,440,000	$.1147	$2,018,000	$.0672

Obviously, for these tonnages the conveyor was the best investment for a lift of more than 200 feet and would become more profitable as the depth of the pit increased.

A similar comparison today would be more favorable to conveyors because of increases in the costs of labor and motor fuel, as well as the effect of inflation on the replacement cost of trucks. Replacement costs for the conveyors would be limited to belting and a few relatively small mechanical and electrical components.

Moving a Mountain to a Cement Plant[6]

In contrast with the above examples of conveyor systems operating over rugged terrain and from deep pit mines, this 4-mile system operates over relatively level Arizona desert. See Figure 1.28. The system hauls the stone, shale, and silica at a substantially lower cost per ton than trucks, even though there are excellent roads between mine and plant.

In addition to its economic advantage over trucks, the conveyor system has interesting design features:

1. The system consists of only two conveyors; one conveyor spans 18,500 feet with a single belt and the other is 2,600 feet long.
2. A special precision accelerating drive accurately controls belt tensions and optimizes load sharing between the two drive pulleys of the longer conveyor. This feature allows the use of a lower-cost belt and safeguards against belt abuse.

FIGURE 1.28. *Belt conveyor in concrete support housing hauls stone and shale at cement plant.*

3. Among other safety features are the prestressed concrete housing that supports the conveyor and devices that provide for automatic shutdown in case of excessive belt tensions or unusual side-tracking of the belt.

Investment Decision—Belt Conveyors versus Trucks

Numerous factors contributed to the selection of each long-distance belt conveyor system described above. However, final investment decisions were based primarily on economics—the lowest cost per ton for the tonnage to be handled during the life of the operation. Actual cost comparisons for several of the systems are given briefly in the text, and further details are contained in the noted published articles. In those cases where the initial cost of a belt conveyor system is higher than that for trucks, the difference is soon overcome by its much lower operating costs, and the financial returns will continue as long as the system is in operation.

These comparisons of cost estimates and operating results, which were made several years ago, would be even more favorable for belt conveyors under present conditions. Since the low operating costs of belt conveyors result largely from their minimal labor requirements and their use of electric energy, the substantial increases in labor rates and fuel prices since these comparisons were made have placed the other forms of transportation at an even greater disadvantage.

Outlined below are some of the factors which must be considered by those who wish to decide whether to invest in a belt conveyor system or a truck operation. The applicable items can be used to suit the reader's specific requirements and circumstances, and costs estimated to obtain the financial information desired. Such information could vary from a brief estimate of

TABLE 1-1. *Owning and Operating Costs—Conveyor Haul*

Annual expenditures in current dollars

	Year									
	1	2	3	4	5	6	7	8	9	N
Capital investment										
Loading arrangement										
Conveyor equipment and structures										
Belting										
Erection of equipment and structures										
Foundations										
Electrical equipment and installation										
Site preparation										
Access road construction										
Lighting										
Repair shop and equipment										
Repair parts storage										
Unloading arrangement										
Maintenance equipment for access road, conveyor movement, cleanup, etc.										
Owning costs per year										
Depreciation										
Interest, taxes, and insurance (% of book value)										
Operating costs per year										
Power										
Maintenance and repair of conveyor installation (labor and material)										
Maintenance of haul road										
Maintenance and repair of auxiliary equipment										
Labor										
Total cost										

owning and operating costs and the anticipated salvage value for a temporary system to highly sophisticated accounting procedures applied to an extensive permanent installation.

An excellent example of the use of sophisticated accounting methods for an investment decision was presented by F. W. Schweitzer and L. G. Dykers.[7] It is based on appropriate cost estimates for a specific set of operating conditions and develops comparisons on present worth, discounted cash flow on investment, and wealth growth rate over a period of 10 years.

For the reader who wishes to compare a particular haulage problem with this example, we include tabular material and a brief discussion of a popular accounting method for this purpose. Dollar amounts are included in the tables simply to clarify the procedure and to illustrate the effects of inflation and depreciation. They *do not* apply to specific conditions, nor are they intended as an accurate comparison of the two haulage methods.

In order to make a true cost comparison between two or more haulage systems, all costs chargeable to each system must be considered, including ancillary facilities required for each system. For example, crushing facilities

TABLE 1-2. *Owning and Operating Costs—Truck Haul*

Annual expenditures in current dollars

	Year									
	1	2	3	4	5	6	7	8	9	N
Capital investment										
Trucks less tires										
Haul road construction										
Haul road maintenance equipment										
Lighting										
Repair shop and equipment										
Repair parts storage building										
Loading arrangement										
Unloading arrangement										
Engineering										
Owning costs per year										
Depreciation										
Interest, taxes, and insurance (% of book value)										
Operating costs per year										
Haul road maintenance										
Maintenance and repair of trucks										
Tires										
Fuel/oil/grease										
Labor										
operators										
mechanics										
(usually included in truck repair)										
Power for lighting										
Total cost										

might be required to reduce lump size for handling on the belt conveyor but would not be necessary for truck haulage. Conversely, a very large shop with extensive special equipment would be required for a fleet of trucks whereas only relatively inexpensive equipment is necessary for servicing the small components of a belt conveyor. Also, the effect of inflation and added investment for additional or replacement equipment and facilities must be considered.

Owning and Operating Costs

Tables 1-1 and 1-2 present suggested methods for determining the capital investment and annual owning and operating costs for truck and conveyor hauls. These tables show the major factors contributing to the costs of each of the two systems. However, for an actual comparison, factors will be added or deducted to suit the particular project. The purpose of this discussion is to point out items other than actual truck or conveyor costs that must be considered in making a valid comparison.

In order to account for inflation and additional investments, the format allows for annual entries over the life of the project and therefore, gives annual costs. These annual figures are important in analyzing present worth, a method described later in this chapter.

As mentioned above, one of the considerations in the investment decision is the owning and operating cost per ton. The annual owning and operating cost is determined for each year over the life of the project. The cost per ton is then determined by dividing the annual owning and operating cost by the tonnage expected to be handled during that year. It is important to note that the annual operating hours have a significant effect on the results of the cost analysis, since the frequency of replacement and annual depreciation are directly affected.

Even though the owning and operating costs analysis indicates an advantage of one system over another, the more economical plan may require an initial investment which is higher than the alternate plan. The worth of this investment may be the determining factor in the investment decision. The following example illustrates one common method of measuring the worth.

Present Worth: Required Rate of Return

Tables 1-3, 1-4, and 1-5 demonstrate the procedure for evaluating the present worth of an investment. The cost figures used in Tables 1-3 and 1-4 are arbitrary and do not represent comparative figures for a specific truck and conveyor haul. To simplify the example, investment tax credit and other tax considerations (mainly depreciation), which would be advisable in an actual analysis, have been ignored. The example assumes 5% annual inflation.

Table 1-3. *Cash Flow—Conveyor System*

In 000's of dollars

	Year									
	1	2	3	4	5	6	7	8	9	10
Capital investment										
Conveyor system	1,900									
Loading system	260					330*				
Belting	700					890*				
(Repair shop and parts storage included in operating costs)										
Access road	50									
Operating costs										
Taxes and insurance	116	101	85	70	54	88	70	53	35	18
Power / Maintenance and repair / Labor	287	301	316	332	349	366	385	404	424	445
Access road maintenance	3	3	3	3	4	4	4	4	4	5
Total cash flow	3,316	405	404	405	407	1,678	459	461	463	468

* For the purpose of this example, these items are replaced at the end of year 5. Anticipated hours of operation and life expectancy of equipment determine the proper entry.

TABLE 1-4. *Cash Flow—Truck System*
In 000's of dollars

	Year 1	2	3	4	5	6	7	8	9	10
Capital investment										
Trucks	1,400					1,787*				
Repair shop and parts storage	50									
Haul road	100									
Loading system	460					588*				
Operating costs										
Taxes and insurance	80	65	49	34	18	98	78	59	39	20
Maintenance of trucks; Maintenance of loaders; Labor; Fuel; Tires	704	739	776	815	856	899	944	991	1,040	1,092
Haul road maintenance	10	10	11	12	12	13	13	14	15	15
Total cash flow	2,804	814	836	861	886	3,385	1,035	1,064	1,094	1,127

* For the purpose of this example, these items are replaced at the end of year 5. Anticipated hours of operation and life expectancy of equipment determine the proper entry.

Tables 1-3 and 1-4 show how the annual expenditures are determined for each system. The present worth of a system is the value of an investment (made today at the rate of return required by the investor) which will produce the annual owning and operating costs for that system over the life of the project. Since each system performs the same function, the system representing the lower investment is more attractive to the investor.

Table 1-5 shows the result using 20% as the required rate of return. In this example, the conveyor system requires a $4,599,000 investment and the truck system, a $6,222,000 investment. Since the conveyor system's present worth is lower than that of the truck system, the conveyor is considered to be

TABLE 1-5. *Present Value Annual Costs Discounted at 20%*
In 000's of dollars

	Belt system			*Truck system*		
Year	*Cash flow*	*Factor*	*Present value*	*Cash flow*	*Factor*	*Present value*
1	$3,316	.833	$2,762	$2,804	.833	$2,336
2	405	.694	281	814	.694	565
3	404	.579	234	836	.579	484
4	405	.482	195	861	.482	415
5	407	.402	164	886	.402	356
6	1,678	.335	562	3,385	.335	1,134
7	459	.279	128	1,035	.279	289
8	461	.233	107	1,064	.233	248
9	463	.194	90	1,094	.194	212
10	468	.162	76	1,127	.162	183
Total			$4,599			$6,222

Difference in present value = 6,222,000 − 4,599,000 = $1,623,000

a more attractive investment. In fact, it could justify an additional capital expenditure of $1,623,000 ($6,222,000 less $4,599,000).

Belt conveyors have long been the primary choice for providing intraplant transportation of bulk materials in every industry in which such materials must be handled. Often, belts are the only practical means for this purpose. By reviewing their uses and advantages under the various conditions described above, the reader will gain a better understanding of their possibilities for application in his operations.

In recent years belt conveyors have gained wider acceptance as a means of transporting bulk materials over relatively long distances because of their lower overall costs per ton and other important benefits. Such long-distance transportation has traditionally been accomplished by rail or truck operations.

This book is published by CEMA to provide the reader with the information necessary to make a technical and economic assessment of belt conveyors as a possible solution to bulk materials handling problems. CEMA members will be glad to aid the reader further by furnishing additional data, estimates, and proposals.

Notes

1. "Conveyors the Key to $6,000,000 Savings," *Construction Equipment Magazine,* August 1974; and "Conveyor Generates Power While Delivering Rock to Dam," *Contractors and Engineers Magazine,* November 1974.

2. Thomas B. Douglas, "Economics of 5½ Mile Transport Conveyor Belt at Ideal Cement Company's Ada, Oklahoma Plant," a paper presented to the American Institute of Mining, Metallurgical, and Petroleum Engineers, St. Louis, Missouri, March 2, 1961.

3. V. F. Malone, "A Study of Long Distance Haulage," *Mining Congress Journal,* October 1964.

4. Robert C. Temps, "Bulk Transportation of Jamaican Bauxite," *Mining Congress Journal,* Vol. 52, No. 10 (1966).

5. J. J. Coile, "In-Pit Crushing and Conveying vs. Truck Haulage," *Mining Congress Journal,* January 1974.

6. "Belt Conveyor Beats Trucks for Moving Bulk Materials," *Material Handling Engineering,* September 1973.

7. F. W. Schweitzer and L. G. Dykers, "Belt Conveyors vs. Truck Haulage: Capital vs. Expense," a paper presented to the Society of Mining Engineers/AIME, Denver, Colorado, September 1976.

Chapter 2

Design Considerations

Contents

The information presented in this book is intended to cover the basic principles of belt conveyor design and to include such formulae, tables and charts as are required to design most belt conveyors. The data are arranged in the order most convenient for the use of an experienced conveyor engineer, but the information is intended to be complete enough for use by engineers less familiar with belt conveyors.

The data are based on industry practice and are reliable when all design considerations are incorporated into the final design. They are applicable to the design of a high percentage of conveyors which are required to operate under reasonably normal conditions. However, some conveyors operate under conditions for which complete data are beyond the scope of this book. These design problems usually require broad experience for a satisfactory solution. A CEMA member should be consulted in such cases, as well as in the design of very large conveyors or complex conveyor systems. Assistance from CEMA members in meeting special design requirements will insure optimum economy and performance.

Belt conveyor design considerations and the location of data pertaining to them are outlined below.

Conveyor Arrangements

Belt conveyors can be arranged to follow an infinite number of profiles or paths of travel. Among these are conveyors which are horizontal, inclined, or declined, or, with the inclusion of concave and convex curves, any combination of these. Also, numerous arrangements are possible for loading to and discharging from the conveyor.

The nomenclature of typical belt conveyor components is illustrated below in Figure 2.1. Various arrangements are illustrated in Figures 2.2

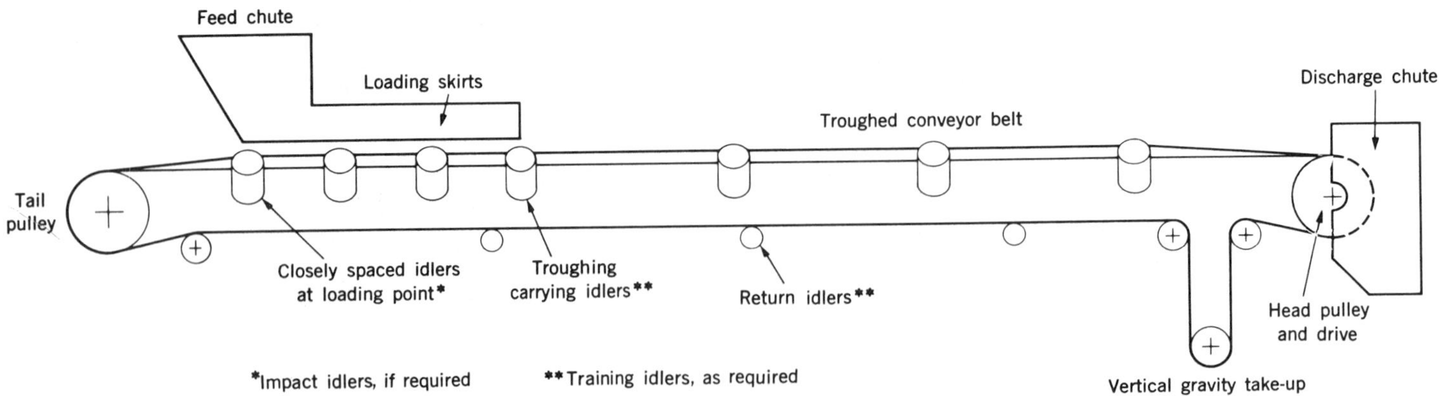

FIGURE 2.1. *Nomenclature of components of a typical belt conveyor.*

through 2.9; loading and discharging arrangements are shown in Figures 2.10 through 2.18. In addition, belt conveyor stackers can be connected to a mainline conveyor to stockpile material several hundred feet on either or both sides into piles of any practical length, as shown in Figure 1.11 in Chapter 1. Or a combination stacker-reclaimer (Figure 1.12) can be used to stockpile material or to reclaim and return it to the main belt conveyor.

Belt conveyors can be designed for practically any desired path of travel, limited only by the strength of belt, angle of incline or decline, or available space. Some of the profiles shown below are more desirable than others. For example, transfers between conveyors should be avoided where possible due to additional wear on the belts at the loading points, dust raised, and possible plugging in the transfer chutes. For these reasons, the arrangement in Figure 2.3 is preferable to those shown in Figures 2.5 and 2.7.

TYPICAL BELT CONVEYOR TRAVEL PATHS

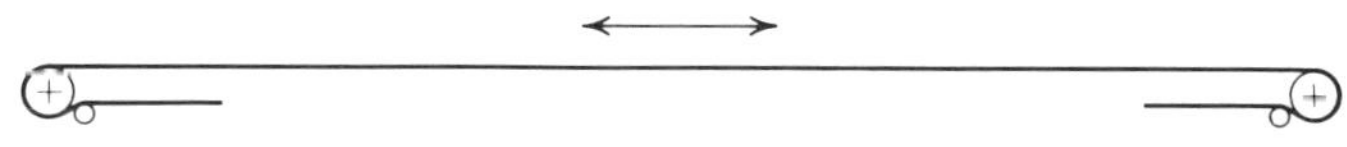

FIGURE 2.2. *Horizontal belt.*

FIGURE 2.3. *Horizontal and ascending path, when space will permit vertical curve and belt strength will permit one belt.*

FIGURE 2.4. *Ascending and horizontal path, when belt tensions will permit one belt and space will permit vertical curve.*

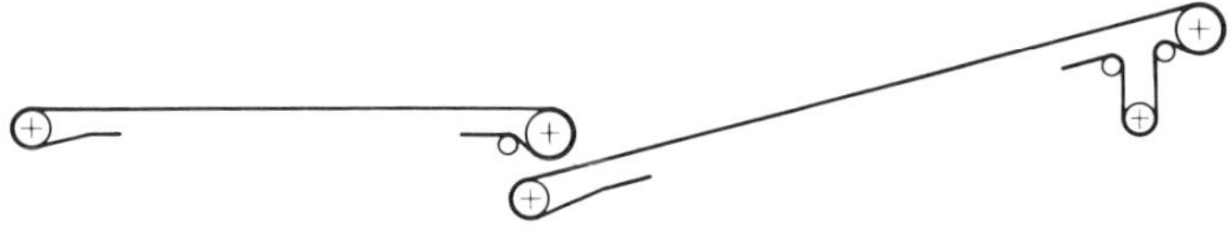

FIGURE 2.5. *Possible horizontal and ascending path, when space will not permit a vertical curve or when the conveyor belt strength requires two belts.*

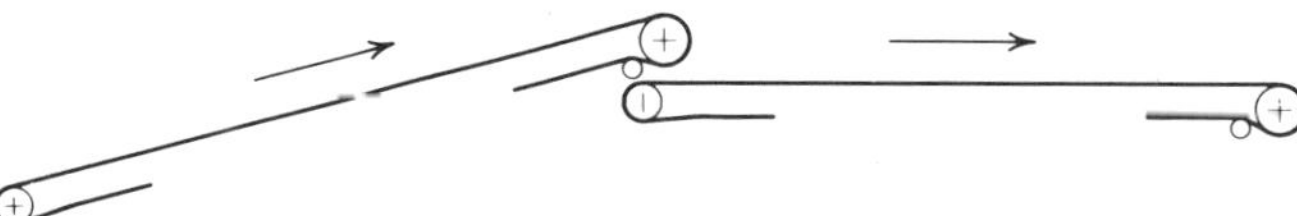

FIGURE 2.6. *Ascending and horizontal path, when advisable to use two conveyor belts.*

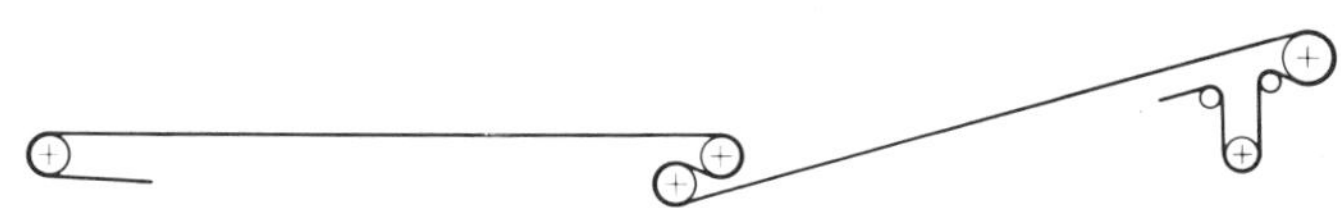

FIGURE 2.7. *Possible horizontal and ascending path, when space will not permit vertical curve but belt strength will permit one belt.*

FIGURE 2.8. *Compound path with declines, horizontal portions, vertical curves, and incline.*

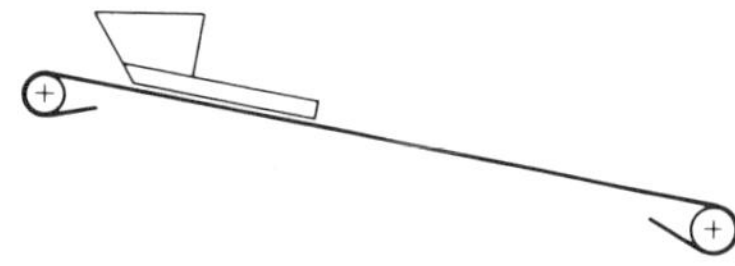

FIGURE 2.9. *Loading can be accomplished, as shown, on minor inclines or declines.*

TYPICAL BELT CONVEYOR LOADING AND DISCHARGING ARRANGEMENTS

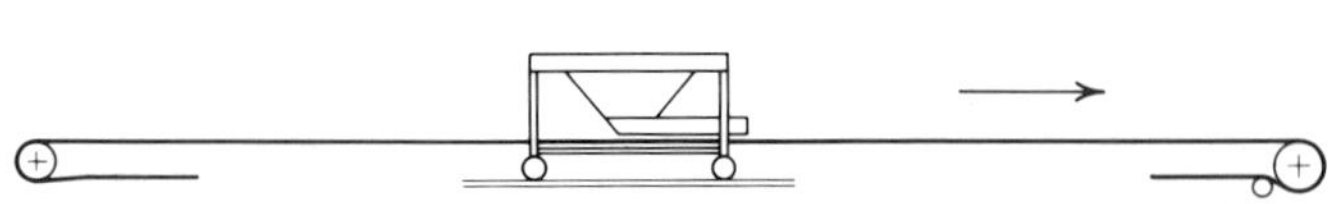

FIGURE 2.10. *Traveling loading chute to receive materials at a number of points along conveyor.*

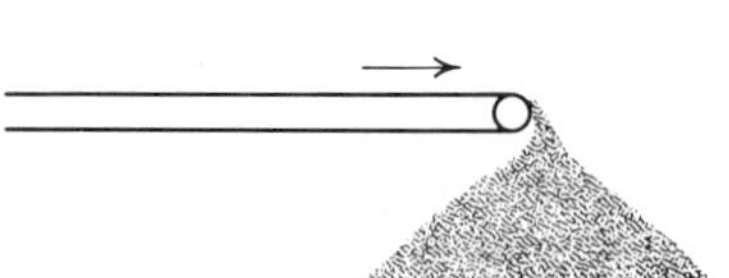

FIGURE 2.11. *Discharge over end pulley to form conical pile.*

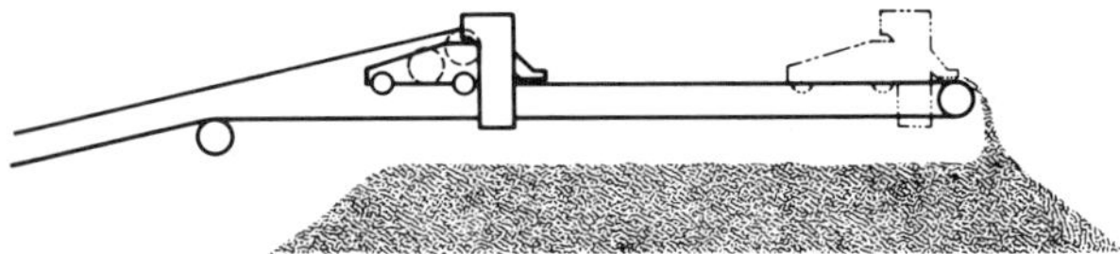

FIGURE 2.12. *Discharge by traveling tripper or through the tripper to the storage pile. See Figure 2.14.*

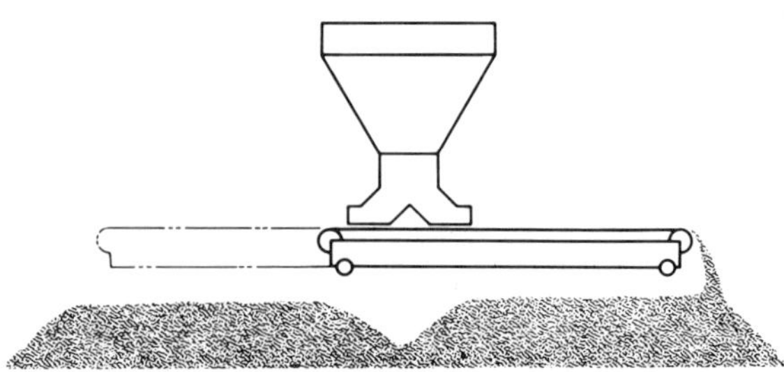

FIGURE 2.13. *Discharge over either end-pulley of a reversible shuttle belt conveyor.*

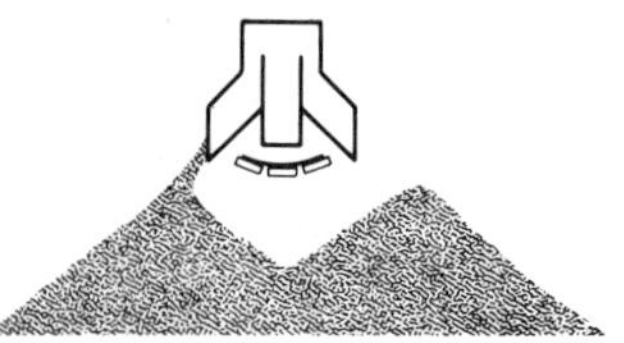

FIGURE 2.14. *Discharge from tripper to one side only, to both sides, or forward again on conveyor belt.*

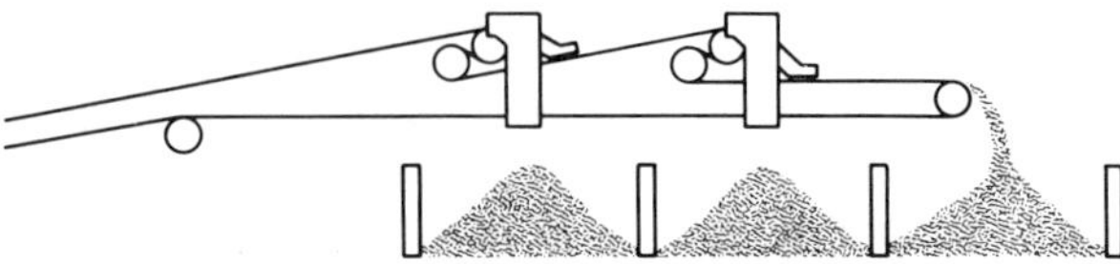

FIGURE 2.15. *Discharge by fixed trippers with or without cross conveyors to fixed piles or bin openings.*

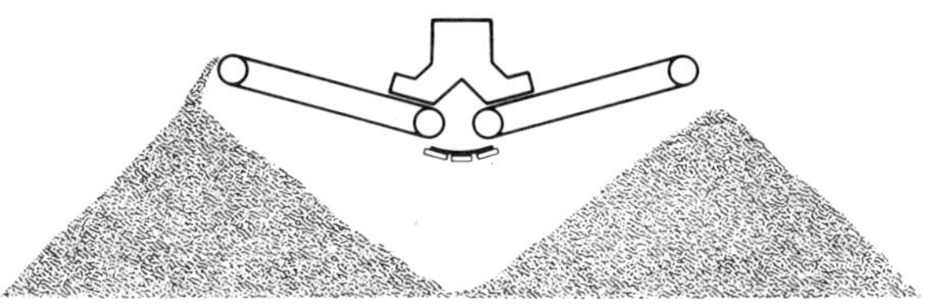

FIGURE 2.16. *Discharge by traveling or stationary trippers to ascending cross conveyors carried by tripper.*

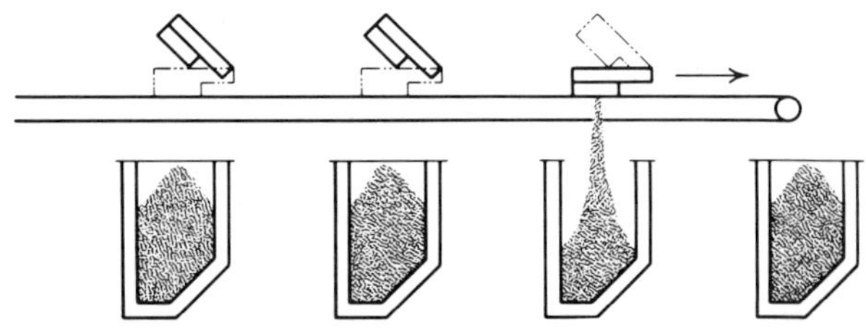

FIGURE 2.17. *Discharge by hinged plows to one or more fixed locations on one or both sides of conveyor plows. Device can be adjusted for proportioned discharge to several places.*

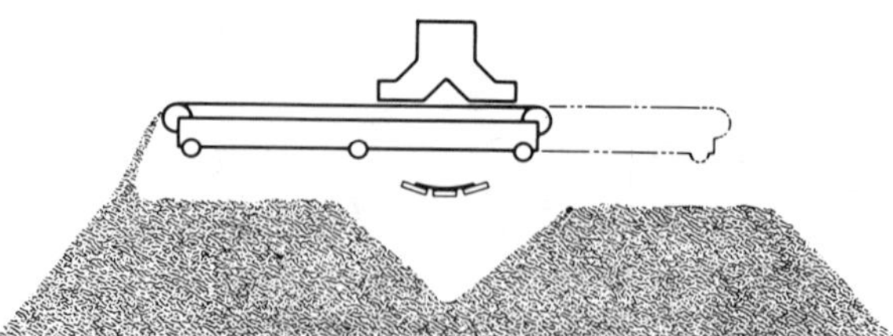

FIGURE 2.18. *Discharge by traveling or stationary tripper carrying reversible shuttle belt.*

Similarly, the profile shown in Figure 2.4 is usually better than that in Figure 2.6.

Figure 2.8 illustrates a profile that can be used on a single, long, overland conveyor, often with several convex and concave curves, as shown in Figure 2.19. On such conveyors, it is essential that belt tensions be determined along the entire length in order to calculate the radii of the curves and locate the maximum belt tension which can occur at an intermediate point.

A few typical arrangements for loading and discharging material are illustrated on page 30.

Figure 1.11 illustrates a traveling stacker stockpiling along a feed conveyor. This type of stacker comprises a traveling tripper with two boom conveyors discharging alternately to each side. The tripper lifts the material from the feed conveyor onto either of the booms for discharge into the piles. The traveling stacker can also be supplied with a single boom fixed for discharge to one side only or with a single, slewing boom which can discharge to both sides of the feed conveyor. See Figure 2.20.

FIGURE 2.19. *This overland conveyor following the natural terrain has an undulating profile with several concave and convex curves.*

FIGURE 2.20. *Traveling winged stacker with slewing boom stockpiles on both sides of the feed conveyor.*

Reclaiming from the elongated piles formed by these stackers is efficiently accomplished by either a combination stacker-reclaimer (Figure 2.21), which utilizes the reversing boom conveyor for both stacking and reclaiming, or by a separate bucket wheel reclaimer, different types of which are shown in Figures 2.22 and 2.23.

FIGURE 2.21. *Bucket wheel of this combination stacker-reclaimer discharges onto the stacker boom conveyor, which is reversed for the reclaim operation.*

FIGURE 2.22. *Bucket wheel reclaimer with wheel at end of boom.*

FIGURE 2.23. *Crawler-mounted bucket wheel reclaimer with wheel mounted on chassis discharges onto a rail-mounted reclaim conveyor with slewing boom.*

Resumes of Chapters 3-13

The following chapters cover various engineering design considerations, including pertinent data, formulae, and applications. Each chapter topic is briefly summarized below.

Characteristics and Conveyability of Bulk Materials (*Chapter 3*)

Belt conveyors are capable of handling an almost unlimited variety of materials. However, their successful performance depends upon a design that is based on a thorough understanding of the characteristics of the material to be handled and careful consideration of its anticipated behavior while being carried on the belt, as well as during delivery to and from it. Some of the characteristics which affect basic design include size of lumps, bulk density, angle of repose, abrasiveness, moisture content, dustiness, stickiness, temperature, and chemical action. The influence of these and other material characteristics on belt conveyor design is discussed in Chapter 3. Most materials which have been handled successfully on belt conveyors are listed and classified in Table 3-3.

Capacities, Belt Widths, and Speeds (*Chapter 4*)

In its simplest terms the rated hourly capacity of a belt conveyor is the total weight of material conveyed in one hour with the belt continuously carrying a uniform cross section of material and traveling at a uniform speed. However, to obtain a desired capacity with successful operation it is important to consider the many factors which influence both belt width and speed.

Width can be determined by the size of lumps, the percentage of lumps to fines, and the angle of repose of the material. Maximum speeds which are acceptable for conveying some materials under certain conditions may not be appropriate for conveying dusty or fragile materials or heavy, sharp-edged ore. Even the cross-sectional area of the load on the belt may vary according to the type of idler used. Design data pertaining to belt widths, speeds, and capacities, including tables and formulae, are covered in Chapter 4.

Belt Conveyor Idlers (*Chapter 5*)

It is important to select the best idler design for a specific condition, since idlers greatly influence belt tensions, power requirements, belt life, capacity, and operational success of a conveyor. Use of the right carrying idler and proper spacing can minimize frictional resistance. Some idlers carry larger cross-sectional loads than others for appropriate materials. Others are needed for such purposes as training the belt, cushioning the belt against heavy lumps, and handling extremely abrasive or sticky materials. The design, selection, and application data for belt conveyor idlers are presented in Chapter 5.

Belt Tension, Power, and Drive Engineering (*Chapter 6*)

Belt tensions not only determine the required strength of the belt, but also influence the design of many other mechanical parts. Furthermore, the power required to drive the conveyor is calculated from the belt tensions.

Factors which contribute to belt tensions and their relation to power requirements, along with appropriate formulae and tables, are covered in Chapter 6. Among the principal factors discussed are the power required to overcome frictional resistance and lift the load, acceleration and deceleration, and drive arrangements. Good design can assure lower belt tension by the proper arrangement of drive pulleys, their lagging, and degrees of belt wrap. Tensions resulting from acceleration can be limited by appropriate electrical controls, as described in Chapter 12.

In order to assist the reader in the use of this data, two different methods for determining tension and power requirements are illustrated. The graphical method, shown on page 130, is used to calculate quickly and conveniently the horsepower required to drive a conveyor. Belt tensions can be determined based on these calculations of horsepower. This method is suitable for use in designing simple, straight-line, horizontal, or inclined conveyors.

The analytical method, described on page 133, provides calculations of all belt tensions as they occur along the conveyor, thus determining magnitudes at intermediate points. This information is necessary to design convex and concave vertical curves, trippers, and remotely located takeups. It is also used to determine the location of maximum belt tension when there is a multiple change of slope in the belt line. The power required to drive or restrain the conveyor is derived from the tension, which represents the amount of pull at the drive pulley necessary to propel or retard the loaded belt. The analytical method is more accurate than the graphical method, and should be used for designing conveyors which exceed the limits set for the former method.

Belt Selection (*Chapter 7*)

Selection of the conveyor belt is a most important design consideration, since the belt constitutes a large portion of the initial cost of a belt conveyor. Also, it is subject to the most wear and contributes substantially to the operating costs of a conveyor.

Selection of the most suitable belt for the required service involves careful consideration of the construction of the belt in conjunction with the idlers and other mechanical components. There are so many belt constructions that it is impractical to include more than a few of them in Chapter 7. Also, the rubber industry is making such rapid strides in the use of synthetic fibers and steel-cord carcasses that improved constructions are frequently available.

Chapter 7 contains information pertaining to the conservative selection of suitable belts for most applications. However, for large, heavily loaded belts or complex conveyor systems, it is advisable to consult a CEMA member to assist in selecting the most economical belt for the required service.

Pulleys and Shafts (*Chapter 8*)

Because conveyor pulleys and shafts form a composite structure in operation, it is accepted engineering practice to consider their design together. Chapter 8 contains tables and formulae to assist the designer in selecting the most effective pulley and shaft combination. Also included is information on special-purpose pulleys, such as those with rubber lagging for better traction, herringbone-grooved lagging for a relatively "non-skid" grip on a wet belt, and wing or slat pulleys for use in handling sticky materials.

Vertical Curves (*Chapter 9*)

Two types of vertical curves may be required in the profile of a belt conveyor, namely, concave and convex. In the concave curve the carrying belt turns upward on a radius whose center is above the belt. If improperly designed, the loaded belt may lift off the idlers and result in excessive spillage. The convex curve turns downward on a radius whose center is beneath the belt. In this case, improper design can mean excessive tensions at the edges of the belt, causing damage to it. Chapter 9 includes tables and formulae for use in the proper design of vertical curves.

Belt Takeups, Cleaners, and Accessories (*Chapter 10*)

In addition to the data required to design a basic belt conveyor, there are numerous accessory items of equipment which must be considered in light of their contribution to the economical and successful operation of the conveyor. For example, belt takeups are necessary for maintaining the proper belt tension for drive pulley traction. Cleaning devices can reduce material cleanup problems and eliminate a source of misalignment which might otherwise result in damage to the belt.

Other accessory devices are available for protecting the belt and/or performing certain process functions. Such devices include belt covers, magnetic pulleys for removing tramp iron from the material being conveyed, tramp iron detectors, and various methods and devices for weighing and/or sampling the conveyed material accurately and continuously.

Conveyor Loading and Discharge (*Chapter 11*)

An extremely important consideration in belt conveyor design is the proper loading of the belt. Improper loading may result in accelerated belt wear, spillage, and/or reduced capacity. Discharge from the belt must be considered in relation to the materials handled, where they are to be delivered, and the trajectory of discharge over the head pulley. Trajectories and chute design are particularly important at transfers from one conveyor to another. Discharge at points other than the head pulley can be accomplished by fixed or movable trippers, plows, etc.

In addition to trippers, Chapter 11 considers auxiliary devices for feeding material to the belt and for receiving the flow from the belt for purposes of stockpiling or otherwise disposing of it.

Motors and Controls (*Chapter 12*)

The electrical equipment must be capable not only of propelling the fully loaded conveyor at its designed speed, but also of controlling starting acceleration within design limits to prevent excessive tension in the belt. Chapter 12 contains tables, charts, and other information on motor types and control devices useful for this purpose. Also included are data on necessary interlocks for two or more conveyors of a system.

A number of important electronic control devices are discussed in Chapter 12, as well as the advantages in using such modern electronic equipment as programmable controllers, computers, and multiplexing.

Operation, Maintenance, and Safety (*Chapter 13*)

Chapter 13 provides guidance on the considerations of operation, maintenance, and safety that are so important to assure dependability, efficient utilization, and reasonable life expectancy of the equipment.

Chapter 3

Characteristics and Conveyability of Bulk Materials

Contents

The successful design of a belt conveyor must begin with an accurate appraisal of the characteristics of the material to be transported. A few important characteristics require definition.

The angle of repose of a material is the angle which the surface of a normal, freely formed pile makes to the horizontal.

The angle of surcharge of a material is the angle to the horizontal which the surface of the material assumes while the material is at rest on a moving conveyor belt. This angle usually is 5° to 15° less than the angle of repose, though in some materials it may be as much as 20° less.

The flowability of a material, as measured by its angle of repose and angle of surcharge, determines the cross-section of the material load which safely can be carried on a belt. It also is an index of the safe angle of incline of the belt conveyor.

The flowability is determined by such material characteristics as: size and shape of the fine particles and lumps, roughness or smoothness of the surface of the material particles, proportion of fines and lumps present, and moisture content of the material.

Table 3-1 illustrates and defines the normal relationship of the foregoing properties and the general characteristics of materials.

TABLE 3-1. *Flowability—Angle of Surcharge—Angle of Repose*

*Very free flowing 1**	*Free flowing 2**	*Average flowing 3**		*Sluggish 4**
5° Angle of surcharge	*10° Angle of surcharge*	*20° Angle of surcharge*	*25° Angle of surcharge*	*30° Angle of surcharge*
5°	10°	20°	25°	30°
0°–19° *Angle of repose*	20°–29° *Angle of repose*	30°–34° *Angle of repose*	35°–39° *Angle of repose*	40°–*up Angle of repose*
Material characteristics				
Uniform size, very small rounded particle, either very wet or very dry, such as dry silica sand, cement, wet concrete, etc.	Rounded, dry polished particles, of medium weight, such as whole grain and beans.	Irregular, granular or lumpy materials of medium weight, such as anthracite coal, cottonseed meal, clay, etc.	Typical common materials such as bituminous coal, stone, most ores, etc.	Irregular, stringy, fibrous, interlocking material, such as wood chips, bagasse, tempered foundry sand, etc.

*Code designations conform to bulk material characteristics chart, Table 3-2.

Consideration should also be given to the weight per cubic foot of the material; its dustiness, wetness, stickiness, abrasiveness; its chemically corrosive action; and its temperature. Some general information concerning these properties of many materials is given in Table 3-2. Table 3-3 gives a list of materials with their physical characteristics and classification code designations. It must be understood that the data given in this table are for average conditions and average materials. Each characteristic may vary in specific instances, especially angles of repose and maximum conveyor inclinations.

Proper consideration must be given to those materials, the characteristics of which vary under various conditions of handling, atmospheric humidity, age, or long storage. In some cases, accuracy may require that carefully conducted tests be run to establish the material characteristics under the required conditions.

Materials or characteristics omitted from Table 3-3 may be roughly appraised by comparison with listed materials of the same general type.

Behavior of Materials on a Moving Belt

Attention must be drawn to the fact that the normal characteristics of materials are considerably influenced by the movement, slope, and speed of the conveyor belt that carries them.

As the conveyor belt passes successively over each carrying idler, the material on it is correspondingly agitated. This agitation tends to work the larger pieces to the surface of the load and the smaller particles or fines to the bottom. It also tends to flatten the material surface slope (i.e., the angle of surcharge) and explains why this angle is less than the angle of repose.

Any difference between the forward velocity of the material as it is being loaded and the conveyor belt that is receiving it must be equalized by the acceleration of the material. This acceleration causes turbulence in the material.

Any vertical velocity of the material as it is being loaded must be absorbed in the resilient construction of the conveyor belt and the impact idlers used under the loading point. In this process, a further increase in material turbulence is produced.

These three influences are emphasized when the conveyor belt is on an incline or decline, and also when the conveyor belt is operated at high speeds. These influences are emphasized even more when the material handled is loose and contains large rounded lumps, such as coarse washed gravel, the tendency of which is to bounce and roll on the conveyor belt.

Effect of Inclines and Declines

The nominal cross section of the material on a horizontal conveyor belt is measured in a plane normal to the belt. On an inclined or declined conveyor belt, gravity necessitates that the actual cross section of the load be considered in a vertical plane. To maintain the total width of the material load on the belt and to maintain unchanged surcharge angles, the cross section of the load possible on an inclined or declined belt must be less than that on a horizontal belt. Referring to the diagram used in the derivation of belt conveyor capacity (Figure 4.2), the area A_b does not change although the area A_s does decrease as the cosine of the conveyor slope.

The total effect is influenced by the surcharge angle at which the material will ride on the conveyor belt. However, in most cases, the actual loss of capacity is less than 3%.

The following three precautions may well be observed: lumps are more likely to roll off the edges of inclined conveyor belts than horizontal ones; for belts of constant slope, the spillage of material is more likely to occur immediately beyond the loading point; materials which aerate excessively, such as some very fine ground cements, or materials in which the proportion of water is so high that a slurry is created, must be carried on inclines and at such a conveyor belt speed that the tendency of the material to slide back is fully offset.

TABLE 3-2. *Material Class Description*

	Material characteristics	*Code*
Size	Very fine—100 mesh and under	A
	Fine—1/8 inch and under	B
	Granular—Under 1/2 inch	C
	Lumpy—containing lumps over 1/2 inch	D
	Irregular—stringy, interlocking, mats together	E
Flowability Angle of Repose	Very free-flowing—angle of repose less than 19°	1
	Free-flowing—angle of repose 20° to 29°	2
	Average flowing—angle of repose 30° to 39°	3
	Sluggish—angle of repose 40° and over	4
Abrasiveness	Nonabrasive	5
	Abrasive	6
	Very abrasive	7
	Very sharp—cuts or gouges belt covers	8
Miscellaneous Characteristics (Sometimes more than one of these characteristics may apply)	Very dusty	L
	Aerates and develops fluid characteristics	M
	Contains explosive dust	N
	Contaminable, affecting use or saleability	P
	Degradable, affecting use or saleability	Q
	Gives off harmful fumes or dust	R
	Highly corrosive	S
	Mildly corrosive	T
	Hygroscopic	U
	Interlocks or mats	V
	Oils or chemical present—may affect rubber products	W
	Packs under pressure	X
	Very light and fluffy—may be wind-swept	Y
	Elevated temperature	Z

Example: A very fine material that is free-flowing, abrasive, and contains explosive dust would be designated: Class A26N

TABLE 3-3. *Material Characteristics and Weight Per Cubic Foot*

Material	Average weight (lbs per cu ft)	Angle of repose (degrees)	Recommended maximum inclination	Code
Alfalfa meal	17	45		B46Y
Alfalfa pellets	41–43	20–29		C25
Alfalfa seed	10–15	29		B26N
Almonds, broken or whole	28–30	30–44		C36Q
Alum, fine	45–50	30–44		B35
Alum, lumpy	50–60	30–44		D35
Alumina	50–65	22	10–12	B27M
*Aluminum chips	7–15	45		E46Y
Aluminum hydrate	18	34	20–24	C35
Aluminum ore (see bauxite)	—			—
Aluminum oxide	70–120	29		A27M
Aluminum silicate	49	30–44		B35S
Aluminum sulphate	54	32	17	D35
Ammonium chloride, crystalline	45–52	30–44		B36S
Ammonium nitrate	45	30–44		*C36NUS
Ammonium sulphate (granular)	45–58	44		*C35TU
Antimony powder	60	30–44		A36
Aplite	70–80	30–44		A35
Arsenic, pulverized	30	20–29		*A26
Arsenic oxide	100–120	30–44		A35R
Asbestos, ore or rock	81	30–44		D37R
Asbestos, shred	20–25	45		E46XY
*Ash, black, ground	105	32	17	*B35
Ashes, coal, dry, 3 inch & under	35–40	45		D46T
Ashes, coal, wet, 3 inch & under	45–50	45		D46T
Ashes, fly	40–45	42	20–25	A37
Ashes, gas-producer, wet	78			D47T
Asphalt, binder for paving	80–85			C45
Asphalt, crushed, ½ inch & under	45	30–44		C35
Bagasse	7–10	45		E45Y
Bakelite & similar plastics (powdered)	35–45	45		B45
Barite	180	30–44		B36
Barium carbonate	72	45		A45
*Barium carbonate filter cake	72	32		A36
Barium hydrate	62–65	43		A36
Barium oxide	150–200			A46
*Bark, wood, refuse	10–20	45	27	E45VY
Barley	37–48	23	10–15	B25N
Basalt	80–103	20–28		B26
Bauxite, ground, dry	68	20–29	20	B26
Bauxite, mine run	80–90	31	17	E37
Bauxite, crushed, 3 inch & under	75–85	30–44	20	D37
Beans, castor, whole	36	20–29	8–10	C25W
Beans, castor, meal	35–40			B15W
Beans, navy, dry	48	29		C25

* May vary considerably—consult a CEMA member.

TABLE 3-3 *continued.*

Material	*Average weight (lbs per cu ft)*	*Angle of repose (degrees)*	*Recommended maximum inclination*	*Code*
Beans, navy, steeped	60	35-40		C35
Beet pulp, dry	12-15			E45
Beet pulp, wet	25-45			E46
Beets, whole	48	50		D45
*Bentonite, crude	35-40	42-44		D36X
Bentonite, 100 mesh & under	50-60	42	20	A36XY
Bones	34-40	45		*C46
Boneblack, 100 mesh & under	20-25	20-29		A25Y
Bonechar	27-40	30-44		B36
Bonemeal	50-60	30-44		B36
Borate of lime	60	30-44		A35
Borax, ½-inch screenings	55-60	30-44		C36
Borax, 3 inch and under	60-70	30-44		D35
Boric acid, fine	55	20-29		B26T
Bran	10-20	30-44		B35NY
Brewer's grain, spent, dry	25-30	45		C45
Brewer's grain, spent, wet	55-60	45		C45T
Bronze chips, dry	30-50	44-57		B47
Buckwheat	37-42	25	11-13	B25N
Calcium carbide (crushed)	70-80	30-44		D36N
Carbon, activated, dry, fine	8-20	20-29		B26Y
Carbon black, pelletized	20-25	25		B25Q
Carbon black, powder	4-7	30-44		*A35Y
Carborundum, 3 inch and under	100	20-29		D27
Casein	36	30-44		B35
Cast iron chips	90-120	45		C46
Caustic soda	88	29-43		A36
Cement, Portland	72-99	30-44	20-23	A36M
Cement, Portland, aerated	60-75			A16M
Cement, rock (see limestone)	100-110			D36
Cement clinker	75-95	30-40	18-20	D37
Chalk, lumpy	75-85	45		D46
*Charcoal	18-25	35	20-25	D36Q
Chrome ore (chromite)	125-140	30-44		D37
Cinders, blast furnace	57	35	18-20	*D37T
Cinders, coal	40	35	20	*D37T
Clay (see also bentonite, diatomaceous earth, fullers earth, kaolin, and Marl)	—			—
Clay, calcined	80-100			B37
Clay, ceramic, dry, fines	60-80	30-44		A35
Clay, dry, fines	100-120	35	20-22	C37
Clay, dry, lumpy	60-75	35	18-20	D36
Clinker, cement (see cement clinker)	—			—
Clover seed	48	28	15	B25N
Coal, anthracite, river, or culm, 1/8 inch and under	60	35	18	B35TY

* May vary considerably—consult a CEMA member.

Table 3-3 *continued.*

Material	*Average weight (lbs per cu ft)*	*Angle of repose (degrees)*	*Recommended maximum inclination*	*Code*
Coal, anthracite, sized	55–60	27	16	C26
Coal, bituminous, mined 50 mesh & under	50–54	45	24	B45T
Coal, bituminous, mined & sized	45–55	35	16	D35T
Coal, bituminous, mined, run of mine	45–55	38	18	D35T
*Coal, bituminous, mined, slack, ½ inch & under	43–50	40	22	C35T
Coal, bituminous, stripping, not cleaned	50–60			D36T
Coal, lignite	40–45	38	22	D36T
Cocoa beans	30–45	30–44		C35Q
Cocoa nibs	35	30–44		C35
Coffee, chaff	20	20–29		B25MY
Coffee, green bean	32–45	30–44	10–15	C35Q
Coffee, ground	25	23	10	B25
*Coffee, roasted bean	22–26			C25PQU
Coffee, soluble	19			B45PQ
Coke, loose	23–35	30–44	18	B37QVT
Coke, petroleum calcined	35–45	30–44	20	D36Y
Coke breeze, ¼ inch and under	25–35	30–44	20–22	C37Y
Compost	30–50			E45ST
Concrete, cinder	90–100		12–30	D46
Copper ore	120–150	30–44	20	*D37
Copper sulfate	75–85	31	17	D36
Cork, granulated	12–15			C45
Corn, cracked	45–50			C35W
Corn, ear	56			
Corn, shelled	45	21	10	C25NW
Corn sugar	31	30–44		B35
Corn germs	21			B35W
Corn grits	40–45	30–44		B35W
Cornmeal	32–40	35	22	B35W
Cottonseed, dry, de-linted	22–40	29	16	C35W
Cottonseed, dry, not de-linted	18–25	35	19	C35W
Cottonseed cake, crushed	40–45	30–44		B35
Cottonseed cake, lumpy	40–45	30–44		D35W
Cottonseed hulls	12	45		*B45Y
Cottonseed meal	35–40	35	22	B35W
Cottonseed meats	40	30–44		B35W
Cracklings, crushed, 3 inch & under	40–50	45		D45
Cryolite, dust	75–90	30–44		A36
Cryolite, lumpy	90–100	30–44		D36
Cullet	80–120	30–44	20	D37Z
Diatomaceous earth	11–14	30–44		A36MY
Dicalcium phosphate	40–50	45		A45

* May vary considerably—consult a CEMA member.

Table 3-3 *continued.*

Material	*Average weight (lbs per cu ft)*	*Angle of repose (degrees)*	*Recommended maximum inclination*	*Code*
Disodium phosphate	25-31	30-44		B36QT
Dolomite, lumpy	80-100	30-44	22	D36
Dolomite, pulverized	46	41		B36
Earth, as excavated—dry	70-80	35	20	B36
Earth, wet, containing clay	100-110	45	23	B46
Ebonite, crushed ½ inch & under	65-70	30-44		C35
Emery	230	20-29		A27
Epsom salts	40-50	30-44		B35
Feldspar, ½-inch screenings	70-85	38	18	B36
Feldspar, 1½- to 3-inch lumps	90-110	34	17	D36
Feldspar, 200 mesh	100	30-44		A37
Ferrous carbonate	85-90	30-44		B36
Ferrous sulfate	50-75			C36
Ferrous sulfide	120-135	20-29		C36
Filter press mud, sugar factory	70			A15
Fish meal	35-40			B45W
Fish scrap	40-50			E45W
Flaxseed	45	21	12	B25NW
Flaxseed meal	25	30-44		B35W
Flour, wheat	35-40	45	21	A45PN
Flue dust, boiler house, dry	35-40	20		A17MTY
Fluorspar, ½-inch screenings	85-105	45		C46
Fluorspar, 1½- to 3-inch lumps	110-120	45		D46
Foundry refuse, old sand cores, etc.	70-100	30-44		D37Z
Fullers earth, dry	30-35	23		B26
Fullers earth, oily	60-65	20-29		B26
Fullers earth, oil filter, burned	40	20-29		B26
Fullers earth, oil filter, raw	35-40	35	20	* B26
Galena (lead sulfide)	240-260	30-44		A36
Glass batch (textile fiber glass)	45-55	0-10		A16LM
Glass batch (wool & container)	80-100	30-44	20-22	D38Z
Gelatin, granulated	32	20-29		C25Q
Glue, ground 1/8 inch and under	40	30-44		B36
Glue, pearl	40	25	11	C25
Glue, vegetable, powdered	40	30-44		
Gluten meal	40	30-44		B35P
Grain, distillery, spent, dry	30	30-44		E35WY
Grain, distillery, spent, wet	40-60	45		C45V
Granite, ½-inch screenings	80-90	20-29		C27
Granite, 1½- to 3-inch lumps	85-90	20-29		D27
Granite, broken	95-100	30-44		D37
Graphite, flake	40	30-44		C35
Graphite, flour	28	20-29		A25
Graphite ore	65-75	30-44		D37
Grass seed	10-12	30-44		B35NY

* May vary considerably—consult a CEMA member.

TABLE 3-3 *continued.*

Material	*Average weight (lbs per cu ft)*	*Angle of repose (degrees)*	*Recommended maximum inclination*	*Code*
Gravel, bank run	90–100	38	20	
Gravel, dry, sharp	90–100	30–44	15–17	D37
Gravel, pebbles	90–100	30	12	D36
Gypsum, ½-inch screenings	70–80	40	21	C36
Gypsum, 1½- to 3-inch lumps	70–80	30	15	D36
Guano, dry	70	20–29		B26
Hominy	37–50	30–44		C35
Hops, spent, dry	35	45		E45
Hops, spent, wet	50–55	45		E45T
Ice, crushed	35–45	19		D16
Ilmenite ore	140–160	30–44		B37
Iron ore	100–200	35	18–20	*D36
Iron ore pellets	116–130	30–44	13–15	D37Q
Iron sponge	100–135	30–44		
Iron sulphate	50–75	30–44		C35
Iron sulfide	120–135	30–44		D36
Kaolin clay, 3 inch and under	63	35	19	D36
Lactose	32	30–44		A35PX
Lead arsenate	72	45		B45R
Lead carbonate	240–260	30–44		A36MR
Lead ores	200–270	30	15	*B36RT
Lead oxides	60–150	45		B45
Lead oxides, pulverized	200–250	30–44		A36
Lead silicate, granulated	230	40		B36
Lead sulfate, pulverized	184	45		B46
Lead sulfide	240–260	30–44		A36
Lignite, air-dried	45–55	30–44		*D35
Lime, ground, 1/8 inch and under	60–65	43	23	B35X
*Lime, hydrated, 1/8 inch & under	40	40	21	B35MX
Lime, hydrated, pulverized	32–40	42	22	A35MXY
Lime, pebble	53–56	30	17	D35
Limestone, agricultural, 1/8 inch & under	68	30–44	20	B36
Limestone, crushed	85–90	38	18	C36X
Linseed cake, pea size	50	30–44		C35W
Linseed meal	27	34	20	B35
Magnesium chloride	33	40		C45
Magnesium sulfate	40–50	30–44		
*Malt, dry, ground, 1/8 inch and under	22	30–44		B35NR
Malt, dry, whole	27–30	20–29		C25N
Malt, wet or green	60–65	45		C45

* May vary considerably—consult a CEMA member.

TABLE 3-3 *continued.*

Material	*Average weight (lbs per cu ft)*	*Angle of repose (degrees)*	*Recommended maximum inclination*	*Code*
Malt, meal	36–40	30–44		B35
Manganese dioxide	80			*
Manganese ore	125–140	39	20	*D37
Manganese oxide	120	30–44		A36
Manganese sulfate	70	30–44		C37
Marble, crushed, ½ inch & under	80–95	30–44		D37
Marl	80	30–44		C37
Meat scraps	50–55	30–44		E35VW
Mica, flakes	17–22	19		B16MY
Mica, ground	13–15	34	23	*B36
Milk, dried, flaked	5–6	30–44		B35MPY
Milk, dry powder	36	45		B45P
Milk, malted	30–35	45		A45PX
Milk, whole, powdered	20	30–44		B35PUXY
Mill scale	100–125	45		E46T
Milo maize	56	30–44		C35N
*Molybdenite, powdered	107	40	25	B35
Molybdenum ore	107	40		B36
Monosodium phosphate	50	30–44		B36
Mustard seed	45–48	20–29		B25N
Nephelene syenite	90–105	30–44		B36
Niacin	35	30–44		B36
Nickel—cobalt sulfate ore	80–150	30–44		*D37T
Oats	26–35	21	10	C25M
Oats, rolled	19–24	30–44		C35NY
Oil cake	48–50	45		D45W
Oxalic acid crystals	60	30–44		B35SU
Oyster shells, ground, under ½ inch	50–60	30–44		C36T
Oyster shells, whole	80	30–44		D36TV
Paper pulp stock	40–60	19		*E15MV
Peanuts, in shells	15–24	30–44		D35Q
Peanuts, shelled	35–45	30–44		C35Q
Peas, dried	45–50			C15NQ
Petroleum coke (see coke)	—			—
Phosphate, acid, fertilizer	60	26	13	B25T
Phosphate, triple super, ground fertilizer	50–55	45	30	B45T
Phosphate rock, broken, dry	75–85	25–29	12–15	D26
Phosphate rock, pulverized	60	40	25	B36
Polyethylene pellets	35	23		B25PQ
Polystyrene pellets	35	23		B25PQ
Polyvinyl chloride	20–30	45		A45KT
Potash (muriate), dry	70	20–29		B27
Potash (muriate), mine run	75	30–44		D37
Potash salts, sylvite, etc.	80	20–29		B25T

* May vary considerably—consult a CEMA member.

TABLE 3-3 *continued.*

Material	*Average weight (lbs per cu ft)*	*Angle of repose (degrees)*	*Recommended maximum inclination*	*Code*
Potassium carbonate	51	20–29		B26
Potassium chloride, pellets	120–130	30–44		C36T
Potassium nitrate	76–80	20–29		C26T
Potassium sulfate	42–48	45		B36X
Pumice, 1/8 inch & under	40–45	45		B47
Pyrites, iron, 2- to 3-inch lumps	135–145	20–29		D26T
Pyrites, pellets	120–130	30–44		C36T
Quartz, ½-inch screenings	80–90	20–29		C27Z
Quartz, 1½- to 3-inch lumps	85–95	20–29		D27Z
Rice, hulled or polished	45–48	19	8	B15
Rice, rough	36	30–44		B35M
Rice grits	42–45	30–44		B35
Rock, crushed	125–145	20–29		D26
Rock, soft, excavated with shovel	100–110	30–44	22	D36
Rubber, pelletized	50–55	35	22	D35
Rubber, reclaim	25–30	32	18	D35
Rye	42–46	23	8	B25N
Rye meal	35–40	19		B15
Safflower cake	50	30–44		D35
Safflower meal	50	30–44		B35
Safflower seed	45	20–29		B25N
Salicylic acid	29			B25U
Salt, common dry, coarse	40–55		18–22	C36TU
Salt, common dry, fine	70–80	25	11	D26TUW
Salt cake, dry, coarse	85	36	21	B36TW
Salt cake, dry, pulverized	60–85	20–29		B26NT
Saltpeter	80	30–44		A35T
Sand, bank, damp	105–130	45	20–22	B47
Sand, bank, dry	90–110	35	16–18	B37
Sand, core	65	41	26	B35X
Sand, foundry, prepared	80–90	30–44	24	B37
Sand, foundry, shakeout	90–100	39	22	D37
Sand, silica, dry	90–100	20–29	10–15	B27
Sandstone, broken	85–90	30–44		D37
Sawdust	10–13	36	22	*B35
Sesame seed, dry	27–41	20–29		B25N
Sewage (sludge)	40–50	20–29		E25TW
Sewage sludge, dried	45–55	30–44		B36
Sewage sludge, moist	55	30–44		B36
Shale, broken	90–100	20–29		D26QZ
Shale, crushed	85–90	39	22	C36
Shellac	80	45		C45
Shellac, powdered or granulated	31			B35PY
Silica gel (silica acid), dry	45	30–44		C37U

* May vary considerably—consult a CEMA member.

TABLE 3-3 *continued.*

Material	*Average weight (lbs per cu ft)*	*Angle of repose (degrees)*	*Recommended maximum inclination*	*Code*
Sinter	100–135	35		*D37
Slag, blast furnace, crushed	80–90	25	10	A27
Slag, furnace, granular, dry	60–65	25	13–16	C27
Slag, furnace, granular, wet	90–100	45	20–22	B47
Slate, crushed, ½ inch & under	80–90	28	15	C36
Slate, 1½- to 3-inch lumps	85–95			D26
Soap beads or granules	15–25	30–44		C35Q
Soda ash, briquettes	50	22	7	C26
Soda ash, heavy	55–65	32	19	B36
Soda ash, light	20–35	37	22	A36Y
Sodium aluminate, ground	72	30–44		B36
Sodium aluminum sulfate	75	30–44		A36
Sodium antimonate, crushed	49	31		C36
Sodium nitrate	70–80	24	11	*D25
Sodium phosphate	50–65	37		B36
Sodium sulfite, dry	96	45		B45X
Sorghum seed	32–52	30–44		B36
Soybeans, cracked	30–40	35	15–18	C36NW
Soybeans, whole	45–50	21–28	12–16	C27NW
Soybean cake, over ½ inch	40–43	32	17	D35W
Soybean flakes, raw	20–26	30–44		C35Y
Soybean meal, cold	40	32–37	16–20	B35
Soybean meal, hot	40	30–44		B35T
Starch	25–50	24	12	*B25
Steel chips, crushed	100–150	30–44		D37WZ
Steel trimmings	75–150	35	18	E37V
Sugar, raw, cane	55–65	45		B46TX
Sugar, refined, granulated, dry	50–55	30–44		B35PU
Sugar, refined, granulated, wet	55–65	30–44		C35X
Sugar, beet pulp, dry	12–15	20–29		C26
Sugar beet pulp, wet	25–45	20–29		C26X
Sugar cane, knifed	15–18	45		E45V
Sulfate, crushed, ½ inch & under	50–60	30–44	20	C35NS
Sulfate, powdered	50–60	30–44	21	B35NW
Sulfate, 3 inch and under	80–85	30–44	18	D35NS
Sunflower seed	19–38	20		C25
Taconite, pellets	116–130	30–44	13–15	D37Q
Talc, ½-inch screenings	80–90	20–29		C25
Talc, 1½- to 3-inch lumps	85–95	20–29		D25
Timothy seed	36	20–29		B25NY
Titanium dioxide	140	30–44		B36
Titanium sponge	60–70	45		E47
Tobacco leaves, dry	12–14	45		E45QV
Tobacco scraps	15–25	45		D45Y
Tobacco stems	15	45		E45Y
Traprock, ½-inch screenings	90–100	30–44		C37

* May vary considerably—consult a CEMA member.

Table 3-3 ***continued.***

Material	Average weight (lbs per cu ft)	Angle of repose (degrees)	Recommended maximum inclination	Code
Traprock, 2- to 3-inch lumps	100–110	30–44		D37
Tricalcium phosphate	21–50	45 +		A45
Trisodium phosphate	60			D36
Trisodium phosphate, granular	60	30–44	11	B35
Trisodium phosphate, pulverized	50	40	25	B35
* Urea prills, dry	43–46	25		B25
Vermiculite, expanded	16	45		C45Y
Vermiculite ore	70–80		20	D36Y
Walnut shells, crushed	35–45	30–44		B37
Wheat	45–48	28	12	C25N
Wheat, cracked	35–45	30–44		B35N
Wheat germ, dry	18–28	20–29		B25
White lead	75–100	30–44		A36MR
Wood chips	10–30	45	27	E45WY
Wood chips, hogged, fuel	15–25	45		D45
Wood shavings	8–15			E45V
Zinc concentrates	75–80			B26
Zinc ore, crushed	160	38	22	*
Zinc ore, roasted	110	38		C36
Zinc oxide, heavy	30–35	45–55		A45X
Zinc oxide, light	10–15	45		A45XY

* May vary considerably—consult a CEMA member.

Chapter 4

Capacities, Belt Widths, and Speeds

Contents

Belt Widths

The width of conveyor belts customarily is expressed in inches. The belt widths treated in this manual—and which are available from conveyor belt manufacturers in the United States—are as follows: 18, 24, 30, 36, 42, 48, 54, 60, 72, 84, and 96 inches.

Generally, for a given speed, the belt width and the belt conveyor capacity increase together. However, the width of the narrower belts may be governed by the size of lumps to be handled. Belts must be wide enough so that any combination of prevailing lumps and finer material does not load the lumps too close to the edge of the conveyor belt. Also, the inside dimensions of loading chutes and the distance between skirtboards must be sufficient to pass various combinations of lumps without jamming.

Lump Size Considerations

The lump size influences the belt specifications and the choice of carrying idlers. There is also an empirical relationship between lump size and belt width.

The recommended maximum lump size for various belt widths is as follows: For a 20° surcharge, with 10% lumps and 90% fines, the recommended maximum lump size is 1/3 the belt width ($b/3$). With all lumps and no fines, the recommended maximum lump size is 1/5 the belt width ($b/5$).

For a 30° surcharge, with 10% lumps and 90% fines, the recommended maximum lump size is 1/6 the belt width ($b/6$). With all lumps and no fines, the recommended maximum lump size is 1/10 the belt width ($b/10$).

Another way to determine belt width for a specific lump size is illustrated in Figure 4.1. This simple chart shows the belt width necessary for a given size lump, for various proportions of lumps and fines, and for various surcharge loadings.

Belt Speeds

Suitable belt conveyor speeds depend largely upon the characteristics of the material to be conveyed, the capacity desired, and the belt tensions employed.

Powdery materials should be conveyed at speeds low enough to minimize dusting, particularly at the loading and discharge points. Fragile

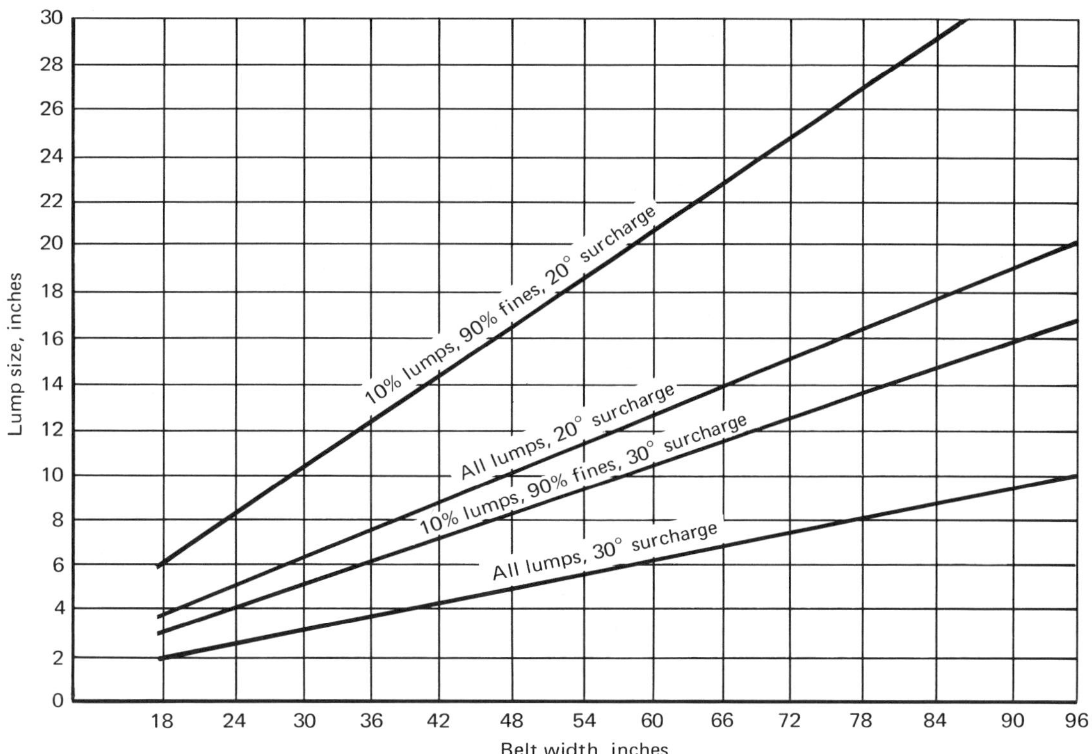

FIGURE 4.1. *Belt width necessary for a given lump size. Fines: no greater than 1/10 maximum lump size.*

materials also limit belt speeds. Low speeds are necessary so that degradation of fragile materials will not occur at the loading and discharge points, as the conveyor belt and the material being carried move over the idlers.

Heavy, sharp-edged materials should be carried at moderate speeds. This is good practice because the sharp edges are likely to wear the belt cover unduly, particularly if the loading velocity of the material in the direction of belt travel is appreciably lower than the belt speed.

General recommendations for *maximum* speeds of belt conveyors are shown in Table 4-1.

Under favorable loading and transfer conditions, for troughed conveyor belts wider than 30 inches, speeds in excess of those specified in Table 4-1 are used for fine materials, damp sand, coal, earth with no large lumps, and crushed stone. An increase in belt speed permits decreases in belt width and tension. However, these benefits must be weighed against the possible disadvantage of increased belt wear, material degradation, windage losses, lump impact on carrying idlers, and generally reduced life of all conveyor components. Consult a CEMA member company when considering operation at these higher speeds.

The design of the loading area and the discharge of the material over the head pulley also must be considered when choosing the belt speed. If the material is dry and fine and the belt velocity is high, dusting of the material may be intolerable. Also, if the material is heavy or contains large lumps, or if the particle edges are angular and sharp, a high velocity of discharge may cause undue wear on the discharge or transfer chutes.

TABLE 4-1. *Recommended Maximum Belt Speeds*

Material being conveyed	*Belt speeds (fpm)*	*Belt width (inches)*
Grain or other free-flowing, nonabrasive material	500 700 800 1000	18 24-30 36-42 48-96
Coal, damp clay, soft ores, overburden and earth, fine-crushed stone	400 600 800 1000	18 24-36 42-60 72-96
Heavy, hard, sharp-edged ore, coarse-crushed stone	350 500 600	18 24-36 Over 36
Foundry sand, prepared or damp; shakeout sand with small cores, with or without small castings (not hot enough to harm belting)	350	Any width
Prepared foundry sand and similar damp (or dry abrasive) materials discharged from belt by rubber-edged plows	200	Any width
Nonabrasive materials discharged from belt by means of plows	200, except for wood pulp, where 300 to 400 is preferable	Any width
Feeder belts, flat or troughed, for feeding fine, nonabrasive, or mildly abrasive materials from hoppers and bins	50 to 100	Any width

Belt Conveyor Capacities

For a given speed, belt conveyor capacities increase as the belt width increases. Also, the capacity of a belt conveyor depends on the surcharge angle and on the inclination of the side rolls of three-roll troughing idlers.

The nominal cross section of the material on a belt is measured in a plane normal to the belt. On an inclined or declined conveyor, the material tends to conform to its surcharge angle as measured in a vertical plane. This decreases the area, A_s, as the cosine of the angle of conveyor slope. See Figure 4.2. However, in most cases, the actual loss of capacity is very small.

Assuming a uniform feed to the conveyor, the cross-sectional area of the load on the conveyor belt is the determinant of the belt conveyor capacity. In this manual, the cross-sectional area is based upon the following two conditions. First, the material load on the troughed belt does not extend to the belt edges. The distance from the edges of the material load to the edges of the belt is set at "standard edge distance," which is defined as $0.055b + 0.9$ inch, where b is the width of the belt in inches. Throughout this manual, standard edge distance is presumed to be in effect unless otherwise specified. Second, the top of the load of the material is the arc of a circle tangent, at the edges of the load, to the surcharge angle of loading.

Troughed Belt Load Areas—Standard Edge Distance

Referring to Figure 4.2, the area of load cross section is divided into two parts. One is the trapezoidal area, A_b; the other is the circular segment area, A_s, which is termed the surcharge area. The sum of these two areas ($A_b + A_s$) equals A_t, which is the total cross-sectional area.

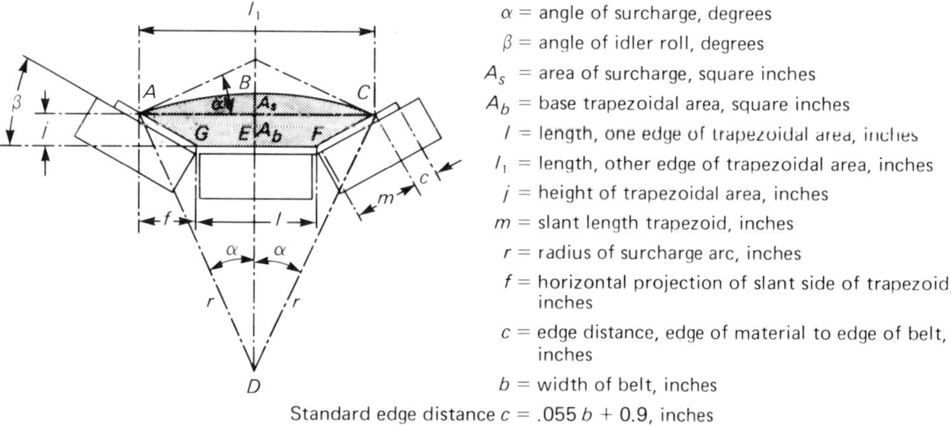

FIGURE 4.2. *Area of load cross section.*

Based on an analysis of the three-equal-roll troughing idlers of eight manufacturers, the length of the flat surface of the center roll averages $0.371b$, where b is the belt width in inches. Graphical full-scale analysis of a 5-ply belt with 1/8-inch and 1/32-inch covers, lying on an average three-equal-roll troughing idler, indicates that the flat distance on the belt-carrying surface over the center idler roll is 1/4-inch greater than center roll length. So:

Trapezoidal area, A_b

1. Area trapezoid ($AECFG$) $A_b = \left(\frac{l + l_1}{2}\right) j$
2. Width belt, $b = l + 2m + 2c$

3. $l_1 = l + 2f$
$f = m \cos \beta$
$l = 0.371b + 0.25$
$c = 0.055b + 0.9$
$b = 0.371b + 0.25 + 2m + 2(0.055b + 0.9)$
$2m = b - 0.481b - 2.05$
$m = 0.2595b - 1.025$
$f = m \cos \beta = (0.2595b - 1.025) \cos \beta$
$2f = 2(0.2595b - 1.025) \cos \beta$
$l_1 = 0.371b + 0.25 + 2(0.2595b - 1.025) \cos \beta$

4. $$\frac{l + l_1}{2} = \frac{0.371b + 0.25 + 0.371b + 0.25 + 2(0.2595b - 1.025) \cos \beta}{2}$$
$$= 0.371b + 0.25 + (0.2595b - 1.025) \cos \beta$$

5. $j = m \sin \beta$
$j = (0.2595b - 1.025) \sin \beta$

6. Area of Trapezoid $A_b = \left(\frac{l + l_1}{2}\right) j$

or
$[0.371b + 0.25 + (0.2595b - 1.025) \cos \beta] \times$
$[(0.2595b - 1.025) \sin \beta]$

Circular segment (surcharge) area, A_s

7. Area whole sector $(ABCD) = \frac{\pi r^2 2\alpha}{360}$

8. Area triangle $(AECD) = \frac{r^2 \sin 2\alpha}{2}$

9. Area segment $(ABCE)$ $A_s = \frac{\pi r^2\, 2\alpha}{360} - \frac{r^2 \sin 2\alpha}{2}$

or
Area $A_s = r^2\left(\frac{\pi\alpha}{180} - \frac{\sin 2\alpha}{2}\right)$

10. $r = \frac{0.5l_1}{\sin \alpha} = \frac{l_1}{2 \sin \alpha}$
$$r = \frac{0.371b + 0.25 + 2(0.2595b - 1.025) \cos \beta}{2 \sin \alpha}$$
$$= \frac{0.1855b + 0.125 + (0.2595b - 1.025) \cos \beta}{\sin \alpha}$$

11. $$A_s = \left(\frac{0.1855b + 0.125 + (0.2595b - 1.025) \cos \beta}{\sin \alpha}\right)^2 \times \left(\frac{\pi\alpha}{180} - \frac{\sin 2\alpha}{2}\right)$$

12. Total Area, A_t (ft²) $= \frac{A_b + A_s}{144}$

Flat Belt Load Areas—Standard Edge Distance

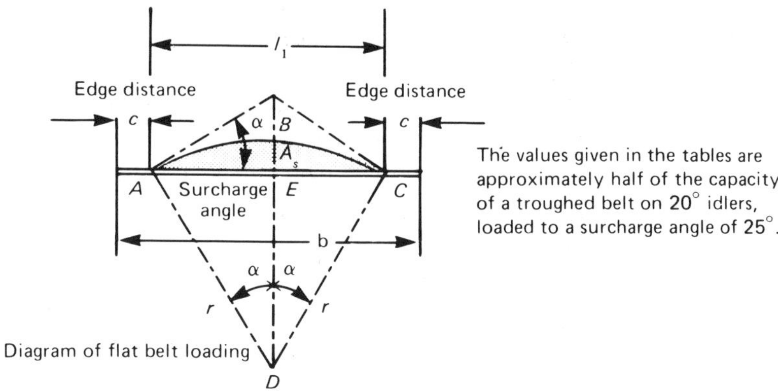

FIGURE 4.3. *Flat belt loading.*

1. Area sector $(ABCD) = \dfrac{2\pi r^2\alpha}{360} = \dfrac{\pi r^2 \alpha}{180}$

   Area triangle $(CDE) = \dfrac{(r\cos\alpha)(r\sin\alpha)}{2} = \dfrac{r^2\sin 2\alpha}{4}$

   Area triangle $(ACD) = \dfrac{2r^2\sin 2\alpha}{4} = \dfrac{r^2\sin 2\alpha}{2}$

   A_s = Area sector $(ABCE) = \dfrac{\pi r^2\alpha}{180} - \dfrac{r^2\sin 2\alpha}{2} = r^2\left(\dfrac{\pi\alpha}{180} - \dfrac{\sin 2\alpha}{2}\right)$

2. $l_1 = 2\,EC = 2\,r\sin\alpha \qquad r = \dfrac{l_1}{2\sin\alpha}$

   For standard edge distance, $c = 0.055b + 0.9$ inch.

   $l_1 = b - 2c = b - 2(.055b + 0.9) = .890b - 1.8$

   $r = \dfrac{.890b - 1.8}{2\sin\alpha} = \dfrac{.445b - 0.9}{\sin\alpha}$

3. $A_s = \left(\dfrac{.445b - 0.9}{\sin\alpha}\right)^2 \times \left(\dfrac{\pi\alpha}{180} - \dfrac{\sin 2\alpha}{2}\right)$

4. For flat belt, Total Area A_t (ft²) $= \dfrac{A_s}{144}$

Belt Conveyor Capacity Tables and Their Use

Troughed and flat belt conveyor capacities are detailed in Tables 4-2 through 4-5. These tables are set up for 20°, 35°, and 45° troughing idler shapes and for flat belts; for various degrees of surcharge angles which correspond to the slumping characteristics of the materials to be conveyed; and for belt speeds of 100 fpm.

To make the best use of these tables, the following eight steps should be taken:

1. Referring to Tables 3-1 and 3-3, determine the surcharge angle of the material. The surcharge angle, on the average, will be 5° to 15° less than the angle of repose.
2. Refer to Table 3-3 to determine the density of the material in pounds per cubic foot (lb/ft³).
3. Choose the idler shape suited to the material and to the conveying problem. Refer to Chapter 5.
4. Refer to Table 4-1, "Recommended Maximum Belt Speeds." Select a suitable conveyor belt speed.
5. Convert the desired tonnage per hour (tph) to be conveyed to the equivalent in cubic ft per hour (ft³/hr).

$$\text{ft}^3/\text{hr} = \frac{\text{tph} \times 2000}{\text{material density}}$$

6. Convert the desired capacity in cubic ft per hour to the equivalent capacity at a belt speed of 100 fpm.

$$\text{Capacity (equivalent)} = (\text{ft}^3/\text{hr}) \times \left(\frac{100}{\text{actual belt speed (fpm)}}\right)$$

7. Using the equivalent capacity so found, refer to Tables 4-2 through 4-5 and find the appropriate belt width.
8. If the material is lumpy, check the selected belt width against the curves in Figure 4.1. The lump size may determine the belt width, in which case the selected belt speed may require revision.

TABLE 4-2. *20° Troughed Belt—Three Equal Rolls Standard Edge Distance = 0.055b + 0.9 Inch*

Belt Width (Inches)	*A_t - Cross Section of Load (Ft^2)*							*Capacity at 100 FPM (Ft^3/Hr)*						
	Surcharge Angle							*Surcharge Angle*						
	0°	5°	10°	15°	20°	25°	30°	0°	5°	10°	15°	20°	25°	30°
18	.089	.108	.128	.147	.167	.188	.209	537	653	769	886	1005	1128	1254
24	.173	.209	.246	.283	.320	.359	.399	1041	1258	1477	1698	1924	2155	2394
30	.284	.343	.402	.462	.522	.585	.649	1708	2060	2414	2772	3137	3511	3897
36	.423	.509	.596	.684	.774	.866	.960	2538	3057	3579	4107	4645	5196	5765
42	.588	.708	.828	.950	1.074	1.201	1.332	3533	4250	4972	5703	6447	7210	7997
48	.781	.940	1.099	1.260	1.424	1.592	1.765	4691	5640	6594	7560	8544	9552	10592
54	1.002	1.204	1.407	1.613	1.822	2.037	2.258	6013	7225	8444	9678	10935	12223	13552
60	1.249	1.501	1.753	2.009	2.270	2.537	2.812	7498	9006	10522	12057	13621	15223	16876
72	1.826	2.192	2.560	2.933	3.312	3.701	4.102	10961	13155	15364	17599	19876	22210	24617
84	2.513	3.014	3.519	4.030	4.551	5.085	5.635	15079	18089	21119	24186	27309	30511	33813
96	3.308	3.967	4.631	5.302	5.986	6.687	7.411	19850	23806	27787	31816	35921	40128	44466

TABLE 4-3. *35° Troughed Belt—Three Equal Rolls Standard Edge Distance = 0.055b + 0.9 Inch*

Belt Width (Inches)	A_t - Cross Section of Load (Ft^2)							Capacity at 100 FPM (Ft^3/Hr)						
	Surcharge Angle							Surcharge Angle						
	0°	5°	10°	15°	20°	25°	30°	0°	5°	10°	15°	20°	25°	30°
18	.144	.160	.177	.194	.212	.230	.248	864	964	1066	1169	1274	1381	1492
24	.278	.309	.341	.373	.406	.440	.474	1668	1857	2048	2241	2438	2640	2847
30	.455	.506	.557	.609	.662	.716	.772	2733	3039	3346	3658	3975	4300	4636
36	.676	.751	.826	.903	.980	1.060	1.142	4058	4508	4961	5419	5886	6364	6857
42	.940	1.044	1.148	1.254	1.361	1.471	1.585	5644	6266	6891	7524	8169	8830	9511
48	1.248	1.385	1.523	1.662	1.804	1.949	2.099	7491	8312	9138	9974	10825	11698	12598
54	1.599	1.774	1.950	2.128	2.309	2.494	2.686	9598	10646	11700	12768	13855	14969	16118
60	1.994	2.211	2.429	2.651	2.876	3.107	3.345	11966	13269	14580	15906	17257	18642	21058
72	2.913	3.229	3.547	3.869	4.197	4.532	4.879	17484	19378	21285	23215	25182	27196	29275
84	4.007	4.440	4.876	5.317	5.766	6.226	6.701	24043	26641	29256	31902	34597	37360	40210
96	5.274	5.842	6.415	6.994	7.584	8.189	8.812	31645	35058	38490	41966	45506	49134	52876

TABLE 4-4. *45° Troughed Belt—Three Equal Rolls Standard Edge Distance = 0.055b + 0.9 Inch*

Belt Width (Inches)	A_t - Cross Section of Load (Ft^2)							Capacity at 100 FPM (Ft^3/Hr)						
	Surcharge Angle							Surcharge Angle						
	0°	5°	10°	15°	20°	25°	30°	0°	5°	10°	15°	20°	25°	30°
18	.170	.184	.199	.214	.230	.245	.262	1021	1109	1198	1289	1380	1475	1572
24	.327	.355	.383	.411	.439	.469	.499	1967	2132	2299	2467	2638	2814	2996
30	.536	.580	.625	.670	.716	.763	.812	3218	3484	3752	4023	4299	4581	4873
36	.795	.860	.926	.992	1.060	1.129	1.200	4775	5165	5558	5955	6360	6775	7204
42	1.106	1.195	1.286	1.377	1.470	1.566	1.664	6636	7175	7717	8265	8824	9397	9987
48	1.467	1.585	1.704	1.825	1.948	2.074	2.204	8803	9514	10229	10953	11690	12445	13224
54	1.879	2.030	2.182	2.336	2.492	2.653	2.819	11276	12182	13094	14017	14957	15921	16915
60	2.342	2.529	2.718	2.909	3.104	3.303	3.509	14053	15179	16312	17458	18626	19823	21059
72	3.420	3.693	3.967	4.245	4.528	4.818	5.117	20524	22160	23807	25473	27171	28910	30705
84	4.702	5.076	5.452	5.832	6.220	6.617	7.027	28216	30458	32713	34997	37322	39706	42165
96	6.188	6.678	7.172	7.671	8.180	8.701	9.239	37128	40071	43032	46029	49081	52210	55437

TABLE 4-5. *Flat Belt Capacity Standard Edge Distance = 0.055b + 0.9 Inch*

Belt Width (Inches)	A_s - Cross Section of Load (Ft^2)							Capacity at 100 FPM (Ft^3/Hr)						
	Surcharge Angle							Surcharge Angle						
	0°	5°	10°	15°	20°	25°	30°	0°	5°	10°	15°	20°	25°	30°
18		.020	.041	.062	.083	.105	.127		123	246	372	498	630	762
24		.039	.077	.117	.157	.198	.241		232	466	702	942	1190	1444
30		.063	.126	.190	.255	.321	.390		376	756	1137	1527	1928	2340
36		.092	.185	.280	.376	.474	.575		555	1113	1677	2253	2844	3450
42		.130	.257	.387	.520	.656	.796		768	1540	2322	3120	3936	4776
48		.169	.340	.512	.688	.868	1.053		1016	2037	3072	4126	5208	6318
54		.216	.434	.654	.879	1.109	1.346		1298	2604	3927	5273	6654	8076
60		.269	.540	.814	1.093	1.380	1.675		1614	3240	4885	6560	8278	10050
72		.392	.786	1.186	1.593	2.010	2.440		2353	4720	7116	9558	12060	14640
84		.538	1.080	1.628	2.186	2.758	3.349		3229	6478	9767	13117	16550	20091
96		.707	1.419	2.139	2.873	3.625	4.400		4243	8514	12835	17238	21750	26404

Chapter 5

Belt Conveyor Idlers

Contents

Idler Requirements

Important requirements for idlers are proper support and protection for the belt and proper support for the load being conveyed.

Belt conveyor idlers for bulk materials are designed to incorporate rolls with various diameters. The rolls are fitted with antifriction bearings and seals, and are mounted on shafts.

Frictional resistance of the idler roll influences belt tension and, consequently, the horsepower requirement. Roll diameter, bearing design, and seal requirements constitute the major components affecting frictional resistance.

This manual does not discuss the relative merits of the various antifriction bearings used, nor the merits of the seals to protect these bearings from dirt and moisture and to retain the lubricant. Each belt conveyor idler manufacturer chooses a particular bearing and seal arrangement. Much ingenuity has been exercised by these idler manufacturers to provide dependable idlers.

Idler Classifications

Selection of the proper roll diameter and size of bearing and shaft is based on the type of service, operating condition, load carried, and belt speed. For ease and accuracy of idler selection, the various idler designs can be grouped into classifications as shown in Table 5-1.

Table 5-1. *Idler Classification*

Classification	*Former series no.*	*Roll diameter (inches)*	*Description*
A4	I	4	Light duty
A5	I	5	" "
B4	II	4	" "
B5	II	5	" "
C4	III	4	Medium duty
C5	III	5	" "
C6	IV	6	" "
D5	NA	5	" "
D6	NA	6	" "
E6	V	6	Heavy duty
E7	VI	7	" "

General Types of Belt Conveyor Idlers

There are two basic types of belt conveyor idlers: carrying idlers, which support the loaded run of the conveyor belt; and return idlers, which support the empty return run of the conveyor belt. See Figures 5.1 through 5.3.

Carrying Idlers. Carrying idlers are of two general configurations. One is used for troughed belts and usually consists of three rolls. The two outer rolls are inclined upward; the center roll is horizontal. The other configuration is used for supporting flat belts. This idler generally consists of a single horizontal roll positioned between brackets which attach directly to the conveyor frame.

Return Idlers. Return idlers usually are horizontal rolls, positioned between brackets which normally are attached to the underside of the support structure on which the carrying idlers are mounted. Two-roll "V" return idlers are also used for better training and higher load ratings.

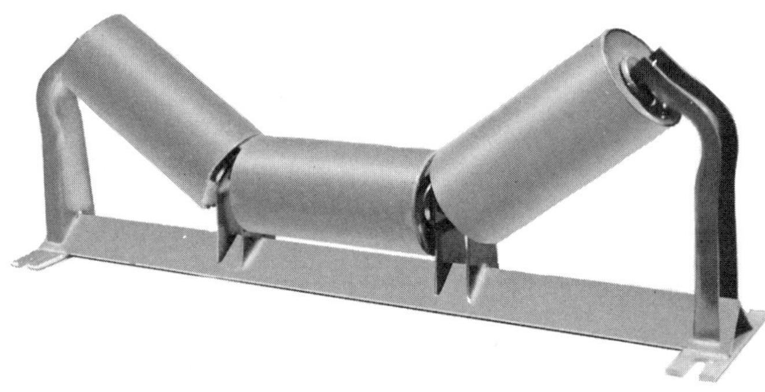

FIGURE 5.1. *35° troughing idler.*

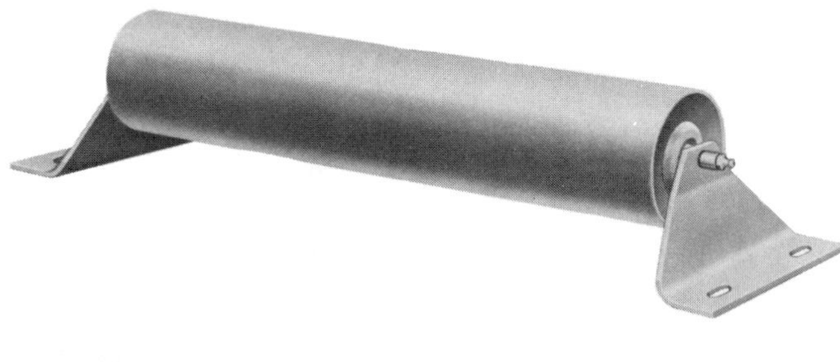

FIGURE 5.2. *Flat belt idler.*

FIGURE 5.3. *Return belt idler.*

Troughing Carrying Idlers

As the capacity tables in Chapter 4 indicate, troughed belts carry far greater tonnages than flat belts, for the same width of belt and belt speed. Troughing carrying idlers are therefore very important components of belt conveyors and belt conveyor systems, warranting detailed discussion.

Idlers with end rolls set at 35° and 45° angles, while affording greater carrying capacity for a given width, necessitate a greater transverse flexibility in the conveyor belt. Such idlers have a shorter history of application than 20° troughing idlers. However, improvements in belt carcass design and materials have contributed to wider acceptance and greater use of 35° and 45° troughing idlers.

Troughing idlers are made in two general styles, in-line and offset. The most commonly used is composed of three in-line rolls of equal length. For a given width of belt, end-roll inclination, and material surcharge angle, the three-equal-length-roll troughing idler forms the belt in the best troughed shape to carry a maximum load cross section.

An offset troughing idler is shown in Figure 5.5. In this idler, the inclined rolls are located in a plane alongside the plane of the horizontal roll. These are popular in the grain industry, where very thin belts are used, and in underground mining, where low head room is a problem. Another in-line troughing idler has a long horizontal roll and two short inclined rolls. While this idler does not form a given belt into a trough for maximum load cross section, it is useful where the load must be spread for manual inspection, picking or sorting. The inclined end rolls turn up the belt edges to prevent or minimize spillage. This is known as a picking belt idler. See Figure 5.6.

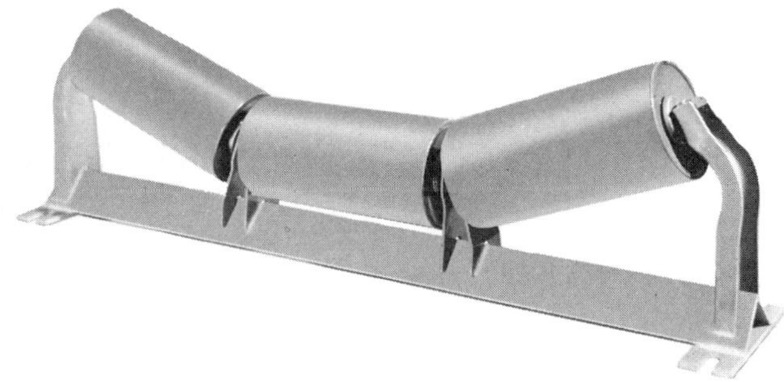

FIGURE 5.4. *20° troughing idler.*

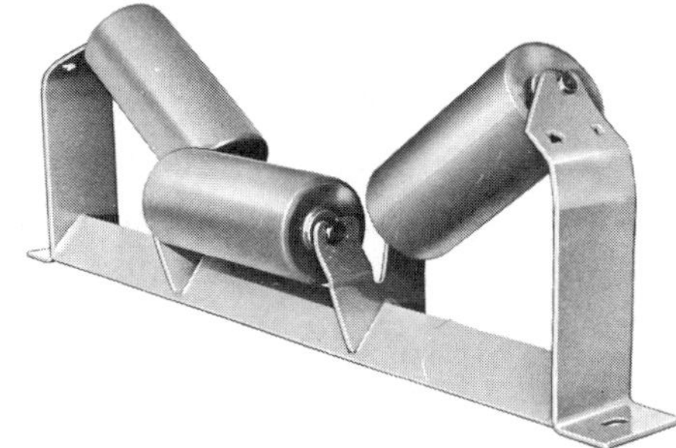

FIGURE 5.5. *35° offset troughing idler.*

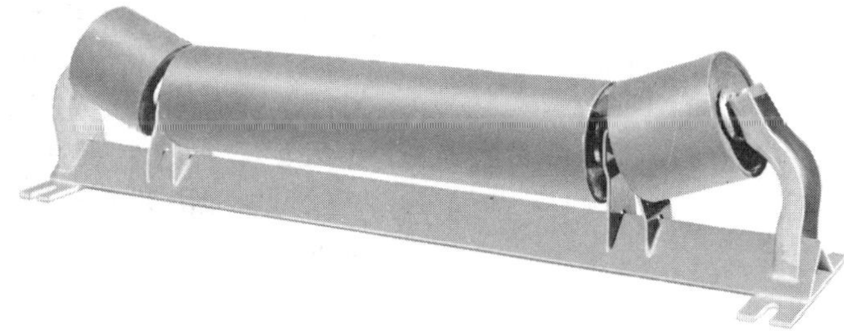

FIGURE 5.6. *20° picking belt idler.*

Impact Idlers

Impact troughing idlers, sometimes referred to as "cushion idlers," have rolls made of a resilient material. They are used at loading points where impact resulting from lump size and weight of the material handled could seriously damage the belt, if it were rigidly supported. One type frequently used consists of a three-roll assembly, each roll being made of spaced, resilient discs. Similar impact idlers are made to support flat belts. See the accompanying illustrations, Figures 5.7 and 5.8. Load ratings of impact idlers are no higher than those of standard idlers. The resilient discs usually are expended in favor of belt protection.

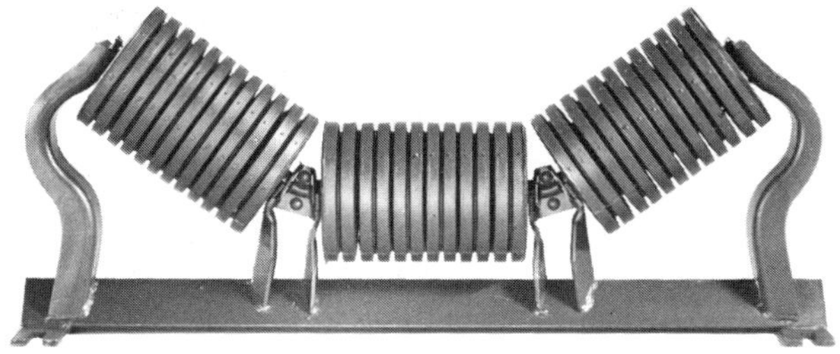

FIGURE 5.7. *35° troughing rubber-cushion impact idler.*

FIGURE 5.8. *Flat-belt rubber-cushion impact idler with fixed shaft.*

Belt Training Idlers, Carrying

Generally speaking, well designed, carefully constructed and maintained belt conveyors will continue to run with proper alignment without the need for special belt training idlers. There are transient conditions, however, that may cause conveyor belts to become misaligned despite all efforts to assure proper installation and maintenance. For this reason, conveyor manufacturers also furnish special belt training idlers, which, if properly maintained, will help to maintain belt alignment in difficult situations.

The usual training idler has the carrying roll frame mounted on a central pivot approximately perpendicular to the conveyor belt. Means are provided to cause the carrying rolls to become skewed with respect to the centerline of the conveyor. As the belt traverses the skewed rolls, they urge the displaced belt to return to the conveyor centerline and, in doing so, the rolls are urged to return to proper alignment. See Figure 5.9.

Fixed guide rolls placed perpendicular to the edge of the conveyor belt are not recommended because continuous contact with the conveyor belt edge accelerates belt edge wear, appreciably reducing belt life. See Figure 5.10.

In general, the greater the belt tensions, the less effective the training idlers.

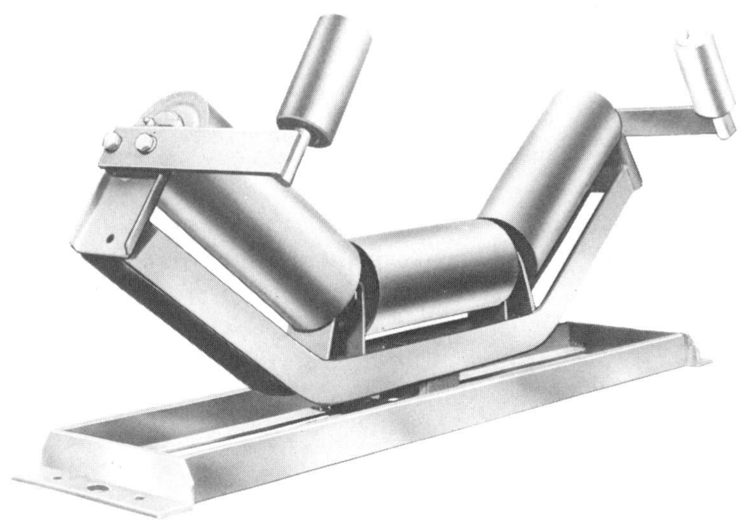

FIGURE 5.9. *35° troughed belt training idler.*

FIGURE 5.10. *Typical fixed guide idler, showing side guide rollers.*

Suspended Idlers

In this type of idler, the rolls (usually 3 or 5) are linked together and suspended from the conveyor frame stringers. This suspended or hanging idler assumes a shape somewhat like that of a catenary and is thus sometimes called a catenary, or garland, idler. It is popular for heavy-duty earth- and ore-moving conveyors. Three-roll idlers are usually used for the carrying run and five-roll idlers are sometimes used at loading points. Resilient discs are not recommended on these idlers at loading points. See Figure 5.11.

The suspended idler is tolerant of both poor alignment and abuse by lumps because of the flexible connection between rolls. It can be furnished with a quick-release type of mounting which permits the entire idler to be

FIGURE 5.11. *Suspended three-roll idler.*

quickly lowered and removed from service (in case of roll failure). Training idlers are generally not used with suspended idlers.

These idlers can be mounted on (suspended from) either rigid frame stringers (channels) or wire-rope stringers.

Return Idlers

These idlers are used to support the return run of the belt. They usually are suspended below the lower flanges of the stringers which support the carrying idlers. It is preferable that return idlers be so mounted that the return run of the belt is visible below the conveyor frame. Figure 5.12 illustrates a typical return idler.

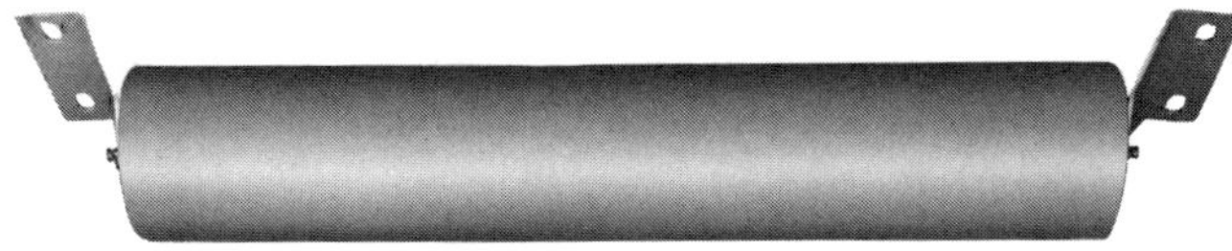

FIGURE 5.12. *Return belt idler.*

Flat Return Idlers. The flat return idler consists of a long single roll, fitted at each end with a mounting bracket. Idler roll length, bracket design, and mounting-hole spacing should allow for adequate transverse belt movement without permitting the belt edges to contact any stationary part of the conveyor or its frame.

Self-Cleaning Return Idlers. An important consideration with return idler applications is the adherence of materials to the carrying surface of the belt. Such material may be abrasive and wear the shell of the return idler rolls. Or, this buildup may be sticky and adhere to the return idler rolls. A large buildup may cause misalignment of the return run of the belt.

Several styles of return idler rolls are available to overcome these difficulties. When sticky materials are a problem, rubber-disc, or rubber-coated helically shaped, self-cleaning return idlers can be used. Disc and helical rolls present very narrow surfaces for adhesion and thus reduce the tendency for material build up. This type of return idler sometimes is erroneously called a "belt cleaning idler." Even though such idlers do "track-off" material adhering to the belt surface on the return run, they do not constitute belt cleaning devices. See Figures 5.13 and 5.14.

FIGURE 5.13. *Rubber-disc return idler.*

FIGURE 5.14. *Helical or spiral self-cleaning return idler.*

On short conveyors, it may be necessary to equip the complete return run with self-cleaning idlers. On long return belt runs, it is necessary to use these idlers only to the point where the material on the belt surface no longer will adhere to and build up on normal return idler rolls. Beyond this point, standard return idlers can be used.

Return Belt Training Idlers. Return belt idlers can be pivotally mounted to train or align the return belt in a manner similar to the training idlers previously described for the carrying run of the belt. See Figure 5.15.

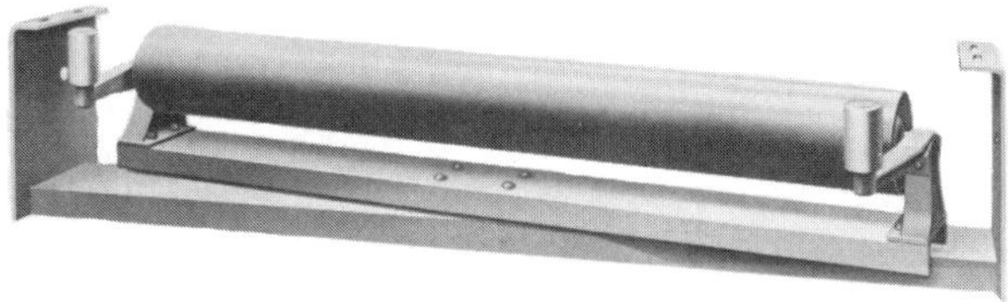

FIGURE 5.15. *Return training idler.*

Two-Roll "V" Return Idlers

With the increased use of heavy, high-tension fabric and steel cable belts, the need for better support and training has resulted in the development of "V" return idlers. A basic "V" return idler consists of two rolls, each tilted at approximately a 10° to 15° angle. These rolls are either of the Garland (suspended) or rigid design. See Figures 5.16 and 5.17. The "V" return idler has some training effect on the belt, while allowing greater idler spacing because of its increased load rating.

"V" return idlers can be supplied in either steel-roll or rubber-disc designs. Field experience has shown that the steel roll is preferable because discs tend to wear out rapidly.

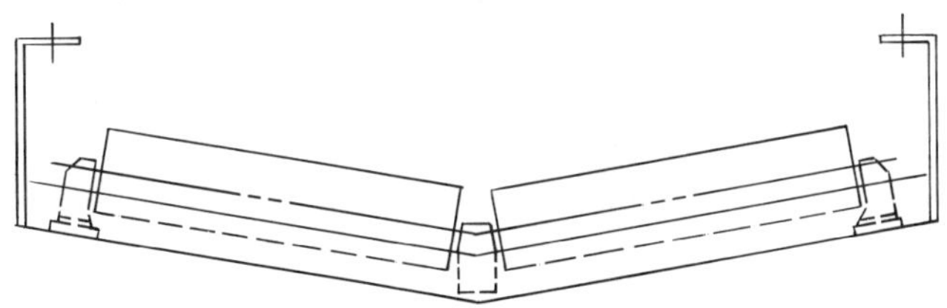

FIGURE 5.16. *Two-roll "V" return idler, rigid design.*

FIGURE 5.17. *Two-roll suspended "V" return idler.*

Idler Spacing

Factors to consider when selecting idler spacing are belt weight, material weight, idler rating, sag, idler life, belt rating, and belt tension.

If too much sag of a loaded troughed belt is permitted between the troughing idlers, the material may spill over the edges of the belt. For the best design, and especially on long-center troughed belt conveyors, the sag between idlers should be limited as described in Chapter 6. Table 5-2 lists

TABLE 5-2. *Suggested Normal Spacing of Belt Idlers (S_i)**

Belt Width (Inches)	*Troughing idlers Weight of material handled, lbs per cu ft*						*Return Idlers*
	30	50	75	100	150	200	
18	5.5 ft	5.0 ft	5.0 ft	5.0 ft	4.5 ft	4.5 ft	10.0 ft
24	5.0 ft	4.5 ft	4.5 ft	4.0 ft	4.0 ft	4.0 ft	10.0 ft
30	5.0 ft	4.5 ft	4.5 ft	4.0 ft	4.0 ft	4.0 ft	10.0 ft
36	5.0 ft	4.5 ft	4.0 ft	4.0 ft	3.5 ft	3.5 ft	10.0 ft
42	4.5 ft	4.5 ft	4.0 ft	3.5 ft	3.0 ft	3.0 ft	10.0 ft
48	4.5 ft	4.0 ft	4.0 ft	3.5 ft	3.0 ft	3.0 ft	10.0 ft
54	4.5 ft	4.0 ft	3.5 ft	3.5 ft	3.0 ft	3.0 ft	10.0 ft
60	4.0 ft	4.0 ft	3.5 ft	3.0 ft	3.0 ft	3.0 ft	10.0 ft
72	4.0 ft	3.5 ft	3.5 ft	3.0 ft	2.5 ft	2.5 ft	8.0 ft
84	3.5 ft	3.5 ft	3.0 ft	2.5 ft	2.5 ft	2.0 ft	8.0 ft
96	3.5 ft	3.5 ft	3.0 ft	2.5 ft	2.0 ft	2.0 ft	8.0 ft

* Spacing may be limited by load rating of idler. See idler load ratings in Tables 5-8—5-12.

suggested normal troughing idler spacing for use in general engineering practice, when the amount of belt sag is not specifically limited. These figures on spacing should be used in conjunction with the information on sag selection in Chapter 6. Spacing is normally varied in 6-inch increments.

Some conveyor systems have been designed successfully utilizing extended idler spacing and/or graduated idler spacing. Extended idler spacing is simply greater than normal spacing. This is sometimes applied where belt tension, sag, belting strength, and idler rating permit. Advantages may be lower idler cost (fewer used) and better belt training.

Graduated idler spacing is greater than normal spacing at high-tension portions of the belt. As the tension along the belt increases, the idler spacing is increased. Usually this type of spacing occurs toward and near the discharge end.

Extended and graduated spacing are not commonly used but if either is employed, care should be taken not to exceed idler rating and sag limits during starting and stopping.

Return Idler Spacing

The suggested normal spacing of return idlers for general belt conveyor work is also given in Table 5-2. For conveyor belts with heavy carcasses, and with a width of 48 inches or more, it is recommended that the return idler spacing be determined by the use of the idler load ratings and by sag considerations.

Carrying Idler Spacing at Loading Points

At loading points, the carrying idlers should be spaced to keep the belt steady and to hold the belt in contact with the rubber edging of the loading skirts along its entire length. Careful attention to the spacing of the carrying idlers at the loading points will minimize material leakage under the skirtboards and, at the same time, will also minimize wear on the belt cover.

Normally, carrying idlers in the loading zone are spaced at half (or less) the normal spacing suggested in Table 5-2. *Caution:* If impact idlers are used at loading zones, impact idler ratings are no higher than standard idler ratings.

Good practice dictates that the spacing of idler rolls under the loading area be such that the major portion of the load engages the belt between idlers.

Troughing Idler Spacing Adjacent to Terminal Pulleys

In passing from the last troughing idler to the terminal pulley, the belt edges are stretched and tension is increased at the outer edges. If the belt edge stress exceeds the elastic limit of the carcass, the belt edge will be stretched permanently and will cause belt training difficulties. On the other hand, if the troughing idlers are placed too far from the terminal pulleys, spillage of the load is likely.

Distance is important in the change (transition) from troughed to flat form. This is especially significant when deeply troughed idlers are used.

Depending on the transition distance, one, two, or more transition-type troughing idlers can be used to support the belt between the last standard troughing idler and the terminal pulley. These idlers can be positioned either at a fixed angle or at an adjustable concentrating angle. Table 5-3 shows recommended transition distances for various troughing angles, belt tensions, and types of belting. In no case should the rating of the idler be exceeded.

TABLE 5-3. ***Recommended Minimum Transition Distances***

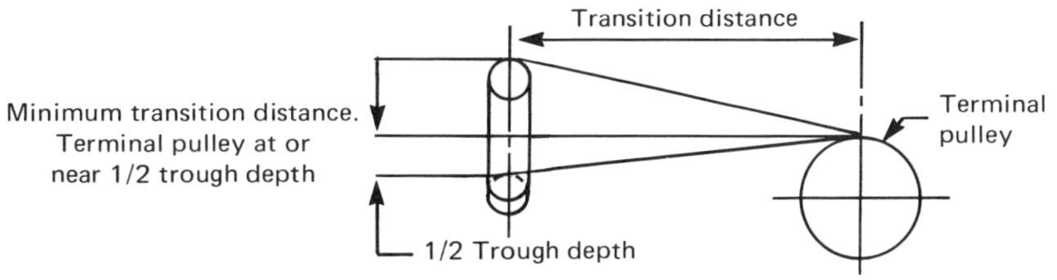

Idler Angle	*% Rated Tension*	*Fabric Belts*	*Steel Cable Belts*
20°	Over 90	.9*b*	2.0*b*
	60 to 90	.8*b*	1.6*b*
	Less than 60	.6*b*	1.0*b*
35°	Over 90	1.6*b*	3.4*b*
	60 to 90	1.3*b*	2.6*b*
	Less than 60	1.0*b*	1.8*b*
45°	Over 90	2.0*b*	4.0*b*
	60 to 90	1.6*b*	3.2*b*
	Less than 60	1.3*b*	2.3*b*

b = Belt width (transition distance will be in the same units as those used for *b*).

TABLE 5-3. *continued. Recommended Minimum Transition Distances*

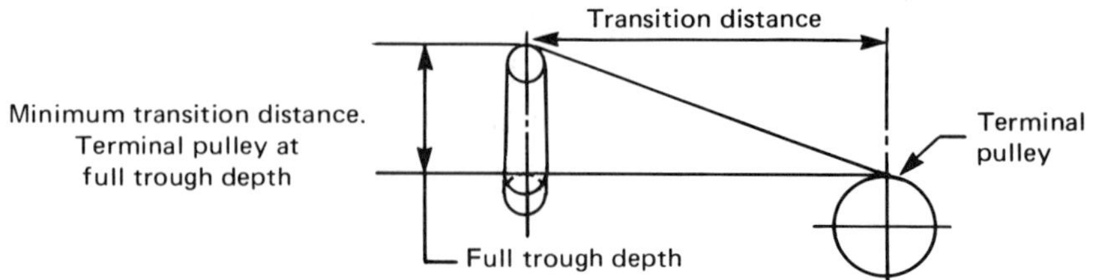

Idler Angle	*% Rated Tension*	*Fabric Belts*	*Steel Cable Belts*
20°	Over 90	1.8*b*	4.0*b*
	60 to 90	1.6*b*	3.2*b*
	Less than 60	1.2*b*	2.8*b*
35°	Over 90	3.2*b*	6.8*b*
	60 to 90	2.4*b*	5.2*b*
	Less than 60	1.8*b*	3.6*b*
45°	Over 90	4.0*b*	8.0*b*
	60 to 90	3.2*b*	6.4*b*
	Less than 60	2.4*b*	4.4*b*

b = Belt width (transition distance will be in the same units as those used for b).

Idler Selection

After the belt width and speed have been determined, it is necessary to select the proper idler classification. This selection is governed by three known conditions: the type of service, the characteristics of the material to be handled, and the belt speed.

Type of Service

A most important consideration is the operating condition under which the idlers will be used. This includes the hours of operation per day, the overall

TABLE 5-4. *K_1 Lump Adjustment Factor*

Maximum Lump Size (Inches)	*Material Weight, lbs/cu. ft. K_1 = 1.1*						
	50	75	100	125	150	175	200
4	1.0	1.0	1.0	1.0	1.1	1.1	1.1
6	1.0	1.0	1.0	1.1	1.1	1.1	1.1
8	1.0	1.0	1.1	1.1	1.1	1.2	1.2
10	1.0	1.1	1.1	1.2	1.2	1.2	1.2
12	1.0	1.1	1.1	1.2	1.2	1.2	1.3
14	1.1	1.1	1.1	1.2	1.2	1.3	1.3
16	1.1	1.1	1.2	1.2	1.3	1.3	1.4
18	1.1	1.1	1.2	1.2	1.3	1.3	1.4

TABLE 5-5 K_2. *Environmental and Maintenance Factors*

Environmental Conditions	*Maintenance*		
	Good	*Fair*	*Poor*
Clean	1.00	1.08	1.11
Moderate	1.06	1.10	1.13
Dirty	1.09	1.12	1.15

TABLE 5-6 K_3. *Service Factor*

Operation	*Factor*
Less than 6 hours per day	0.8
6 to 9 hours per day	1.0
10 to 16 hours per day	1.1
Over 16 hours per day	1.2

TABLE 5-7. K_4 *Belt Speed Correction Factor*

Belt Speed (fpm)	*Roll Diameter, Inches*			
	4	5	6	7
100	0.80	0.80	0.80	0.80
200	0.83	0.80	0.80	0.80
300	0.90	0.85	0.83	0.81
400	0.95	0.91	0.88	0.85
500	0.99	0.95	0.92	0.88
600	1.03	0.98	0.95	0.92
700	1.05	1.01	0.98	0.95
800	—	1.04	1.00	0.97
900	—	1.06	1.03	1.00
1000	—	—	1.05	1.02

life expectancy of the conveyor system and the environment in which the idlers will operate. Lists of service factors based on collective field experience are given in Tables 5-5 and 5-6.

Type of Material Handled

The characteristics of the material to be handled have a direct bearing on the idler selection. The weight of the material governs the idler load and spacing, and lump size modifies the effect of weight by introducing an impact factor. Table 5-4 combines the unit weight and the lump size in a group of empirical factors. Note that, in the table, "lump size" means the largest lump which may occasionally be carried rather than the "average" lump.

The proper selection of return belt idlers is just as important as the selection of carrying idlers. In fact, operating conditions are often more severe for the return idler. The return belt idler contacts the "dirty" side of the belt, resulting in abrasive wear of the idler roll surface. Materials build up on the roll and increase its effective diameter. Because the build up is never uniform, and usually is less at the belt edges, the clean sections of the return roll travel at a slower surface speed than that of the belt. This results in relative slippage, thereby accelerating wear of both the belt cover and the surface of the roll. Thus, the life of the roll shell is usually shorter on return belt idlers than on carrying idlers.

In the selection of return belt idlers, where the only "material" handled is the belt itself, unit weight for the belt, W_b, must be ascertained. This can be estimated accurately enough by referring to tables on belt weights.

Another point to consider is the fact that manufacturers of idlers customarily supply larger rolls with thicker metal walls. These rolls have a potentially longer life than do the smaller rolls with thinner walls.

For the most severe abrasive conditions, covered idler rolls will give longer wear life. A rubber covering can yield a life about four to eight times the life of a steel roll of the same outside diameter. Special materials other than rubber can also be used.

When an idler is subjected to corrosive material (e.g., salt or chemicals), special care is required in the idler selection. The roll can be made of rubber-covered steel or steel covered with any durable, corrosion-resisting material. Normally, the brackets and frame also must be coated with an appropriate resistant material. For such special applications, the conveyor manufacturer should be consulted.

Idler Selection Procedure

To select the proper classification of idler, it is first necessary to determine the idler adjusted load. This is the load handled by the idler but with multiplying factors for lump size, environment, and service factors.

$$\text{Actual Idler Load} = IL = (Wb + Wm)\,Si$$
$$\text{Adjusted Load} = AL$$

Where
- AL = $(IL \times K_1 \times K_2 \times K_3 \times K_4) + IML$
- Wb = Belt Weight (lbs per ft) (See Table 6-1)
- Wm = Material Weight (lbs per ft) (See page 79)
- Si = Idler Spacing (ft) (See Table 5-2)
- K_1 = Lump Adjustment Factor (See Table 5-4)
- K_2 = Environmental and Maintenance Factor (see Table 5-5)
- K_3 = Service Factor (see Table 5-6)
- K_4 = Belt Speed Correction Factor (see Table 5-7)
- IML = Force Due to Idler Height Deviation

Compute adjusted load, AL, from above information. If AL is less than actual idler load, IL, let AL equal IL. Do not use a value for adjusted load that is less than actual idler load.

Using AL, select the proper idler from the idler load rating Tables 5-8 through 5-12.

If the product of $K_1 \times K_2 \times K_3 \times K_4$ is less than 1, a value of 1 should be used.

Forces Due to Idler Height Deviation

When an idler is higher than adjacent idlers, a component of belt tension will add load to that idler. The amount of height deviation can vary with the installation and type of idler. These additional forces should be considered in the idler selection. Accuracy of the installation will determine the magnitude of these forces.

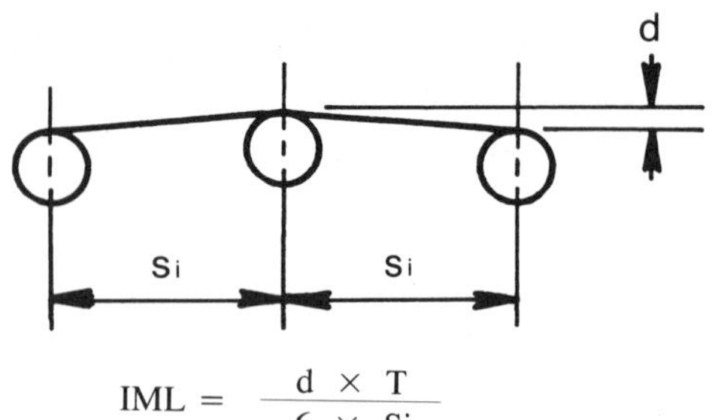

$$IML = \frac{d \times T}{6 \times Si}$$

IML = Idler Misalignment Load (lbs) Due to Belt Tension Only

T = Belt Tension (lbs)

d = Misalignment (inches)

Si = Idler Spacing (feet)

IML does not include material or belt weight.

Table 5-8. *Load Ratings for CEMA A Idlers, lbs*

Belt Width (inches)	*Trough Angle*			
	20°	35°	45°	Return
18	300	300	300	150
24	300	300	289	125
30	300	280	270	100
36	275	256	248	75

Table 5-9. *Load Ratings for CEMA B Idlers, lbs*

Belt Width (inches)	*Trough Angle*			
	20°	35°	45°	Return
18	410	410	410	220
24	410	410	410	190
30	410	410	410	165
36	410	410	396	155
42	390	363	351	140
48	380	353	342	130

Table 5-10. *Load Ratings for CEMA C Idlers, lbs*

Belt Width (inches)	*Trough Angle*			
	20°	35°	45°	Return
18	900	900	900	475
24	900	900	900	325
30	900	900	900	250
36	900	837	810	200
42	850	791	765	150
48	800	744	720	125
54	750	698	675	*
60	700	650	630	*

* Use CEMA D return idlers

Table 5-11. *Load Ratings for CEMA D Idlers, lbs.*

Belt Width (inches)	*Trough Angle*			
	20°	35°	45°	Return
24	1200	1200	1200	600
30	1200	1200	1200	600
36	1200	1200	1200	600
42	1200	1200	1200	500
48	1200	1200	1200	425
54	1200	1116	1080	375
60	1150	1070	1035	280
72	1050	977	945	155

Table 5-12. *Load Ratings for CEMA E Idlers, lbs.*

Belt Width (inches)	*Trough Angle*			
	20°	35°	45°	Return
36	1800	1800	1800	1000
42	1800	1800	1800	1000
48	1800	1800	1800	1000
54	1800	1800	1800	925
60	1800	1800	1800	850
72	1800	1800	1800	700
84	1800	1674	1620	550
96	1750	1628	1575	400

Idler Rating

Idler life is determined by a combination of many factors, such as seals, bearings, shell thickness, maintenance, environment, and load density. While bearing life is often used as an indicator of idler life, it must be recognized that the effect of other variables (e.g., seal effectiveness) may be more important in determining idler life than the bearings. Nevertheless, since bearing rating is the only variable for which laboratory tests have provided standard values, CEMA uses this factor as a guide for establishing idler rating.

The term useful life (*BU*) will be used instead of the common terms for bearing life, *B*-10 or *L*-10 *BU* is longer than *B*-10 or *L*-10, and represents the statistical point in hours where a minimum of 90% of the bearings will still be functional with no increase in torque or noise. *L*-10 represents the statistical point in hours where a minimum of 90% of the bearings will complete or exceed before fatigue failure. Fatigue failure is reached when pitting or spalling of the rolling element or race way reaches .01 sq. in. Useful life (*BU*) is three times *L*-10.

Tables 5-8 through 5-12 show the load ratings for CEMA equal-length-roll idlers A, B, C, D, and E. These ratings are based on 90,000 hours minimum *BU* bearing life at 500 rpm. Note that these load ratings are minimum ratings for CEMA-rated idlers. Actual figures on load ratings supplied by manufacturers may be higher.

Belt Alignment

A belt conveyor must be designed, constructed, and maintained so that the belt consistently runs centrally on its mechanical system of idlers and pulleys. To accomplish this, the following conditions must prevail:

1. All idlers must be in line, square, and level transversely.
2. All pulleys must be in line, with the pulley shafts parallel to one another and at 90° to the center line of the belt.
3. The material must be loaded centrally on the belt.
4. The belt must be straight and properly spliced.
5. The supporting structure must be straight and level transversely.

If, after these conditions have been met, the conveyor belt persistently runs to one side, certain corrective measures can make the belt run centrally. Some of the idlers can be skewed so that the horizontal roll of the idler is at a slight angle to the center line of the belt. The brackets which support the idler roll assembly generally have slotted holes so that such movement of the idler is possible. (*Caution:* This does not apply to reversible belt conveyors.)

Some or all of the troughing idlers should be tilted not more than 2° from the vertical, in the direction of the belt travel. (*Caution:* Troughing idlers which have this tilt built in should not be additionally tilted. Also, in the case of reversible belt conveyors, idlers should not be tilted, as the misalignment of the belt would be accentuated when it runs in the reverse direction.)

Training idlers can be installed to replace either troughing or return idlers without difficulty. They should be used only in an area where other corrective measures are not adequate and should normally be at least 50 feet away from terminals or bend pulleys. Training idlers should not be used on vertical curves where the radius of the curve is less than 800 feet.

Belt training idlers must be selected and installed correctly for the direction of belt travel. Reversible belt training idlers are available for reversible belt conveyors. The aligning action of belt training idlers depends upon free movement of the conveyor belt and of the training idler, so proper cleanliness and maintenance are essential for satisfactory results.

Table 5-13. *Average Weight (lbs) of Troughing Idler Rotating Parts—Steel Rolls*

Belt Width (inches)	*CEMA Idler Class*										
	A4	A5	B4	B5	C4	C5	C6	D5	D6	E6	E7
18	12.7	16.2	15.0	19.2	14.5	19.1	26.7				
24	15.8	21.2	18.3	24.2	17.5	23.2	32.6	23.2	32.6		
30	18.9	25.0	21.8	28.3	20.5	26.8	38.0	26.8	38.0		
36	22.0	28.6	25.3	33.0	23.5	31.3	43.6	31.3	43.6	64.8	81.8
42			30.8	38.1	26.5	35.2	49.2	35.2	49.2	73.3	91.7
48			32.9	41.6	29.5	39.3	54.8	39.3	54.8	81.9	101.3
54						45.9	62.3	45.9	62.3	93.6	121.8
60						50.1	68.3	50.1	68.3	102.2	132.7
72								57.9	77.8	119.4	154.5
84										132.0	164.0
96										145.3	173.0

Table 5-14. *Average Weight (lbs) of Return Idler Rotating Parts—Steel Rolls*

Belt Width (inches)	*CEMA Idler Class*										
	A4	A5	B4	B5	C4	C5	C6	D5	D6	E6	E7
18	11.9	15.5	13.1	16.3	12.2	16.6	21.6				
24	15.6	19.2	16.3	20.9	15.2	20.1	27.1	20.9	30.1		
30	18.5	23.2	19.5	24.5	18.2	24.0	32.3	25.8	35.4		
36	21.9	27.1	22.7	28.5	21.2	28.0	37.6	30.1	40.5	59.0	70.0
42			26.0	33.0	24.6	32.1	43.3	34.3	47.2	67.4	80.1
48			27.4	36.1	27.6	36.1	48.4	38.7	54.4	75.6	89.9
54								43.4	60.8	83.2	99.9
60								49.2	68.1	92.2	109.4
72								55.1	74.9	109.4	129.0
84										114.0	136.2
96										122.0	149.8

Table 5-15. *WK^2 (Lb.-In.^{2}) Average for Three-Equal-Roll Troughing Idlers*

Belt Width (inches)	*CEMA Idler Class*										
	A4	A5	B4	B5	C4	C5	C6	D5	D6	E6	E7
18	39	83	43	90	45	90	179				
24	49	106	53	116	56	116	224	116	224		
30	60	128	65	140	67	140	269	140	269		
36	71	149	78	161	79	161	313	161	313	446	801
42			90	185	91	185	358	185	358	502	876
48			104	207	105	207	400	207	400	564	1017
54						223	447	223	447	628	1127
60						245	493	245	493	689	1234
72								266	538	811	1451
84										985	1598
96										1114	1804

Table 5-16. *WK^2 (Lb.-In.^{2}) Average for Single Steel Return Idlers*

Belt Width (inches)	*CEMA Idler Class*										
	A4	A5	B4	B5	C4	C5	C6	D5	D6	E6	E7
18	39	84	41	85	42	85	165				
24	49	105	50	106	51	106	210	106	210		
30	60	127	61	128	61	128	254	128	254		
36	70	151	72	152	72	152	312	152	312	419	750
42			83	174	83	174	348	174	348	479	857
48			94	196	94	196	385	196	385	539	963
54								218	429	597	1070
60								234	473	659	1177
72								256	513	779	1391
84										933	1581
96										1074	1737

Chapter 6

Belt Tension, Power, And Drive Engineering

Contents

Basic power requirements.

Belt tension calculations.

CEMA horsepower formula.

Drive pulley relationships.

Drive arrangements.

Maximum and minimum belt tensions.

Tension relationships and belt sag between idlers.

Acceleration and deceleration forces.

Analysis of acceleration and deceleration forces.

Design considerations.

Conveyor horsepower determination—graphical method.

Examples of belt tension and horsepower calculations—6 problems.

Belt conveyor drive equipment.

Backstops.

Brakes.

Brakes and backstops in combination.

Devices for acceleration, deceleration, and torque control.

Brake requirement determination (deceleration calculations).

The earliest application engineering of belt conveyors was, to a considerable extent, dependent upon empirical solutions that had been developed by various manufacturers and consultants in this field. The belt conveyor engineering analysis, information, and formulae presented in this manual represent recent improvements in the concepts and data which have been developed over the years, using the observations of actual belt conveyor operation and the best mathematical theory.

Horsepower and tension formulae, incorporating successively all the factors affecting the total force needed to move the belt and its load, are presented here in a manner which permits the separate evaluation of the effect of each factor. These formulae represent the consensus of all CEMA member companies.

In recent years, CEMA member companies have developed computer programs capable of complete engineering analysis of the most complex and extensive belt conveyor systems. These programs are more comprehensive and include more extensive analysis and calculations than can be included in this manual. Although the programs are treated as proprietary information, each CEMA member company welcomes an opportunity to assist in the proper application of belt conveyor equipment. One advantage of using computer programs is the speed and accuracy with which they provide information for alternate conveyor designs.

Basic Power Requirements

The horsepower, hp, required at the drive of a belt conveyor, is derived from the pounds of the effective tension, T_e, required at the drive pulley to propel or restrain the loaded conveyor at the design velocity of the belt V, in fpm:

$$\text{hp} = \frac{T_e \times V}{33{,}000} \qquad \textbf{(1)}$$

To determine the effective tension, T_e, it is necessary to identify and evaluate each of the individual forces acting on the conveyor belt and contributing to the tension required to drive the belt at the driving pulley. T_e is the final summarization of the belt tensions produced by forces such as:

1. The gravitational load to lift or lower the material being transported.
2. The frictional resistance of the conveyor components, drive, and all accessories while operating at design capacity.
3. The frictional resistance of the material as it is being conveyed.
4. The force required to accelerate the material continuously as it is fed onto the conveyor by a chute or a feeder.

The basic formula for calculating the effective tension, T_e, is:

$$T_e = LK_t(K_x + K_yW_b + 0.015\ W_b) + W_m(LK_y \pm H) + T_p + T_{am} + T_{ac} \quad \textbf{(2)}$$

Belt Tension Calculations

The following symbols will be used to assist in the identification and evaluation of the individual forces that cumulatively contribute to T_e and that are therefore components of the total propelling belt tension required at the drive pulley:

A_i = belt tension, or force, required to overcome frictional resistance and rotate idlers, lbs (see page 81)

C_1 = friction modification factor for regenerative conveyor

H = vertical distance that material is lifted or lowered, ft

K_t = ambient temperature correction factor (see Figure 6.1)

K_x = factor used to calculate the frictional resistance of the idlers and the sliding resistance between the belt and idler rolls, lbs per ft (see equation 3, page 81).

K_y = carrying run factor used to calculate the combination of the resistance of the belt and the resistance of the load to flexure as the belt and load move over the idlers (see equation 4, page 82, and Table 6-2). For return run use constant .015 in place of K_y. See T_{yr}

L = length of conveyor, ft

Q = tons per hour conveyed, tph, short tons of 2,000 lbs

S_i = troughing idler spacing, ft

T_{ac} = total of the tensions from conveyor accessories, lbs:

$$T_{ac} = T_{sb} + T_{pl} + T_{tr} + T_{bc}$$

T_{am} = tension resulting from the **force** to **accelerate** the **material** continuously as it is **fed** onto the belt, lbs

T_b = tension resulting from the **force** needed to **lift** or **lower** the **belt**, lbs (see page 105):

$$T_b = \pm H \times W_b$$

T_{bc} = tension resulting from **belt pull** required for **belt-cleaning devices** such as belt scrapers, lbs.

T_e = **effective** belt tension at drive, lbs

T_m = tension resulting from the **force** needed to **lift** or **lower** the **conveyed material**, lbs:

$$T_m = \pm H \times W_m$$

T_p = tension resulting from **resistance** of belt to **flexure** around **pulleys** and the **resistance** of **pulleys** to rotation on their bearings, total for all pulleys, lbs

T_{pl} = tension resulting from the **frictional resistance** of **plows**, lbs

T_{sb} = tension resulting from the **force** to overcome **skirtboard friction**, lbs

T_{tr} = tension resulting from the additional **frictional resistance** of the **pulleys** and the flexure of the belt over units such as **trippers**, lbs

T_x = tension resulting from the **frictional resistance** of the **carrying** and **return idlers**, lbs:

$$T_x = L \times K_x \times K_t$$

T_{yb} = total of the tensions resulting from the **resistance** of the **belt** to **flexure** as it rides over both the **carrying** and **return idlers**, lbs:

$$T_{yb} = T_{yc} + T_{yr}$$

T_{yc} = tension resulting from the **resistance** of the **belt** to **flexure** as it rides over the **carrying idlers**, lbs:

$$T_{yc} = L \times K_y \times W_b \times K_t$$

T_{ym} = tension resulting from the **resistance** of the **material** to **flexure** as it rides with the **belt** over the **carrying idlers**, lbs:

$$T_{ym} = L \times K_y \times W_m$$

T_{yr} = tension resulting from the **resistance** of the **belt** to **flexure** as it rides over the **return idlers**, lbs:

$$T_{yr} = L \times 0.015 \times W_b \times K_t$$

V = design belt speed, fpm

W_b = weight of belt in pounds per foot of belt length. When the exact weight of the belt is not known, use the **average** estimated belt weight (see Table 6-1)

W_m = weight of material, lbs per ft of belt length:

$$W_m = \frac{Q \times 2{,}000}{60 \times V} = \frac{33.33 \times Q}{V}$$

Three multiplying factors, K_t, K_x, and K_y, are used in calculations of three of the components of the effective belt tension, T_e.

K_t—Ambient Temperature Correction Factor

Idler rotational resistance and the flexing resistance of the belt increase in cold weather operation. In extremely cold weather the proper lubricant for idlers must be used to prevent excessive resistance to idler rotation.

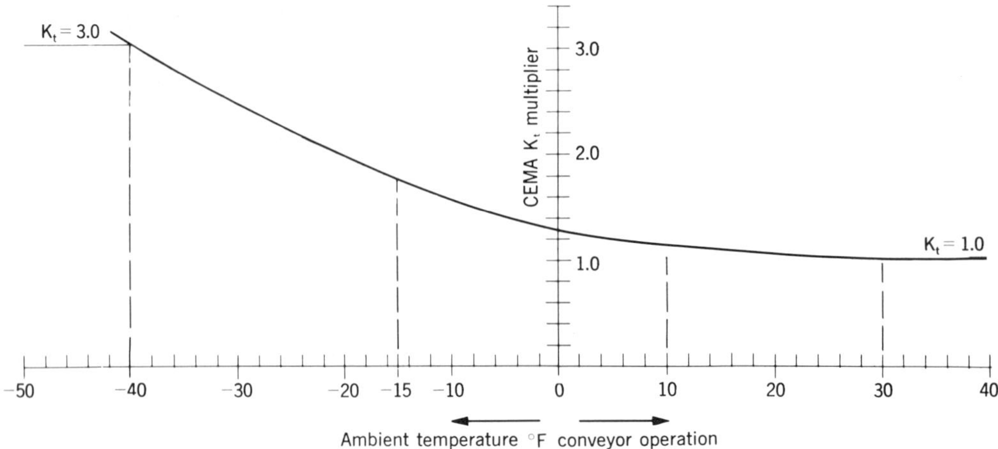

Operation at temperatures below −15°F involves problems in addition to horsepower considerations. Consult conveyor manufacturer for advice on special belting, greasing, and cleaning specifications and necessary design modifications.

FIGURE 6.1. *Variation of temperature correction factor, K_t, with temperature*

K_t is a multiplying factor that will increase the calculated value of belt tensions to allow for the increased resistances that can be expected due to low temperatures. Figure 6.1 provides values for factor K_t.

K_x—Idler Friction Factor

The frictional resistance of idler rolls to rotation and sliding resistance between the belt and the idler rolls can be calculated by using the multiplying factor K_x. K_x is a force in lbs per foot of conveyor length to rotate the idler rolls, carrying and return, and to cover the sliding resistance of the belt on the idler rolls. The K_x value required to rotate the idlers is calculated using equation (3).

The resistance of the idlers to rotation is primarily a function of bearing, grease, and seal resistance. A typical idler roll, equipped with antifriction bearings and supporting a load of 1,000 lbs will require a turning force at the idler roll periphery of from 0.5 to 0.7 lbs to overcome the bearing friction. The milling or churning of the grease in the bearings and the bearing seals will require additional force. This force, however, is generally independent of the load on the idler roll.

Under normal conditions, the grease and seal friction in a well-lubricated idler will vary from 0.1 to 2.3 lbs per idler, depending upon the type of idler, the seals, and the condition of the grease.

Sliding resistance between the belt and idler rolls is generated when the idler rolls are not exactly at 90° to the belt movement. After initial installation, deliberate idler misalignment is often an aid in training the belt. Even the best installations have a small requirement of this type. However, excessive idler misalignment results in an extreme increase in frictional resistance and should be avoided.

TABLE 6-1. *Estimated Average Belt Weight, Multiple- And Reduced-Ply Belts, lbs/ft*

Belt width inches (b)	*Material carried, lbs/ft³*		
	30-74	*75-129*	*130-200*
18	3.5	4	4.5
24	4.5	5.5	6
30	6	7	8
36	9	10	12
42	11	12	14
48	14	15	17
54	16	17	19
60	18	20	22
72	21	24	26
84	25	30	33
96	30	35	38

1. Steel-cable belts—increase above value by 50%
2. Actual belt weights vary with different constructions, manufacturers, cover gauges, etc. Use the above values for estimating. Obtain actual values from the belt manufacturer whenever possible.

Some troughing idlers are designed to operate with a small degree of tilt in the direction of belt travel, to aid in belt training. This tilt results in a slight increase in sliding friction which must be considered in the horsepower formula.

Values of K_x can be calculated from the equation:

$$K_x = 0.00068(W_b + W_m) + \frac{A_i}{S_i}, \text{ lbs tension per foot of belt length} \quad (3)$$

A_i = 1.5 for 6-inch dia. idler rolls, CEMA C6, D6

A_i = 1.8 for 5-inch dia. idler rolls, CEMA A5, B5, C5, D5

A_i = 2.3 for 4-inch dia. idler rolls, CEMA A4, B4, C4

A_i = 2.4 for 7-inch dia. idler rolls, CEMA E7

A_i = 2.8 for 6-inch dia. idler rolls, CEMA E6

For regenerative declined conveyors, $A_i = 0$.

The A_i values tabulated above are averages and include frictional resistance to rotation for both the carrying and return idlers. Return idlers are based on single roll type. **If 2 roll vee return idlers are used, increase A_i value by 5%.** In the case of long conveyors or very high belt speed (over 1000 fpm) refer to CEMA member companies for more specific values of A_i.

K_y—Factor for Calculating the Force of Belt and Load Flexure Over the Idlers

Both the resistance of the belt to flexure as it moves over idlers and the resistance of the load to flexure as it rides the belt over the idlers develop belt-tension forces. K_y is a multiplying factor used in calculating these belt tensioning forces.

Table 6-2 gives values of K_y for carrying idlers as they vary with differences in the weight per foot of the conveyor belt, W_b; load, W_m; idler spacing, S_i; and the percent of slope or angle that the conveyor makes with the horizontal. When applying idler spacing, S_i, other than specified in Table 6-2, use Table 6-3 to determine a corrected K_y value.

Example 1 For a conveyor whose length is 800 ft, and $(W_b + W_m) = 150$ lbs per ft, having a slope of 12%, the K_y value (Table 6-2) is .017. This K_y value is correct only for the idler spacing of 3.0 ft. If a 4.0-ft idler spacing is to be used, using Table 6-3 and the K_y reference values at the top of the table, the K_y of .017 lies between .016 and .018. Through interpolation and using the corresponding K_y values for 4.0-ft spacing, the corrected K_y value is .020.

Example 2 For a conveyor whose length is 1000 ft, and $(W_b + W_m) = 125$ lbs per ft, with a slope of 12%, the K_y value (Table 6-2) is .0165. This value is correct only for 3.5-ft spacing. If 4.5-ft spacing is needed, Table 6-3 shows that .0165 lies between .016 and .018 (reference K_y). Through interpolation and using the corresponding K_y values for 4.5-ft spacing, the corrected K_y value is .0194.

K_y values in Tables 6–2 and 6–3 are applicable for conveyors up to 3,000 feet long with a single slope and a 3% maximum sag of the belt between the troughing and between the return idlers. The return idler spacing is 10 feet nominal and loading of the belt is uniform and continuous.

Equation (4) provides K_y values for the carrying idlers of belt conveyors whose length, number of slopes, and/or average belt tensions exceed the limitations specified above for the conveyors covered by Tables 6–2 and 6–3. This equation is applicable for conveyors in which the average belt tension is 16,000 lbs or less. To determine the K_y factor for use in calculating conveyors of this class, it is necessary, first, to assume a tentative value for the average belt tension. The graphical method for determining conveyor horsepower (pages 130 through 133) may be of assistance in estimating this initial tentative value of average belt tension.

After estimating the average belt tension and selecting an idler spacing, refer to Table 6–4 to obtain values for A and B for use in the following equation:

$$K_y = (W_m + W_b) \times A \times 10^{-4} + B \times 10^{-2} \qquad (4)$$

By using equation (4), an initial value for K_y can be determined and an initial average belt tension can be subsequently calculated. The comparison of this calculated average belt tension with the original tentative value will determine the need to select another assumed belt tension. Recalculate K_y and calculate a second value for the average belt tension. The process should be repeated until there is reasonable agreement between the estimated and final calculated average belt tensions.

There are no tabulated K_y values or mathematical equations to determine a K_y for conveyors having an average belt tension exceeding 16,000 lbs. A reasonably accurate value that can be used for calculations is K_y equals 0.016. It is suggested that this value for K_y be considered a minimum, subject to consultation with a CEMA member company on any specific applications.

The force that results from the resistance of the belt to flexure as it moves over the idlers for the return run is calculated in the same manner as the resistance to flexure for the carrying run, except a constant value of 0.015

TABLE 6–2. *Factor K_y Values*

Conveyor length (ft)	$W_b + W_m$ (lbs per ft)	Percent slope 0 / Approx. degrees 0	3 / 2	6 / 3.5	9 / 5	12 / 7	24 / 14	33 / 18
250	20	0.035	0.035	0.034	0.031	0.031	0.031	0.031
	50	0.035	0.034	0.033	0.032	0.031	0.028	0.027
	75	0.035	0.034	0.032	0.032	0.030	0.027	0.025
	100	0.035	0.033	0.032	0.031	0.030	0.026	0.023
	150	0.035	0.035	0.034	0.033	0.031	0.025	0.021
	200	0.035	0.035	0.035	0.035	0.032	0.024	0.018
	250	0.035	0.035	0.035	0.035	0.033	0.021	0.018
	300	0.035	0.035	0.035	0.035	0.032	0.019	0.018
400	20	0.035	0.034	0.032	0.030	0.030	0.030	0.030
	50	0.035	0.033	0.031	0.029	0.029	0.026	0.025
	75	0.034	0.033	0.030	0.029	0.028	0.024	0.021
	100	0.034	0.032	0.030	0.028	0.028	0.022	0.019
	150	0.035	0.034	0.031	0.028	0.027	0.019	0.016
	200	0.035	0.035	0.033	0.030	0.027	0.016	0.014
	250	0.035	0.035	0.034	0.030	0.026	0.017	0.016
	300	0.035	0.035	0.034	0.029	0.024	0.018	0.018
500	20	0.035	0.033	0.031	0.030	0.030	0.030	0.030
	50	0.034	0.032	0.030	0.028	0.028	0.024	0.023
	75	0.033	0.032	0.029	0.027	0.027	0.021	0.019
	100	0.033	0.031	0.029	0.028	0.026	0.019	0.016
	150	0.035	0.033	0.030	0.027	0.024	0.016	0.016
	200	0.035	0.035	0.030	0.027	0.023	0.016	0.016
	250	0.035	0.035	0.030	0.025	0.021	0.016	0.015
	300	0.035	0.035	0.029	0.024	0.019	0.018	0.018
600	20	0.035	0.032	0.030	0.029	0.029	0.029	0.029
	50	0.033	0.030	0.029	0.027	0.026	0.023	0.021
	75	0.032	0.030	0.028	0.026	0.024	0.020	0.016
	100	0.032	0.030	0.027	0.025	0.022	0.016	0.016
	150	0.035	0.031	0.026	0.024	0.019	0.016	0.016
	200	0.035	0.031	0.026	0.021	0.017	0.016	0.016
	250	0.035	0.031	0.024	0.020	0.017	0.016	0.016
	300	0.035	0.031	0.023	0.018	0.018	0.018	0.018
800	20	0.035	0.031	0.030	0.029	0.029	0.029	0.029
	50	0.032	0.029	0.028	0.026	0.025	0.021	0.018
	75	0.031	0.029	0.026	0.024	0.022	0.016	0.016
	100	0.031	0.028	0.025	0.022	0.020	0.016	0.016
	150	0.034	0.028	0.023	0.019	0.017	0.016	0.016
	200	0.035	0.027	0.021	0.016	0.016	0.016	0.016
	250	0.035	0.026	0.020	0.017	0.016	0.016	0.016
	300	0.035	0.025	0.018	0.018	0.018	0.018	0.018

Idler spacing: The above values of K_y are based on the following idler spacing (for other spacing, see Table 6-3):

$(W_b + W_m)$, lbs per ft	S_i, ft	$(W_b + W_m)$, lbs per ft	S_i, ft
Less than 50	4.5	100 to 149	3.5
50 to 99	4.0	150 and Above	3.0

TABLE 6-2 *cont'd.* *Factor K_y Values*

Conveyor length (ft)	$W_b + W_m$ (lbs per ft)	Percent slope						
		0	3	6	9	12	24	33
		Approximate degrees						
		0	2	3.5	5	7	14	18
	50	0.031	0.028	0.026	0.024	0.023	0.019	0.016
	75	0.030	0.027	0.024	0.022	0.019	0.016	0.016
	100	0.030	0.026	0.022	0.019	0.017	0.016	0.016
1000	150	0.033	0.024	0.019	0.016	0.016	0.016	0.016
	200	0.032	0.023	0.017	0.016	0.016	0.016	0.016
	250	0.033	0.022	0.017	0.016	0.016	0.016	0.016
	300	0.033	0.021	0.018	0.018	0.018	0.018	0.018
	50	0.029	0.026	0.024	0.022	0.021	0.016	0.016
	75	0.028	0.024	0.021	0.019	0.016	0.016	0.016
	100	0.028	0.023	0.019	0.016	0.016	0.016	0.016
1400	150	0.029	0.020	0.016	0.016	0.016	0.016	0.016
	200	0.030	0.021	0.016	0.016	0.016	0.016	0.016
	250	0.030	0.020	0.017	0.016	0.016	0.016	0.016
	300	0.030	0.019	0.018	0.018	0.018	0.018	0.018
	50	0.027	0.024	0.022	0.020	0.018	0.016	0.016
	75	0.026	0.021	0.019	0.016	0.016	0.016	0.016
	100	0.025	0.020	0.016	0.016	0.016	0.016	0.016
2000	150	0.026	0.017	0.016	0.016	0.016	0.016	0.016
	200	0.024	0.016	0.016	0.016	0.016	0.016	0.016
	250	0.023	0.016	0.016	0.016	0.016	0.016	0.016
	300	0.022	0.018	0.018	0.018	0.018	0.018	0.018
	50	0.026	0.023	0.021	0.018	0.017	0.016	0.016
	75	0.025	0.021	0.017	0.016	0.016	0.016	0.016
	100	0.024	0.019	0.016	0.016	0.016	0.016	0.016
2400	150	0.024	0.016	0.016	0.016	0.016	0.016	0.016
	200	0.021	0.016	0.016	0.016	0.016	0.016	0.016
	250	0.021	0.016	0.016	0.016	0.016	0.016	0.016
	300	0.020	0.018	0.018	0.018	0.018	0.018	0.018
	50	0.024	0.022	0.019	0.017	0.016	0.016	0.016
	75	0.023	0.019	0.016	0.016	0.016	0.016	0.016
	100	0.022	0.017	0.016	0.016	0.016	0.016	0.016
3000	150	0.022	0.016	0.016	0.016	0.016	0.016	0.016
	200	0.019	0.016	0.016	0.016	0.016	0.016	0.016
	250	0.018	0.016	0.016	0.016	0.016	0.016	0.016
	300	0.018	0.018	0.018	0.018	0.018	0.018	0.018

Idler spacing: The above values of K_y are based on the following idler spacing (for other spacing, see Table 6-3):

$(W_b + W_m)$, lbs per ft	S_i, ft	$(W_b + W_m)$, lbs per ft	S_i, ft
Less than 50	4.5	100 to 149	3.5
50 to 99	4.0	150 and Above	3.0

is used in place of K_y. The resistance of the belt flexure over idler rolls is a function of the belt construction, cover thickness and indentation by the idler rolls, type of rubber compound, idler roll diameter, temperature, and other factors. The belt flexing resistance increases at lower temperatures.

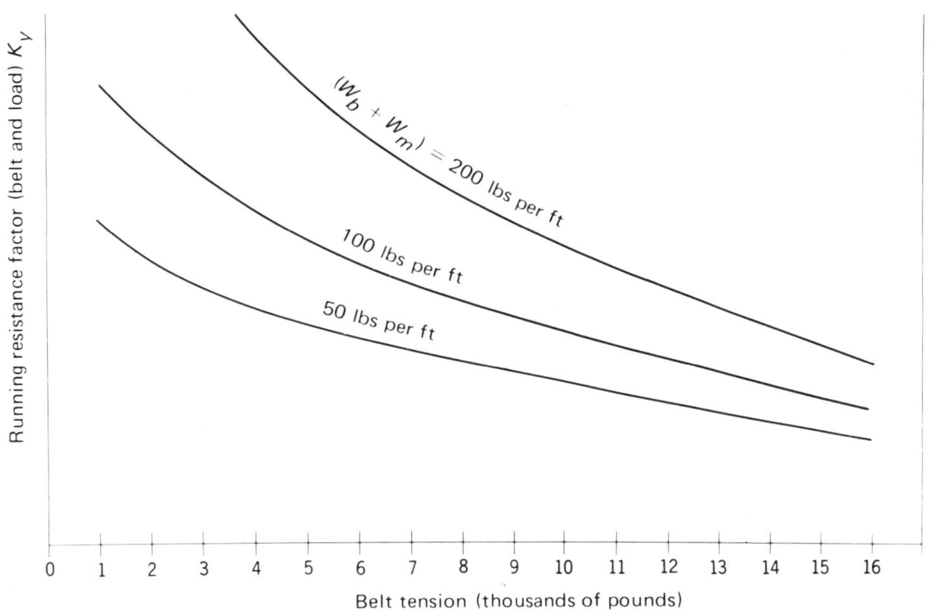

FIGURE 6.2. *Effect of belt tension on resistance of material to flexure over idler rolls.*

TABLE 6–3. ***Corrected Factor K_y Values When Other than Tabular Carrying Idler Spacings Are Used***

$W_b + W_m$ (lbs per ft)	S_i, (ft)	*Reference values of K_y for interpolation*									
		0.016	0.018	0.020	0.022	0.024	0.026	0.028	0.030	0.032	0.034
Less Than 50	3.0	0.016	0.016	0.016	0.0168	0.0183	0.0197	0.0212	0.0227	0.0242	0.0257
	3.5	0.016	0.0160	0.0169	0.0189	0.0207	0.0224	0.0241	0.0257	0.0274	0.0291
	4.0	0.016	0.0165	0.0182	0.0204	0.0223	0.0241	0.0259	0.0278	0.0297	0.0316
	4.5	**0.016**	**0.018**	**0.020**	**0.022**	**0.024**	**0.026**	**0.028**	**0.030**	**0.032**	**0.034**
	5.0	0.0174	0.0195	0.0213	0.0236	0.0254	0.0273	0.0291	0.0031	0.0329	0.0348
50 To 99	3.0	0.016	0.0162	0.0173	0.0186	0.0205	0.0221	0.0239	0.026	0.0274	0.029
	3.5	0.016	0.0165	0.0185	0.0205	0.0222	0.024	0.0262	0.0281	0.030	0.0321
	4.0	**0.016**	**0.018**	**0.020**	**0.022**	**0.024**	**0.026**	**0.028**	**0.030**	**0.032**	**0.034**
	4.5	0.0175	0.0193	0.0214	0.0235	0.0253	0.0272	0.0297	0.0316	0.0335	0.035
	5.0	0.0184	0.021	0.023	0.0253	0.027	0.029	0.0315	0.0335	0.035	0.035
100 To 149	3.0	0.016	0.0164	0.0186	0.0205	0.0228	0.0246	0.0267	0.0285	0.0307	0.0329
	3.5	**0.016**	**0.018**	**0.020**	**0.022**	**0.024**	**0.026**	**0.028**	**0.030**	**0.032**	**0.034**
	4.0	0.0175	0.0197	0.0213	0.0234	0.0253	0.0277	0.0295	0.0312	0.033	0.035
	4.5	0.0188	0.0213	0.0232	0.0253	0.0273	0.0295	0.0314	0.033	0.0346	0.035
	5.0	0.0201	0.0228	0.0250	0.0271	0.0296	0.0316	0.0334	0.035	0.035	0.035
150 To 199	**3.0**	**0.016**	**0.018**	**0.020**	**0.022**	**0.024**	**0.026**	**0.028**	**0.030**	**0.032**	**0.034**
	3.5	0.0172	0.0195	0.0215	0.0235	0.0255	0.0271	0.0289	0.031	0.0333	0.0345
	4.0	0.0187	0.0213	0.0235	0.0252	0.0267	0.0283	0.0303	0.0325	0.0347	0.035
	4.5	0.0209	0.023	0.0253	0.0274	0.0289	0.0305	0.0323	0.0345	0.035	0.035
	5.0	0.0225	0.0248	0.0272	0.0293	0.0311	0.0328	0.0348	0.035	0.035	0.035
200 To 249	**3.0**	**0.016**	**0.018**	**0.020**	**0.022**	**0.024**	**0.026**	**0.028**	**0.030**	**0.032**	**0.034**
	3.5	0.0177	0.0199	0.0216	0.0235	0.0256	0.0278	0.0295	0.031	0.0327	0.0349
	4.0	0.0192	0.0216	0.0236	0.0256	0.0274	0.0291	0.0305	0.0322	0.0339	0.035
	4.5	0.021	0.0234	0.0253	0.0276	0.0298	0.0317	0.0331	0.0347	0.035	0.035
	5.0	0.0227	0.0252	0.0274	0.0298	0.0319	0.0338	0.035	0.035	0.035	0.035

To use this table to correct the value of K_y for idler spacing other than shown in bold type, apply the procedure in the two examples on page 82.

TABLE 6–4. ***A and B Values for Equation $K_y = (W_m + W_b) \times A \times 10^{-4} + B \times 10^{-2}$***

Average belt tension, lbs	*Idler Spacing, ft*									
	3.0		3.5		4.0		4.5		5.0	
	A	B	A	B	A	B	A	B	A	B
1,000	2.1500	1.565	2.1955	1.925	2.2000	2.250	2.2062	2.584	2.1750	2.910
2,000	1.8471	1.345	1.6647	1.744	1.6156	1.982	1.5643	2.197	1.5429	2.331
3,000	1.6286	1.237	1.4667	1.593	1.4325	1.799	1.4194	1.991	1.4719	2.091
4,000	1.4625	1.164	1.3520	1.465	1.3295	1.659	1.3250	1.825	1.3850	1.938
5,000	1.2828	1.122	1.1926	1.381	1.1808	1.559	1.1812	1.714	1.2283	1.839
6,000	1.1379	1.076	1.0741	1.318	1.0625	1.472	1.0661	1.627	1.0962	1.761
7,000	1.0069	1.039	0.9448	1.256	0.9554	1.404	0.9786	1.549	1.0393	1.657
8,000	0.9172	0.998	0.8552	1.194	0.8643	1.337	0.8875	1.472	0.9589	1.583
9,000	0.8207	0.958	0.8000	1.120	0.7893	1.272	0.8339	1.388	0.8911	1.507
10,000	0.7241	0.918	0.7362	1.066	0.7196	1.216	0.7821	1.314	0.8268	1.430
11,000	0.6483	0.885	0.6638	1.024	0.6643	1.167	0.7375	1.238	0.7768	1.340
12,000	0.5828	0.842	0.5828	0.992	0.6232	1.100	0.6750	1.180	0.7411	1.242
13,000	0.5207	0.798	0.5241	0.938	0.5732	1.040	0.6179	1.116	0.6821	1.169
14,000	0.4690	0.763	0.4810	0.897	0.5214	0.996	0.5571	1.069	0.6089	1.123
15,000	0.4172	0.718	0.4431	0.841	0.4732	0.935	0.5179	1.006	0.5607	1.063
16,000	0.3724	0.663	0.3966	0.780	0.4232	0.875	0.4589	0.958	0.5054	1.009

A minimum K_y value of .016 should be used when tensions exceed 16,000 lbs. Refer to page 81 for further explanation.

The resistance of the material load to flexure over idler rolls is a function of belt tension, type of material, shape of the load cross section, and idler spacing. Measurements indicate that the most important factor is belt tension, because this controls the amount of load flexure. Figure 6.2 shows this relationship for a typical idler spacing.

For a given weight per foot of belt and load, the running resistance, in lbs per lb of load, decreases with increases in belt tension. For a given belt tension, running resistance, in lbs per lb of load, increases with increases in the amount of load. However, the running resistance is not proportional to the weight of the load.

Information similar to that in Figure 6.2 has been developed by analyzing a series of field tests on belt conveyors of different widths carrying different materials. Many investigators, both in the United States and abroad, have analyzed similar series of field tests and have obtained similar results. Although the exact expressions differ, all investigators agree that changes in belt tension affect the force required to flex the material over idler rolls to a substantially greater degree than changes in the material handled. The latter does have a noticeable effect, and thus appears to be of less importance in the overall calculation.

Compilation of Components of T_e

The preceding pages describe the methods and provide the data for calculating factors K_t, K_x, and K_y. These factors must be evaluated as the first step to calculating certain components of belt tension that will be summarized to determine the effective tension, T_e, required at the driving pulley.

The procedures for calculating the belt tension components are as follows:

1. T_x—from the frictional resistance of the carrying and return idlers, lbs

   $T_x = L \times K_x \times K_t$ (References: K_x—page 80, K_t—page 79)

2. T_{yb}—from the resistance of the belt to flexure as it moves over the idlers, lbs

   T_{yc}—for carrying idlers, $= L \times K_y \times W_b \times K_t$

   T_{yr}—for return idlers, $= L \times 0.015 \times W_b \times K_t$

   $T_{yb} = T_{yc} + T_{yr}$

   $T_{yb} = L \times K_y \times W_b \times K_t + L \times 0.015 \times W_b \times K_t$

   $= L \times W_b \times K_t(K_y + 0.015)$

   (References: K_y—page 81, K_t—page 79)

3. T_{ym}—from resistance of the material to flexure as it rides the belt over the idlers, lbs

   $T_{ym} = L \times K_y \times W_m$

   (Reference: K_y—page 81)

4. T_m—from force needed to lift or lower the load (material), lbs

   $T_m = \pm H \times W_m$

5. T_p—from resistance of belt to flexure around pulleys and the resistance of pulleys to rotate on their bearings, lbs

   Pulley friction arises from two sources. One source is the resistance of the belt to flexure over the pulleys, which is a function of the pulley diameter and the belt stiffness. The belt stiffness depends upon the ambient temperature and the belt construction.

   The other source of pulley friction is the resistance of the pulley to rotate, which is a function of pillow block bearing friction, lubricant, and seal friction. The pillow block bearing friction depends upon the load on the bearings, but the lubricant and seal frictions generally are independent of load.

   Since the drive pulley friction does not affect belt tension, it is not introduced into the mathematical calculation for belt tension; however, it must be included when determining the total horsepower at the motor shaft.

   Table 6-5 provides conservative values for the pounds of belt tension required to rotate each of the pulleys on a conveyor. However, if a more precise value for belt tension to rotate pulleys is desired refer to **Appendix C** page 338. Examples of belt tension and horsepower calculations shown in this book use values from Table 6-5.

   T_p = total of the belt tensions required to rotate each of the pulleys on the conveyor.

Table 6-5. ***Belt Tension to Rotate Pulleys***

Location of pulleys	*Degrees wrap of belt*	*Pounds tension at belt line*
Tight side	150° to 240°	200 lbs per pulley
Slack side	150° to 240°	150 lbs per pulley
All other pulleys	less than 150°	100 lbs per pulley

Note: Double the above values for pulley shafts which are not operating in antifriction bearings.

6. T_{am}—from force to accelerate the material continuously as it is fed onto the belt.

When material is discharged from chutes or feeders to a belt conveyor, it cannot be assumed that the material is moving in the direction of belt travel, at belt speed, although this may be the case in some instances. Normally, the material loaded onto the belt is traveling at a speed considerably lower than belt speed. The direction of material flow may not be fully in the direction of belt travel. Therefore, the material must be accelerated to the speed of the belt in the direction of belt travel, and this acceleration requires additional effective tension.

The belt tension T_{am} can be derived from the basic equation $F = MV_c$, where:

$$T_{am} = F = MV_c$$

M = mass of material accelerated per second, slugs

W = weight of material accelerated

$$= \frac{Q \times 2000}{3600}, \text{ lbs. per second}$$

Q = tph

g = 32.2 ft/sec²

$$M = \frac{W}{g} = \frac{Q \times 2000}{3600 \times 32.2}$$

V_c = velocity change, fps

$$= \frac{V - V_o}{60}$$

V = design belt speed, fpm

V_o = initial velocity of material as it is fed onto belt, fpm

$$T_{am} = \frac{Q \times 2000}{3600 \times 32.2} \times \frac{V - V_o}{60}$$

$$= 2.8755 \times 10^{-4} \times Q \times (V - V_o)$$

The graph in Figure 6.3 provides a convenient means of estimating the belt tension, T_{am}, for accelerating the material as it is fed onto the belt.

7. T_{ac}—from the resistance generated by conveyor accessories.

Conveyor accessories such as trippers, stackers, plows, belt-cleaning equipment, and skirtboards usually add to the effective tension, T_e. The additional belt tension requirements may come from frictional losses caused by the accessory. If the accessory lifts the conveyed material a force will be added to belt tension.

T_{tr}—from trippers and stackers.

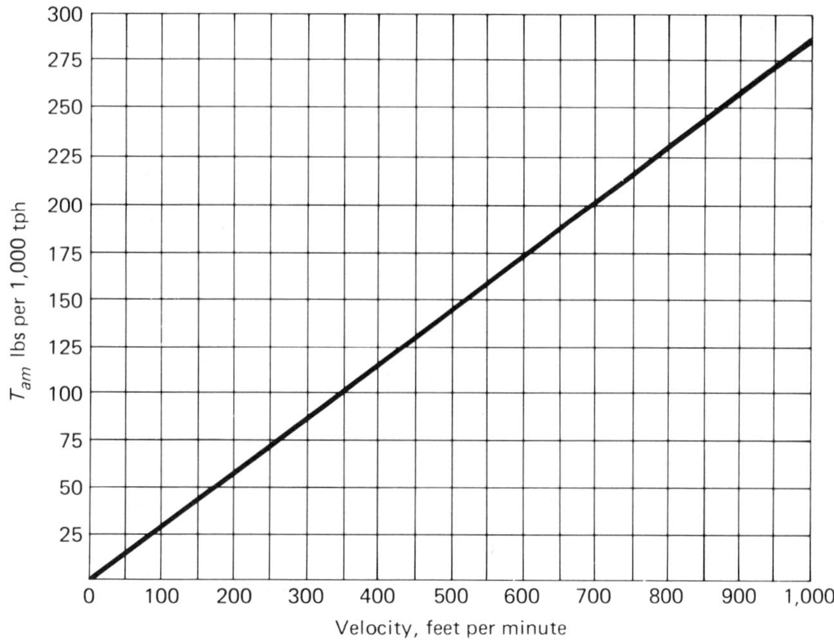

To use this chart:

- Enter chart at belt velocity and read T_{am} per 1,000 tph.
- Again enter chart at material velocity in direction of belt travel and read T_{am} per 1,000 tph. This may be positive, zero, or negative.
- Subtract the second T_{am} reading from the first T_{am} reading and convert the difference from 1,000 tph to the value for the actual tonnage. This will be the T_{am} desired, lbs.

FIGURE 6.3. *Effective tension required to accelerate material as it is fed onto a belt conveyor.*

The additional belt pull to flex the belt over the pulleys and rotate the pulleys in their bearings can be calculated from Table 6-5 or Tables C-1 and C-2.

The force needed to lift the material over the unit can be calculated from the formula $T_m = H \times W_m$, lbs.

Frictional resistance of the idlers, belt, and material should be included with that of the rest of the conveyor.

T_{pl}—from frictional resistance of plows.

The use of a plow on a conveyor will require additional belt tension to overcome both the plowing and frictional resistances developed.

While a flat belt conveyor may be fitted with a number of plows to discharge material at desired locations, seldom is more than one plow in use at one time on one run of the belt conveyor. However, when proportioning plows are used—with each plow taking a fraction of the load from the belt—two or even three separate plows may be simultaneously in contact with the carrying run of the belt.

To approximate the amount of additional belt pull that normally will be required by well-adjusted, rubber-shod plows, the values given in Table 6-6 can be used.

Table 6-6. *Discharge Plow Allowance*

Type of plow	*Additional belt pull per plow, at belt line (lbs per inch belt width)*
Full vee or single slant plow removing all material from belt	5.0
Partial vee or single slant plow removing half material from belt	3.0

T_{bc}—from belt-cleaning devices.

Belt scraper cleaning devices add directly to the belt pull. The additional belt pull required for belt cleaning devices can vary from 2 to 14 lbs per inch width of scraper blade contact. This wide variance is due to the different types of cleaning blades and single cleaner system vs. multiple cleaner systems that are available. In lieu of data on specific cleaning system being used, use 5 lbs per inch width of scraper blade contact for **each blade or scraper device in contact with the belt.**

Rotary brushes and similar rotating cleaning devices do not impose appreciable belt pull, if independently driven and properly adjusted. If such devices are driven from the conveyor drive shaft, suitable additional power should be incorporated in the drive to operate them. Consult a CEMA member company for horsepower required.

T_{sb}—From skirtboard friction.

The force required to overcome skirtboard friction is normally larger per foot of skirtboarded conveyor than the force to move the loaded belt over the idlers. In some cases, this force can be significant. When the total conveyor length is many times that portion of the length provided with the skirtboards, the additional power requirements for the skirtboards is relatively small, and in some cases negligible. However, if a large portion of the conveyor is equipped with skirtboards, the additional belt pull required may be a major factor in the effective tension, T_e, required to operate the conveyor.

When the spacing of the skirtboards is two-thirds the width of the belt, the depth of the material rubbing on the skirtboards will not be more than 10% of the belt width, providing no more than a 20° surcharge load is carried on 20° troughing idlers.

Once the cross section of the load on the belt conveyor has been determined, the skirtboard friction can be calculated by determining the total pressure of the material against the skirtboard, then multiplying this value by the appropriate coefficient of friction of the material handled. The pressure of the material against the skirtboard can be calculated by assuming that the wedge of material contained between a vertical skirtboard and the angle of repose of the material is supported equally by the skirtboard and the belt.

This results in the following formula for conveyors whereon the material assumes its natural surcharge angle:

$$P = \frac{L_b d_m h_s^2}{288} \frac{(1 - \sin \phi)}{(1 + \sin \phi)}$$

where: P = total force against one skirtboard, lbs
L_b = skirtboard length, ft, one skirtboard
d_m = apparent density of the material, lbs per cu ft
h_s = depth of the material touching the skirtboard, in
ϕ = angle of repose of material, degrees.

Combining the apparent density, coefficient of friction, the angle of repose, and the constant into one factor for one type of material, the formula can be expressed:

$$C_s = \frac{2d_m}{288}\left(\frac{1 - \sin\phi}{1 + \sin\phi}\right)$$

$$T = (C_s)(L_b h_s^2) = C_s L_b h_s^2$$

where: T = belt tension to overcome skirtboard friction of two parallel skirtboards, lbs
C_s = factor for the various materials in Table 6-7.

Table 6-7. ***Skirtboard Friction Factor, C_s***

Material	*Factor* C_s	*Material*	*Factor* C_s	*Material*	*Factor* C_s
Alumina, pulv., dry	0.1210	Coke, ground fine	0.0452	Limestone, pulv., dry	0.1280
Ashes, coal, dry	0.0571	Coke, lumps and fines	0.0186	Magnesium chloride, dry	0.0276
Bauxite, ground	0.1881	Copra, lumpy	0.0203	Oats	0.0219
Beans, Navy, dry	0.0798	Cullet	0.0836	Phosphate rock, dry, broken	0.1086
Borax	0.0734	Flour, wheat	0.0265	Salt, common, dry fine	0.0814
Bran, granular	0.0238	Grains, wheat, corn or rye	0.0433	Sand, dry, bank	0.1378
Cement, Portland, dry	0.2120	Gravel, bank run	0.1145	Sawdust, dry	0.0086
Cement clinker	0.1228	Gypsum, ½" screenings	0.0900	Soda ash, heavy	0.0705
Clay, ceramic, dry fines	0.0924	Iron Ore, 200 lbs per cu ft	0.2760	Starch, small lumps	0.0623
Coal, anthracite, sized	0.0538	Lime, burned, 1/8"	0.1166	Sugar, granulated dry	0.0349
Coal, bituminous, mined	0.0754	Lime, hydrated	0.0490	Wood chips, hogged fuel	0.0095

To this skirtboard friction must be added 3 lbs for every linear foot of each skirtboard, to overcome friction of the rubber skirtboard edging, when used, with the belt. Then,

$$T_{sb} = T + 2L_b \times 3$$
$$= C_s L_b h_s^2 + 2L_b \times 3$$
$$T_{sb} = L_b(C_s h_s^2 + 6)$$

Summary of Components of T_e

Once they have been compiled, the components of belt tension can be summarized to determine the effective tension, T_e, required at the driving pulley.

T_e equals the total of the following:

$$
\begin{array}{lll}
& T_x\text{, idler friction} & = L \times K_x \times K_t \\
+ & T_{yc}\text{, belt flexure, carrying idlers} & = L \times K_y \times W_b \times K_t \\
+ & T_{yr}\text{, belt flexure, return idlers} & = L \times 0.015 \times W_b \times K_t \\
& \text{Subtotal (A)} & = LK_t(K_x + K_yW_b + 0.015W_b) \\
+ & T_{ym}\text{, material flexure} & = L \times K_y \times W_m \\
+ & T_m\text{, lift or lower} & = \pm H \times W_m \\
& \text{Subtotal (B)} & = W_m(LK_y \pm H) \\
+ & T_p\text{, pulley resistance} & \\
+ & T_{am}\text{, accelerated material} & \\
+ & T_{ac}\text{, accessories } (T_{tr} + T_{pl} + T_{bc} + T_{sb}) & \\
& \text{Subtotal (C)} & = T_p + T_{am} + T_{ac}
\end{array}
$$

$$
\begin{aligned}
T_e &= \Sigma \text{ Subtotals (A), (B), and (C)} \\
&= LK_t(K_x + K_yW_b + 0.015W_b) + W_m(LK_y \pm H) + T_p + T_{am} + T_{ac}
\end{aligned}
$$

CEMA Horsepower Formula

Equation 1, page 77, provides the means for calculating the horsepower required by a belt conveyor having an effective tension, T_e, at the drive pulley and a design velocity, V, of the belt, as follows:

$$hp = \frac{T_e \times V}{33{,}000}$$

Combining equations (1) and (2) on pages 77-78, the horsepower load can be expressed as follows:

$$hp = \frac{V}{33{,}000}\,[LK_t(K_x + K_yW_b + 0.015W_b) + W_m(LK_y \pm H) + T_p + T_{am} + T_{ac}]$$

The motor which will drive a fully loaded belt conveyor without becoming overheated may not be able to accelerate the loaded conveyor from rest to the design speed. To insure adequate starting capabilities, the following conditions must exist. First, the locked rotor torque of the motor should exceed the sum of the torque required to lift the material, plus approximately twice the torque required to overcome total conveyor friction, despite any possible voltage deficiencies that may exist during the acceleration period. This may not be true for long, horizontal conveyors or for declined conveyors.

Second, the motor speed-torque curve should not drop below a line drawn from the locked rotor torque requirement to the torque of the running horsepower requirement at full speed. This is further explained in Chapter 12, "Motors and Controls."

For examples illustrating the use of the equations in determining the effective belt tension, T_e, at the drive pulley and the horsepower to operate the belt conveyor, refer to the two problems on pages 134 through 140.

It is also possible to arrive at a close approximation of the horsepower required to operate a belt conveyor by means of a graphical solution. This method, used under proper circumstances, is quick and relatively simple. Generally, a graphical solution will provide a somewhat conservative value of required horsepower. However, it must be recognized that it is impractical to incorporate all elements of belt conveyor design into a simple graphical solution. Therefore, the graphs should be used based on a complete understanding of all aspects of the analytical method of calculating belt conveyor tension and horsepower, in order to allow for adjustment of the results to account for unusual situations. It is recommended that final design be based on calculations made by the analytical method. The graphical method of designing belt conveyors is described on pages 130-133.

Drive Pulley Relationships

The force required to drive a belt conveyor must be transmitted from the drive pulley to the belt by means of friction between their two surfaces. The force required to restrain a downhill regenerative conveyor is transmitted in exactly the same manner. In order to transmit power, there must be a difference in the tension in the belt as it approaches and leaves the drive pulley. This difference in tensions is supplied by the driving power source. Figures 6.4 and 6.5 illustrate typical arrangements of single pulley drives.

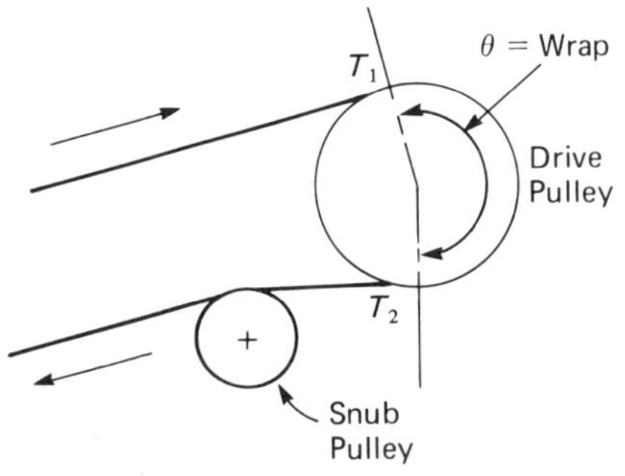

FIGURE 6.4. *Inclined or horizontal conveyor, pulley driving belt.*

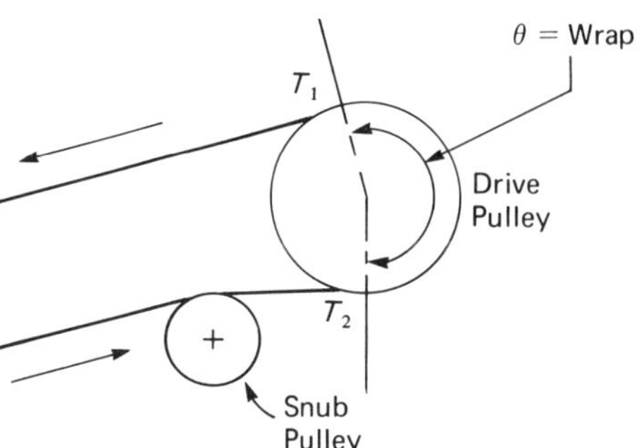

FIGURE 6.5. *Declined conveyor. Lowering load with regeneration, belt driving pulley.*

It should be noted that if power is transmitted from the pulley to the belt, the approaching portion of the belt will have the larger tension, T_1, and the departing portion will have the smaller tension, T_2. If power is transmitted from the belt to the pulley, as with a regenerative declined conveyor, the reverse is true. Wrap is used here to refer to the angle or arc of contact the belt makes with the pulley's circumference.

Wrap Factor, C_w

The wrap factor, C_w, is a mathematical value used in the determination of the effective belt tension, T_e, that can be dependably developed by the drive pulley. The T_e that can be developed is governed by the coefficient of friction existing between the pulley and the belt, wrap, and the values of T_1 and T_2.

The following symbols and formulas are used to evaluate the drive pulley relationships:

$T_e = T_1 - T_2$ = effective belt tension, lbs

T_1 = tight-side tension at pulley, lbs

T_2 = slack-side tension at pulley, lbs

e = base of naperian logarithms = 2.718

f = coefficient of friction between pulley surface and belt surface (0.25 rubber surfaced belt driving bare steel or cast iron pulley; 0.35 rubber surfaced belt driving rubber lagged pulley surface). Values apply to normal running calculations.

θ = wrap of belt around the pulley, radians (one degree = 0.0174 radians)

C_w = wrap factor (see Table 6–8)

$$= \frac{T_2}{T_e} = \frac{1}{e^{f\theta} - 1}$$

It should be noted that the wrap factors do not determine T_2 but only establish its safe minimum value for a dry belt. A wet belt and pulley will substantially reduce the power which can be transmitted from the one to the other because of the lower coefficient of friction of the wet surfaces. Various expedients, such as grooving the lagging on the pulley, lessen this problem. However, the best solution is to keep the driving side of the belt dry. If this is impractical, increasing the wrap is helpful, or providing some means of increasing the slack side tension, T_2. This can be done, for example, by increasing the counterweight in a gravity takeup.

Wrap Factor with Screw Takeup

When a screw takeup is used, Table 6–8 indicates an increased wrap factor. This increased wrap factor is necessary to provide sufficient slack side tension, T_2, to drive the conveyor in spite of the amount of stretch in the conveyor belt, for which the screw takeup makes no automatic provision.

TABLE 6–8. *Wrap Factor, C_w (Rubber surfaced belt)*

Type of pulley drive	θ Wrap	Automatic takeup		Manual takeup	
		Bare pulley	Lagged pulley	Bare pulley	Lagged pulley
Single no snub	180°	0.84	0.50	1.2	0.8
Single with snub	200°	0.72	0.42	1.0	0.7
	210°	0.66	0.38	1.0	0.7
	220°	0.62	0.35	0.9	0.6
	240°	0.54	0.30	0.8	0.6
Dual	380°	0.23	0.11	0.5	0.3
	420°	0.18	0.08	—	—

Note: For wet belts and smooth lagging use bare pulley factor.
For wet belts and grooved lagging, use lagged pulley factor.
If wrap is unknown, assume the following:

Type of Drive	*Assumed Wrap*
Single—no snub	180°
Single—with snub	210°
Dual	380°

Wrap θ (Arc of Contact)

So far, it has been shown that the relationship between the values known as T_1, the tight side tension (and generally the tension for which the belt must be designed and built), and T_2, the slack side tension (the minimum value which must be available for driving the belt successfully), is influenced by the angle of wrap of the belt around the drive pulley and by the coefficients of friction established by the belt and pulley surfaces as they make contact. It has been indicated that the coefficient of friction may vary when driving a rubber-surfaced belt by a bare steel or cast iron pulley, or by a rubber-lagged pulley surface.

The angle of wrap of the belt around the drive pulley can be varied by the use of a snub pulley or, for larger angles of wrap, by supplying power, under the proper conditions, to more than one drive pulley.

The wrap limits for various types of pulley drives can be determined from Table 6–9.

TABLE 6–9. *Wrap Limits*

Type of pulley drive	Wrap limits*	
	From	To
Single—no snub	180°	180°
Single—with snub	180°	240°
Dual	360°	480°

*The above wrap angles apply to either bare or lagged pulleys.

For most cases, the belt will have an angle of wrap around the drive pulley of about 180° to 240°. Often, it will be necessary to arrange a drive which uses an angle of wrap greater than 180°. This is accomplished by the appropriate positioning of a snub pulley, which can extend the angle of wrap to 240°. However, its use is subject to the following limitations: (1) Snub pulley diameter is limited by the belt specifications; (2) In order to thread through a new or replacement belt between the pulleys, a suitable clearance between pulley rims should be allowed; (3) The departing direction of the belt from the snub pulley (plus clearance for belt fasteners, etc.) must be below the deck plate or on the underside of the carrying idlers. These limitations will be found to restrict a snubbed drive to an angle of wrap not exceeding 240°, in most cases. If a greater angle of wrap is necessary, it may be necessary to use a dual-pulley drive.

Dual-Pulley Drives

A dual-pulley drive uses two or more separate motors, one or more driving the primary drive pulley and one or more driving the secondary drive pulley. Table 6-8, Wrap Factor C_w shows the major increase in wrap that becomes available when using a dual-pulley drive. This increase in available wrap can often provide for lower maximum belt tension and a more efficient and lower cost conveyor design.

In any such system where two drive pulleys are involved, the secondary pulley starts out with a certain value of T_2. Contingent on its angle of wrap and the applicable friction coefficient, the secondary pulley can produce a value, T_3, such that:

$$T_1 - T_3 = T_{ep} \text{ (primary)}$$
$$T_3 - T_2 = T_{es} \text{ (secondary)}$$
$$T_{ep} \text{ (primary)} + T_{es} \text{ (secondary)} = T_e \text{ (total for conveyor)}$$

The value, T_3, for the secondary pulley, clearly is the only value available to be used as the slack side tension in the preceding primary drive. This value of T_3, added to the T_{ep} for the primary pulley, yields T_1. The sum of the secondary T_{es} and the primary T_{ep} yields a total T_e which the combined two-pulley drive can produce.

For the maximum efficiency of a two-pulley or dual drive, as described above, it is evident that the proportionate size of the two motors employed must be related appropriately to the angles of wrap and the coefficients of friction at the respective pulleys.

The ratio of tight side tension/slack side tension for each of the drive components, when multiplied together, gives the constant which would apply to the combined or total drive. Or, putting this another way, T_1/T_3 multiplied by T_3/T_2 will equal T_1/T_2, provided the drive conditions are the same for both pulleys. However, if the primary drive utilizes the clean side of the belt while the secondary drive is permitted to operate on the carrying or dirty side of the belt, the friction coefficient and wrap factor for the secondary pulley will vary and the tension relationship should be investigated.

For any conveyor drive which utilizes more than one drive pulley, a snub pulley arrangement is preferable, so that both pulleys drive on the same clean side of the belt.

The following symbols and formulas will be of assistance in evaluating the drive pulley relationships for dual pulley drives:

T_3 = belt tension between the primary and secondary drive pulleys
C_{ws} = wrap factor for the secondary drive pulley
C_{wp} = wrap factor for the primary drive pulley
C_w = the combined wrap factor for both drive pulleys
T_{es} = effective tension on the secondary drive pulley
T_{ep} = effective tension on the primary drive pulley

$$T_2 = T_e C_w$$

$$T_1 = T_2 + T_e$$

$$T_1 = T_{ep} + T_3$$

$$T_3 = T_2 + T_{es}$$

$$C_{wp} = \frac{T_3}{T_{ep}}$$

$$C_{ws} = \frac{T_3}{T_{es}} - 1$$

$$T_e = T_{es} + T_{ep}, \text{ whence}$$

$$T_2 = (T_{es} + T_{ep})(C_w)$$

$$T_{es} = \frac{T_2}{C_{ws}} \qquad \textit{by definition}$$

$$T_{ep} = \frac{T_3}{C_{wp}} \qquad \textit{by definition}$$

$$T_3 = T_2 + T_{es}, \text{ whence}$$

$$T_{ep} = \frac{T_2 + T_{es}}{C_{wp}}$$

Substituting:

$$T_2 = \left[\frac{T_2}{C_{ws}} + \frac{(T_2 + T_{es})}{C_{wp}}\right] C_w$$

$$= \left(\frac{T_2}{C_{ws}} + \frac{T_2}{C_{wp}} + \frac{T_{es}}{C_{wp}}\right) C_w$$

$$T_2 = \left(\frac{T_2 C_{wp} + T_2 C_{ws} + T_{es} C_{ws}}{C_{ws} C_{wp}}\right) C_w$$

solving for C_w, and noting that $T_{es} C_{ws} = T_2$,

$$C_w = \frac{T_2 C_{ws} C_{wp}}{T_2 C_{wp} + T_2 C_{ws} + T_2}, \text{ and because } T_{es} C_{ws} = T_2,$$

$$C_w = \frac{C_{ws}C_{wp}}{C_{wp} + C_{ws} + 1}$$

For example, if the angles of wrap of the primary and secondary drive pulleys are 180° and 220°, respectively, the factors are as follows for lagged pulleys (see Table 6-8):

C_{ws} = 0.35 for 220° angle of wrap
C_{wp} = 0.50 for 180° angle of wrap
C_w = 0.095 for 400° total angle of wrap by interpolating between 380° and 420°, or:

$$C_w = \frac{C_{wp}C_{ws}}{1 + C_{wp} + C_{ws}} = \frac{(0.5)(0.35)}{1 + 0.35 + 0.50} = 0.0945$$

The tensions exterior to a dual-pulley drive are the same as those for a single-pulley drive.

A part of the effective tension, T_e, is taken on the primary drive pulley and a part on the secondary drive pulley. Using two motors, the ratio of T_{ep} to T_{es} is the ratio of the horsepower ratings of the two motors.

For example, if the total calculated horsepower is 250, this could be supplied, allowing for drive losses, by using a 200-hp primary drive and a 75-hp secondary drive, with a drive efficiency of 90%.

The primary pulley would take

$$(200/275)(250) = 182 \text{ hp}$$

The secondary pulley would take

$$(75/275)(250) = 68 \text{ hp}$$

If the belt velocity, V, is 400 fpm, then

$$T_{ep} = \frac{(182)(33{,}000)}{400} = 15{,}000 \text{ lbs}$$

$$T_{es} = \frac{(68)(33{,}000)}{400} = 5{,}625 \text{ lbs}$$

and

$$\frac{T_{ep}}{T_{es}} = \frac{15{,}000}{5{,}625} = 2.67$$ (see Problem 1, page 134.)

Drive Arrangements

The final selection and design of a conveyor drive arrangement is influenced by many factors, including the performance requirements, the preferred physical location, and relative costs of components and installation.

Figures 6.6A through 6.7F illustrate some of the drive combinations that have been furnished. Other arrangements may be better suited to a particular conveyor in a particular location. CEMA member companies can assist in final recommendations.

Note that the illustrated arrangements that are on a substantially downhill run are usually regenerative and are so indicated in the title.

Figure 6.6. *Single-Pulley Drive Arrangements*

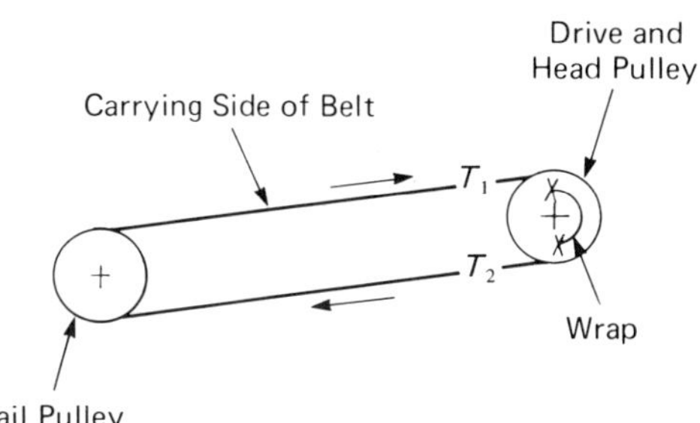

6.6A. *Single-pulley drive at head end of conveyor without snub pulley.*

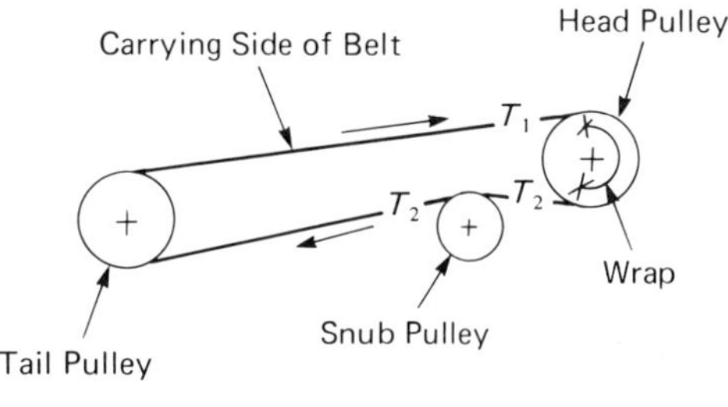

6.6B. *Single-pulley drive at head end of conveyor with snub pulley.*

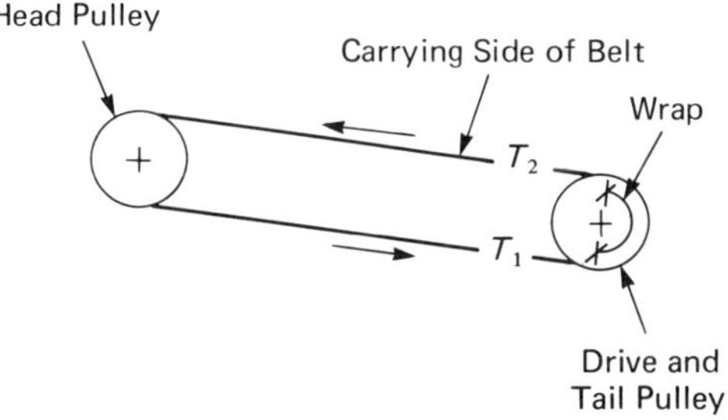

6.6C. *Single-pulley drive at tail end without snub pulley. Used when head end drive cannot be applied.*

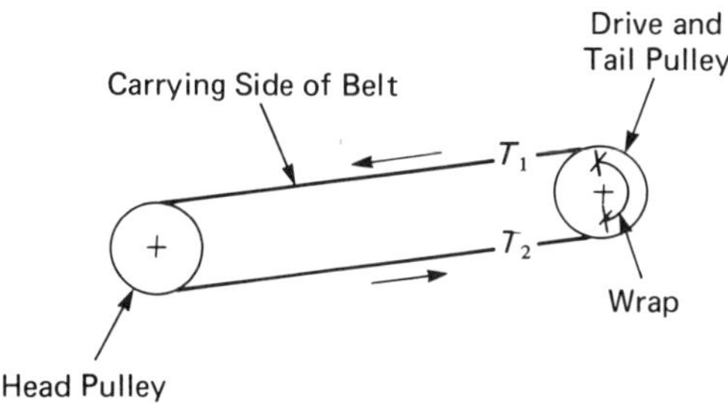

6.6D. *Single-pulley drive at tail end of conveyor without snub pulley; regenerative.*

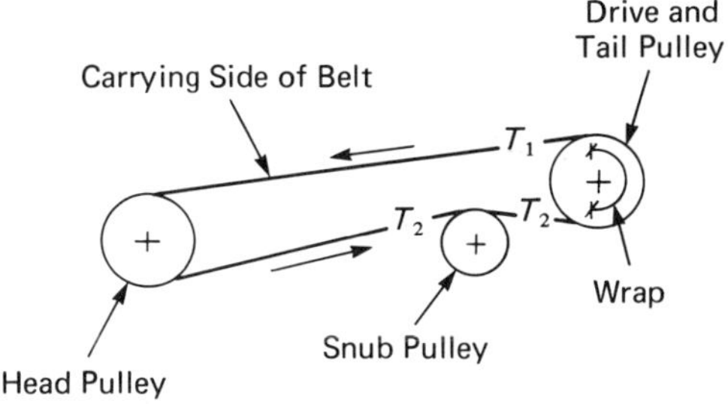

6.6E. *Single-pulley drive at tail end of conveyor with snub pulley; regenerative.*

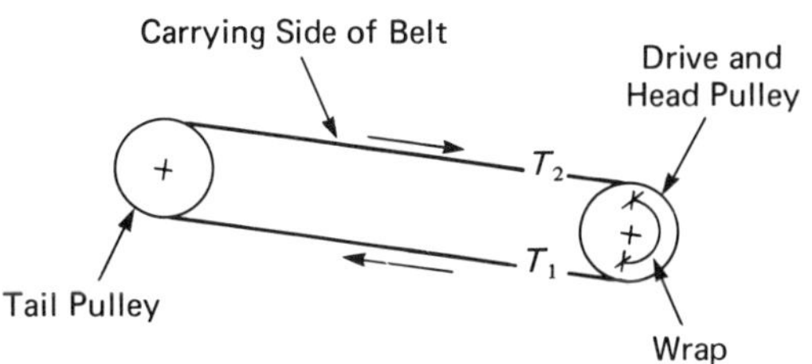

6.6F. *Single-pulley drive at head end of conveyor without snub pulley; regenerative.*

FIGURE 6.6 *continued.* ***Single-Pulley Drive Arrangements***

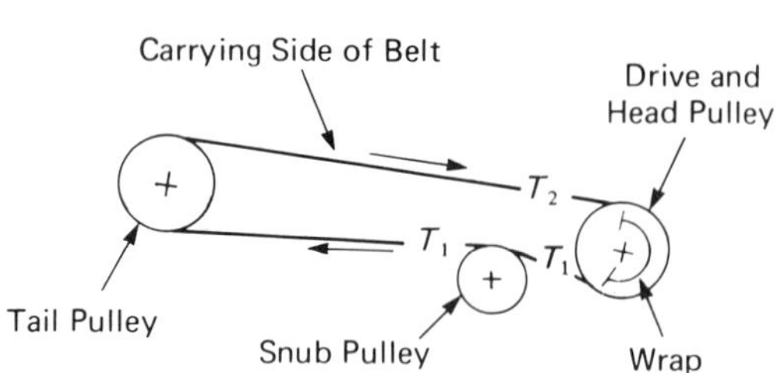

6.6G. *Single-pulley drive at head end of conveyor with snub pulley; regenerative.*

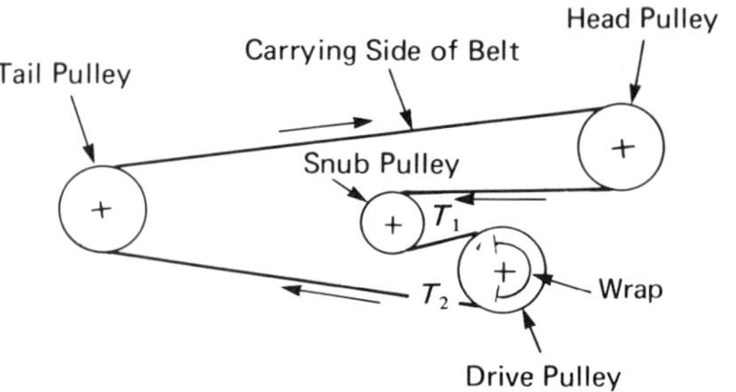

6.6H. *Single-pulley drive on return run.*

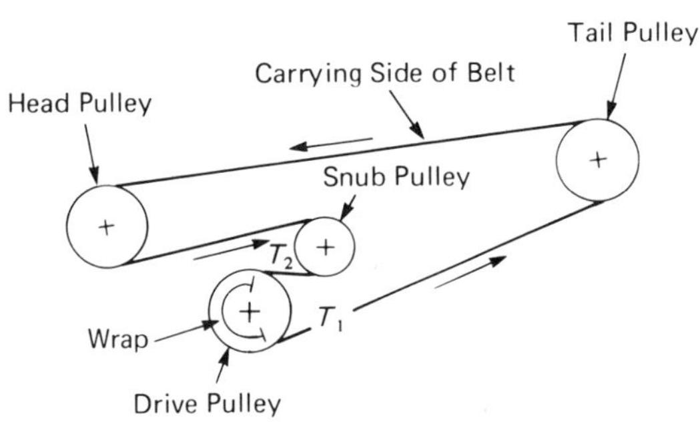

6.6I. *Single-pulley drive on return run; regenerative.*

FIGURE 6.7. ***Dual-Pulley Drive Arrangements***

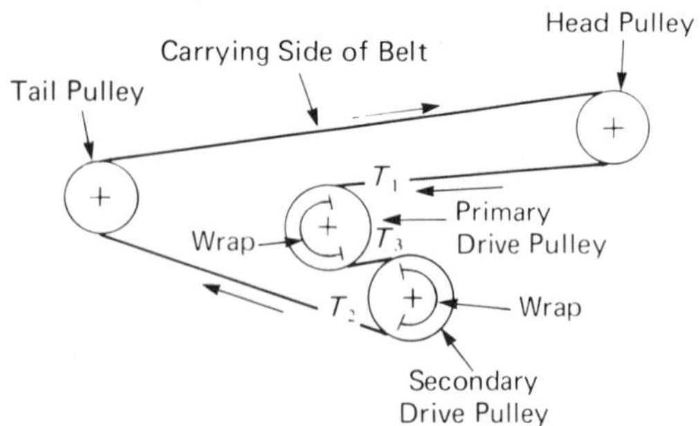

6.7A. *Dual-pulley drive on return run..*

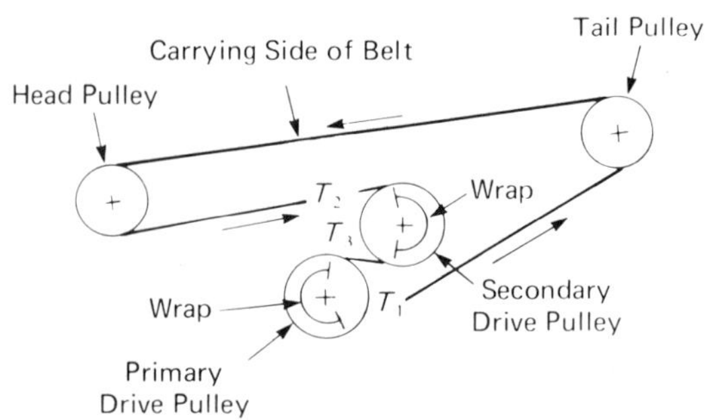

6.7B. *Dual-pulley drive on return run; regenerative.*

Figure 6.7 *con't.* *Dual-Pulley Drive Arrangements*

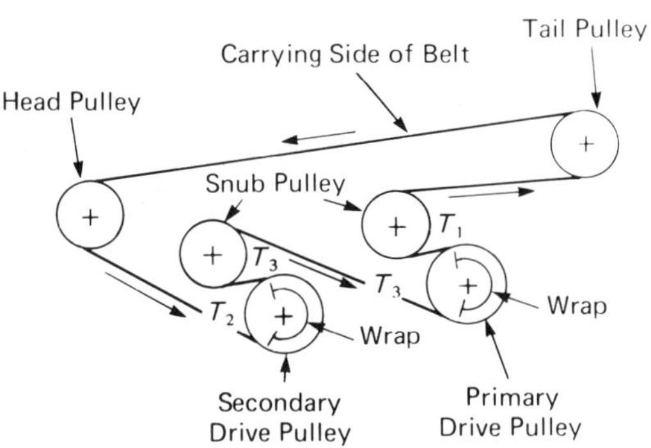

6.7C. *Dual-pulley drive on return run; regenerative.*

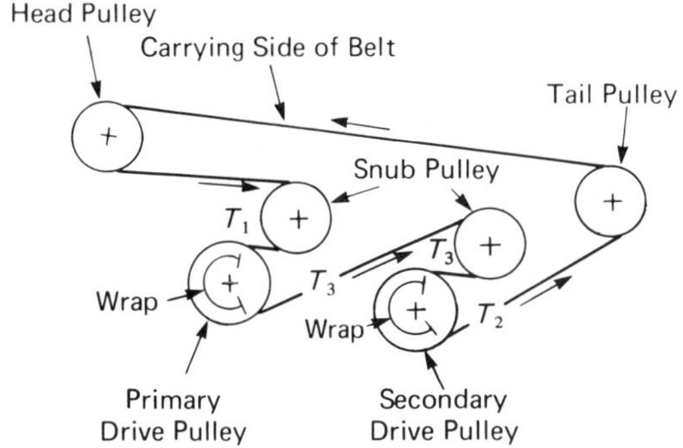

6.7D. *Dual-pulley drive on return run. Drive pulleys engage clean side of belt.*

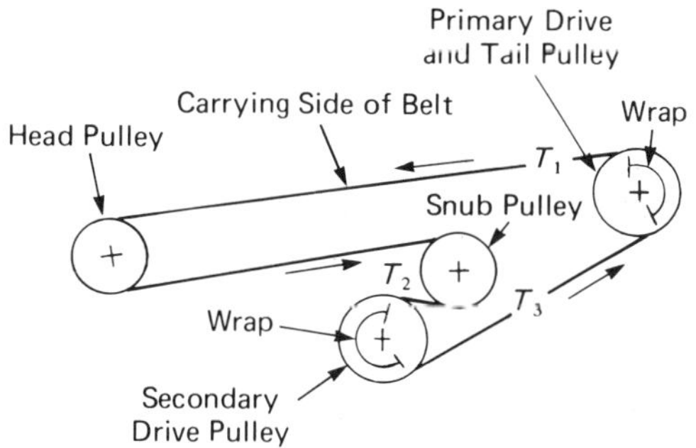

6.7E. *Dual-pulley drive with primary drive on tail pulley of conveyor; regenerative.*

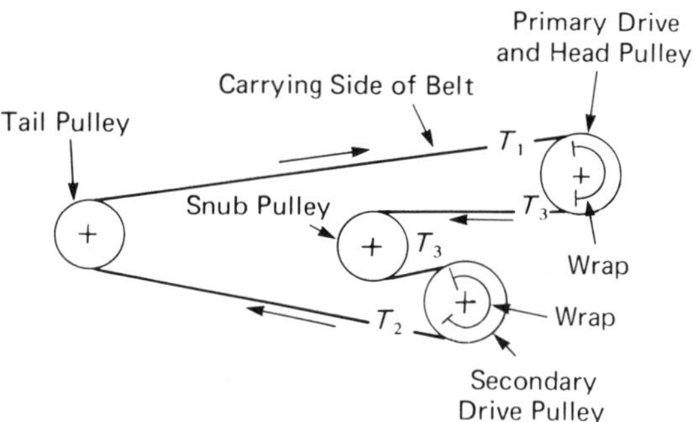

6.7F. *Dual-pulley drive with primary drive on head pulley of conveyor.*

Maximum and Minimum Belt Tensions

For the illustrated common conveyor profiles and drive arrangements, minimum and maximum tensions will be discussed and procedures given for calculating the belt tension at any point in the conveyor. The applicable formulas are indicated with the various profile and drive arrangements where single-pulley drives are involved. The tensions involved in multiple-pulley drives are treated separately.

Maximum Belt Tension

Operating Maximum Belt Tension. The operating maximum belt tension is defined as the maximum belt tension occurring when the belt is conveying the design load from the loading point continuously to the point of discharge. Operating maximum tension usually occurs at the discharge point on horizontal or inclined conveyors and at the loading point on regenerative declined conveyors. On compound conveyors, the operating maximum belt tension frequently occurs elsewhere. Because the operating maximum belt tension must be known to select a belt, its location and magnitude must be determined. For details on belt tensions, refer to Figures 6.8 through 6.16.

Conveyors having horizontal and lowering, or horizontal and elevating, sections can have maximum tensions at points other than a terminal pulley. In this case, belt tensions can be calculated by considering the horizontal and sloping sections as separate conveyors.

Temporary Operating Maximum Belt Tension. A temporary operating maximum belt tension is that maximum tension which occurs only for short periods. For example, a conveyor with a profile which contains an incline, a decline, and then another incline, may generate a higher operating tension when only the inclines are loaded and the decline is empty. These temporary operating maximum belt tensions should be considered in the selection of the belt and the conveyor machinery.

Starting and Stopping Maximum Tension

The starting torque of an electric motor may be more than 2½ times the motor full-load rating. Such a torque transmitted to a conveyor belt could result in starting tensions many times more than the chosen operating tension. To prevent progressive weakening of splices and subsequent failure, such starting maximum tensions should be avoided. Refer to Chapter 12. Likewise, if the belt is brought to rest very rapidly, especially on decline conveyors, the inertia of the loaded belt may produce high tensions.

The generally recommended maximum for starting belt tension is 150% of the allowable belt working tension. On conveyors with tensions under 75 lbs per ply inch or the equivalent, the maximum can be increased to as high as 180%. For final design allowances, conveyor equipment or rubber belt manufacturers should be consulted.

Minimum Belt Tension, T_{min}

For conveyors which do not overhaul the drive, the minimum belt tension on the carrying run will usually occur at the tail (feed) end. For conveyors which do overhaul their drive, the minimum belt tension will usually occur at the head (discharge) end. The locations and magnitude of minimum belt tensions are given in connection with the conveyor profiles and drives shown in Figures 6.8 through 6.16.

It will be seen that the minimum tension is influenced by the T_2 tension required to drive, without slippage of the belt on the pulley, and by the T_0 tension required to limit the belt sag at the point of minimum tension. The minimum tension is calculated both ways and the larger value used. If T_0 to limit belt sag is larger than the T_{min} produced by the T_2 tension necessary to drive the belt without slippage, a new T_2 tension is calculated, using T_0 and considering the slope tension, T_b, and the return belt friction, T_{yr}. Formulas for calculating T_2, having T_0, T_b, and T_{yr}, are given for each of the conveyor profiles and drive arrangements.

Tension Relationships and Belt Sag Between Idlers

Chapter 5, "Belt conveyor idlers," presents the basic facts on the subject of idler spacing. A major requirement, noted in Chapter 5, is that the sag of the belt between idlers must be limited to avoid spillage of conveyed material over the edges of the belt. The sag between idlers is closely related to the weight of the belt and material, the idler spacing, and the tension in the belt.

Graduated Spacing of Troughing Idlers

For belt conveyors with long centers, it is practical to vary the idler spacing so as to equalize the catenary sag of the belt as the belt tension increases.

The basic equation for the sag in a catenary can be written:

$$\text{Sag, ft} = \frac{WS_i^2}{8T}$$

where W = weight, $(W_b + W_m)$, lbs per ft, of belt and material
S_i = idler spacing, ft
T = tension in belt, lbs

The basic sag formula can also be expressed as a relation of belt tension, T, idler spacing, S_i, and the weight per foot of belt and load, $(W_b + W_m)$, in the form:

y = vertical drop (sag) between idlers, ft

$$y = \frac{S_i^2 (W_b + W_m)}{8T}$$

Experience has shown that when a conveyor belt sags more than 3% of the span between idlers, load spillage often results. For 3% sag the equation becomes:

$$\frac{S_i^2 (W_b + W_m)}{8T} = \frac{3\, S_i}{100}$$

While pure catenary equations are used, the allowable percent sag takes into account such factors as stiffness of the belt carcass, strength of the belt span due to the "channel" shape of a troughed belt, etc.

Simplifying for minimum tension to produce various percentages of belt sag yields the following formulas:

For 3% sag, $T_0 = 4.2\, S_i (W_b + W_m)$

For 2% sag, $T_0 = 6.25\, S_i (W_b + W_m)$

For 1½% sag, $T_0 = 8.4\, S_i (W_b + W_m)$

See Table 6-10 for recommended belt sag percentages for various full load conditions.

The graduated spacing should be calculated to observe the following limitations: (1) A maximum of 3% sag should be maintained when belt is operating with a normal load. (2) A maximum of 4.5% sag should be maintained when the loaded belt is standing still. (3) The idler spacing should not exceed twice the suggested normal spacing of the troughing idlers listed in Table 5-2. (4) The load on any idler should never exceed the idler load ratings given in Chapter 5.

Moreover, the number of spacing variations must be based on practical considerations, such as the number of different stringer sections in the conveyor support structure, so that the fabrication cost of the support structure does not become excessive. Usually, the spacing of troughing idlers is varied in 6-inch increments.

Limiting the calculated belt sag to 3% of the idler spacing, at any point on the conveyor, will generally prevent spillage of the material from conveyor belts operating over 20° troughing idlers.

When handling lumpy material on belts operating on 35° (or deeper) troughing idlers, belt tension should be increased to reduce the percent of sag. Deep-troughed conveyor belts normally carry a relatively large cross-sectional loading and corresponding heavy weight of material per foot of length. Therefore, the material exerts a greater pressure against the side of the trough, tending to cause greater transverse belt flexure. The purpose of increasing the minimum belt tension in belts operating on idlers of greater than 20° troughing angle is to keep this transverse belt flexure to an acceptable minimum and thus prevent spillage.

Similarly, when frequent surge loads are encountered or a substantial percentage of large lumps is expected, the material weight per foot of conveyor will be increased. Consideration of increased minimum belt tension at, or closely adjacent to, the loading points is recommended.

Table 6–10. ***Recommended Belt Sag Percentages for Various Full Load Conditions***

Idler troughing angle	*Material*		
	All fines	*One-half the maximum lump size**	*Maximum lump size**
20°	3%	3%	3%
35°	3%	2%	2%
45°	3%	2%	1½%

*See Figure 4.1

Note: Reduced load cross sections will permit an increase in the sag percentage resulting in a decreased minimum tension. Such a choice may lead to a more economical belt selection, as the maximum tension correspondingly will be reduced.

Slack Side Tension, T_2. The minimum tension required to drive the belt without slippage is the product of T_e and C_w. However, the value to be used for minimum belt tension on the carrying run is either T_o (calculated as above) plus or minus the tension T_b and plus or minus the return belt friction T_{yr}, or the minimum tension to drive without slippage $T_e \times C_w$. By rearranging and substituting terms,

$$T_2 = T_0 \pm T_b \pm T_{yr}$$

or, by the above definition,

$$T_2 = T_e C_w$$

Use the larger value of T_2.

Tension, T_b. The weight of the carrying and/or return run belt for a sloped conveyor is carried on the pulley at the top of the slope. This must be considered in calculating the T_2 tension, as indicated above.

$$T_b = HW_b$$

where W_b = weight of belt, lbs per ft
H = net change in elevation, ft

Return Belt Friction Tension, T_{yr}. The return belt friction is the belt tension resulting from the empty belt moving over the return run idlers:

$$T_{yr} = 0.015\ LW_b\ K_t$$

where L = length, ft, of conveyor to center of terminal pulleys
K_t = temperature correction factor as defined on pages 79–80. For temperatures above 32°F, $K_t = 1.0$

Belt Tensions for Conveyors of Marginal Decline. Allowances made for frictional losses in a conveyor are intended as conservative assumptions. When a declined conveyor is involved, such allowances should be discounted when considered in conjunction with maximum possible regenerative tensions (or horsepower).

Belt Tensions for Typical Conveyors. When calculating tensions at any point in these conveyor profiles, the portions of the conveyor on zero slope, inclines, or declines should be considered as separate conveyors.

Belt Tension at Any Point, X, on Conveyor Length. In order to understand clearly the formulas for evaluating the belt tension at any point, X, on the belt conveyor length, it is necessary to establish the following nomenclature:

L_x = distance, ft, from tail pulley to point X along the conveyor

H_x = vertical distance, ft, from tail pulley to point X

T_{cx} = belt tension, lbs, at point X on the carrying run

T_{rx} = belt tension, lbs, at point X on the return run

T_t = belt tension, lbs, at tail pulley

T_{hp} = belt tension, lbs, at head pulley

T_{wcx} = tension, lbs, at point X on the carrying run, resulting from the weight of belt and material carried

T_{fcx} = tension, lbs, at point X on the carrying run, resulting from friction

T_{wrx} = tension, lbs, at point X on the return run, resulting from the weight of the empty belt

T_{frx} = tension, lbs, at point X on the return run, resulting from friction

$T_{wcx} = H_x (W_b + W_m)$

$T_{fcx} = L_x [K_t K_x + K_y W_b] + L_x K_y W_m$

$T_{wrx} = H_x W_b$

$T_{frx} = 0.015\ L_x W_b K_t$

Formulas for T_{cx} and T_{rx} are given for all the belt conveyor profiles and drives in Figures 6.8 through 6.16. Nondriving pulley frictions have been omitted.

Analysis of Belt Tensions

In addition to calculation of the effective belt tension, *Te*, which occurs at the drive pulley, a designer must consider the belt tension values that occur at other points of the conveyor's belt path.

Figures 6.8 through 6.16 illustrate various possible conveyor layouts and profiles and the appropriate tension analysis. Some of these examples are more commonly applied than others; the order of presentation is not intended to infer preference of design. Many of these diagrams illustrate the takeup, *TU*, in alternate locations. It is most unusual for a conveyor to employ more than one takeup; a preferred single location should be chosen.

FIGURE 6.8. *Head Pulley Drive—Horizontal or Elevating*.*

$T_e = T_1 - T_2$
$T_2 = C_w \times T_e$ or $T_2 = T_t + T_b - T_{yr}$
Use the larger value of T_2

$T_t = T_0$ or
$T_t = T_2 - T_b + T_{yr}$
Use the larger value of T_t
$T_t = T_{min}$ $T_1 = T_{max}$

$T_{cx} = T_t + T_{wcx} + T_{fcx}$
$T_{rx} = T_t + T_{wrx} - T_{frx}$

Takeup on return run or at tail pulley

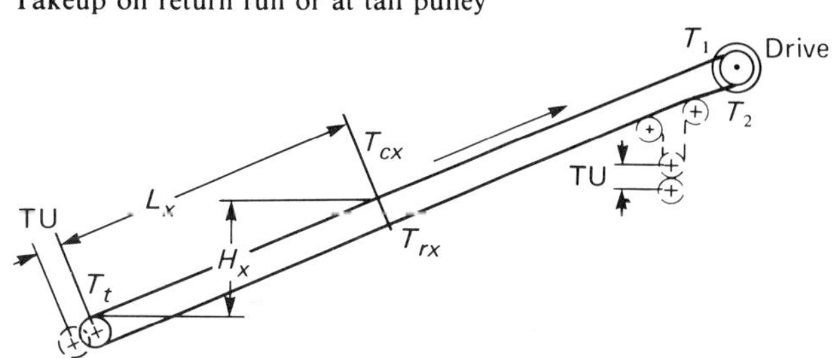

6.8A. *Inclined conveyor with head pulley drive.*

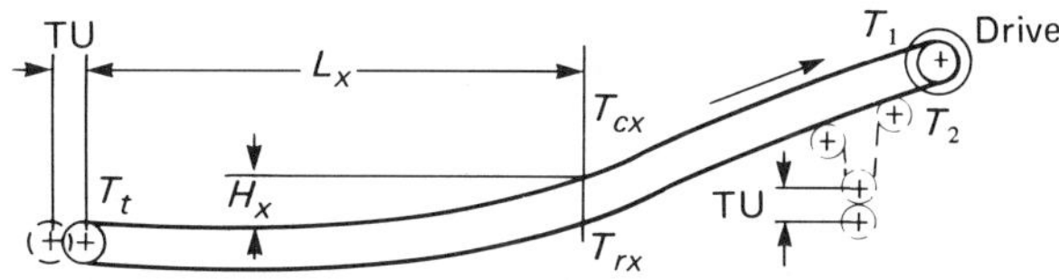

6.8B. *Horizontal belt conveyor with concave vertical curve, and head pulley drive.*

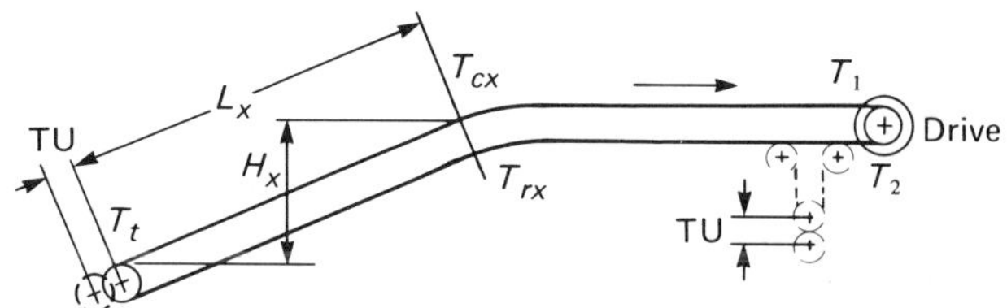

6.8C. *Horizontal belt conveyor with convex vertical curve , and head pulley drive.*

*Note: Two takeups are shown only to illustrate alternative. Two automatic takeups cannot function properly on the same conveyor.

FIGURE 6.9. ***Head Pulley Drive—Lowering without Regenerative Load****

$T_e = T_1 - T_2$
$T_2 = C_w \times T_e$ or $T_2 = T_0 - T_e$
Use the larger value of T_2

$T_1 = T_e + T_2$
$T_t = T_2 + T_b + T_{yr}$
$T_{max} = T_t$ or T_1
$T_{min} = T_t$ or T_1
$T_{cx} = T_t - T_{wcx} + T_{fcx}$
$T_{rx} = T_t - T_{wrx} - T_{frx}$

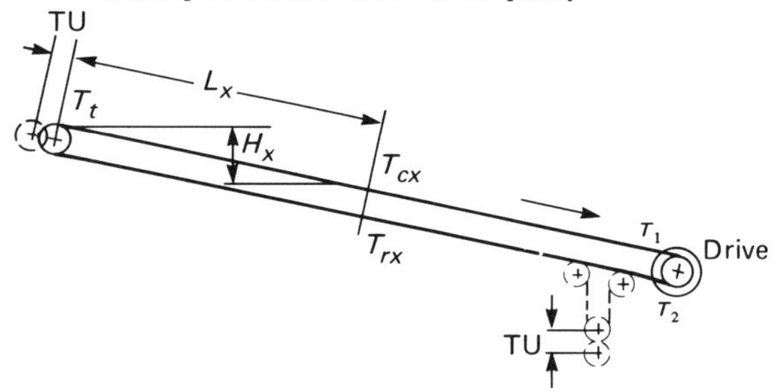

6.9A. *Declined belt conveyor with head pulley drive. Lowering without regenerative load.*

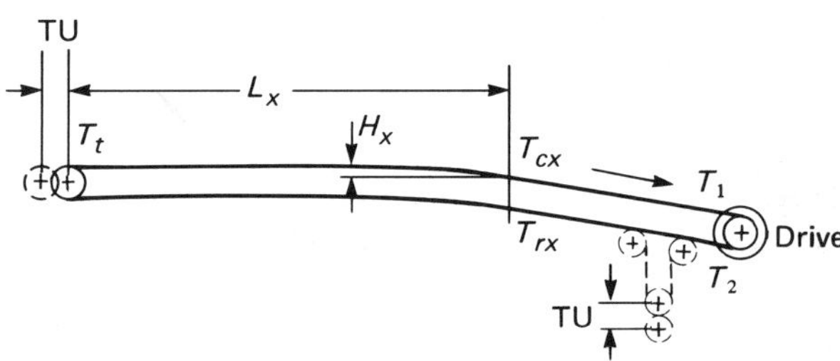

6.9B. *Conveyor with convex vertical curve, head pulley drive. Lowering without regenerative load.*

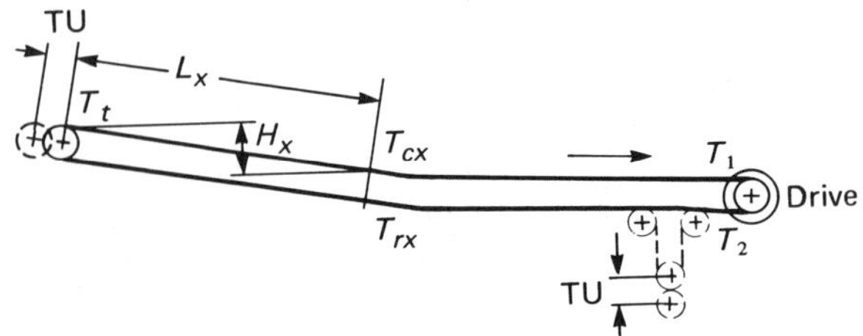

6.9C. *Conveyor with concave vertical curve, head pulley drive. Lowering without regenerative load.*

**Note: Two takeups are shown only to illustrate alternatives. Two automatic takeups cannot function properly on the same conveyor.*

FIGURE 6.10. *Head Pulley Drive—Lowering with Regenerative Load*

$T_e = T_1 - T_2$
$T_2 = C_w \times T_e$ or $T_2 = T_0$
Use the larger value of T_2

$T_t = T_{max} = T_1 + T_b + T_{yr}$
$T_{min} = T_2$
$T_{cx} = T_t - T_{wcx} + T_{fcx}$
$T_{rx} = T_t - T_{wrx} - T_{frx}$

Takeup on return run not recommended to avoid driving through the takeup

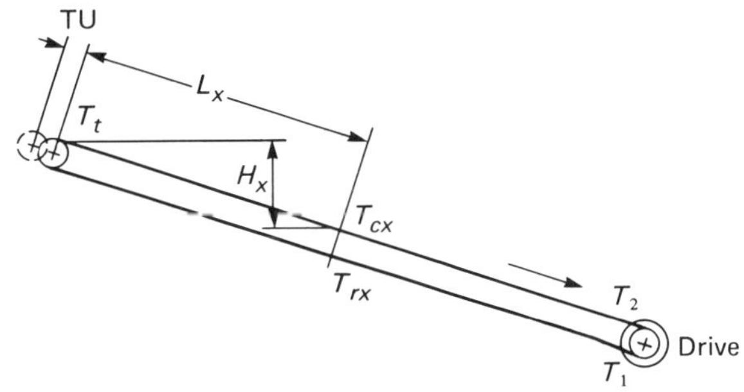

6.10A. *Declined belt conveyor with head pulley drive. Lowering with regenerative load.*

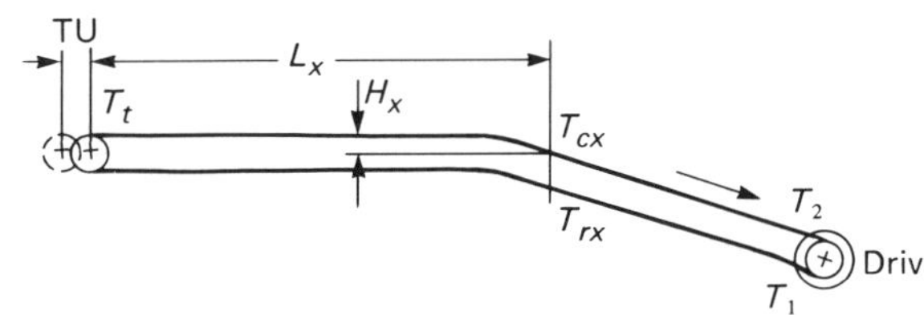

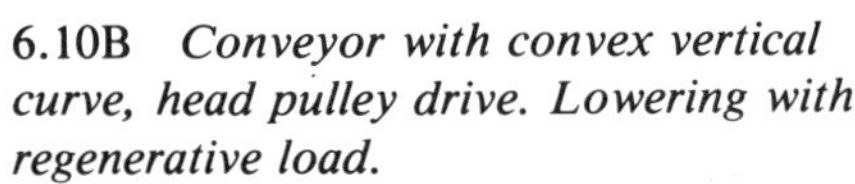
6.10B *Conveyor with convex vertical curve, head pulley drive. Lowering with regenerative load.*

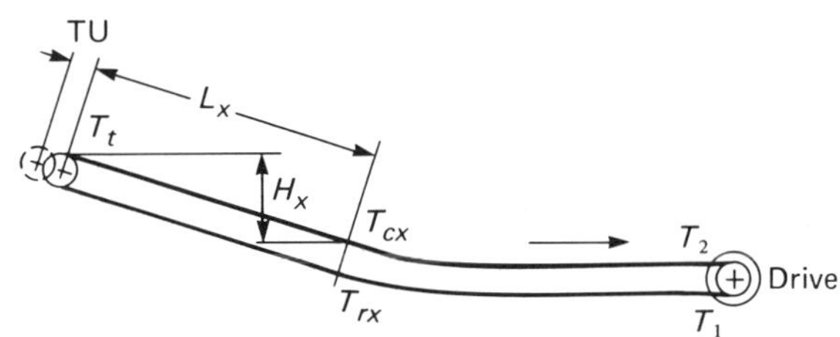

6.10C. *Conveyor with concave vertical curve, head pulley drive. Lowering with regenerative load.*

FIGURE 6.11. ***Tail Pulley Drive—Horizontal or Elevating***

$T_e = T_1 - T_2$
$T_2 = C_w \times T_e$ or $T_2 = T_0$
Use the larger value of T_2

$T_t = T_2$
$T_{min} = T_2$
$T_{hp} = T_1 - T_{yr} + T_b$
$T_{max} = T_1$
$T_{cx} = T_2 - T_{wcx} + T_{fcx}$
$T_{rx} = T_1 - T_{wrx} - T_{frx}$

Takeup on return run not recommended. For arrangement which is preferred to above, drive is on return run, Figure 6.15A, page 114.

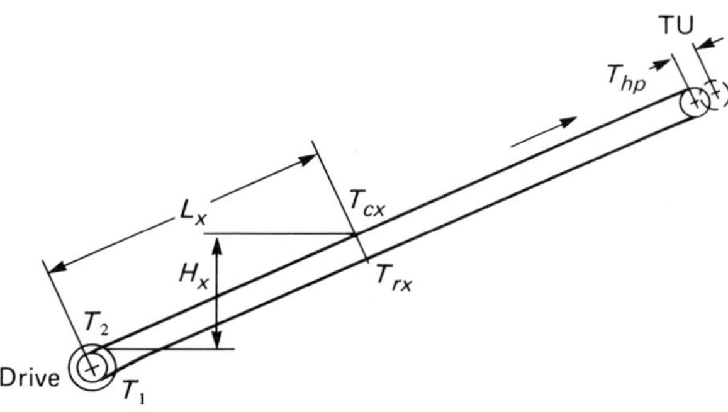

6.11A. *Inclined conveyor with tail pulley drive.*

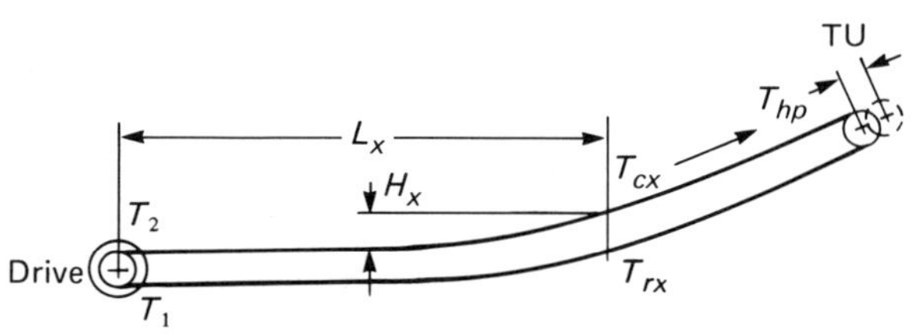

6.11B. *Horizontal belt conveyor with concave vertical curve, and tail pulley drive.*

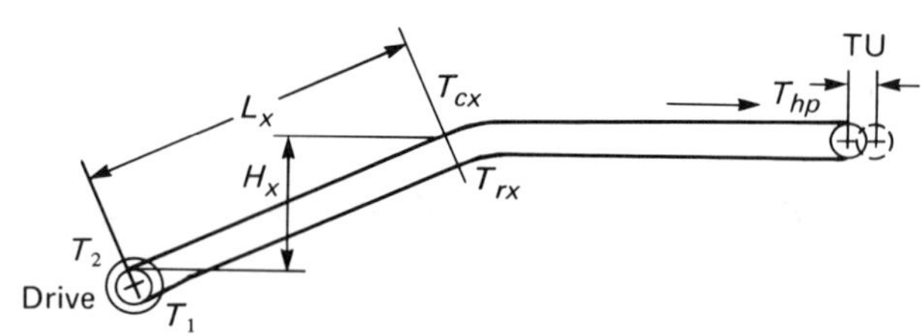

6.11C. *Horizontal belt conveyor with convex vertical curve, and tail pulley drive.*

FIGURE 6.12. ***Tail Pulley Drive—Lowering without Regenerative Load***

$T_e = T_1 - T_2$

$T_2 = C_w \times T_e$ or $T_2 = T_{hp} + T_b + T_{yr} - T_e$

Use the larger value of T_2

$T_{hp} = T_0$ or $T_{hp} = T_1 - T_b - T_{yr}$

Use the larger value of T_{hp}

$T_{hp} = T_{min}$

$T_1 = T_e + T_2 = T_{max}$

$T_{cx} = T_2 - T_{wcx} + T_{fcx}$

$T_{rx} = T_1 - T_{wrx} - T_{frx}$

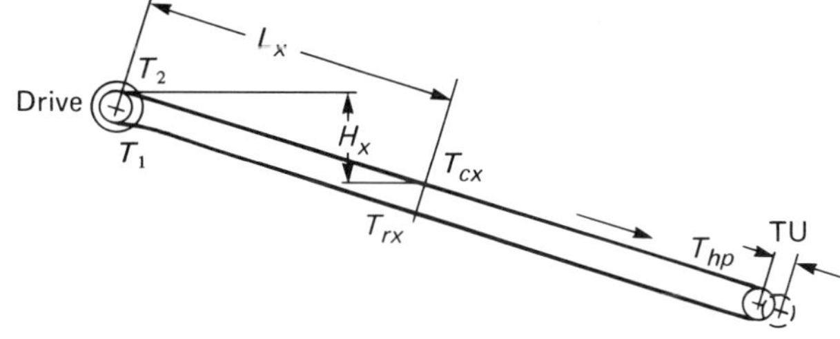

6.12A. *Declined belt conveyor with tail pulley drive. Lowering without regenerative load.*

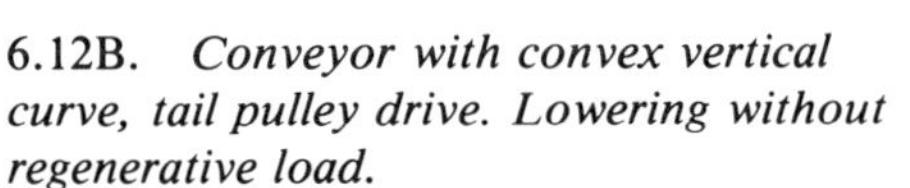

6.12B. *Conveyor with convex vertical curve, tail pulley drive. Lowering without regenerative load.*

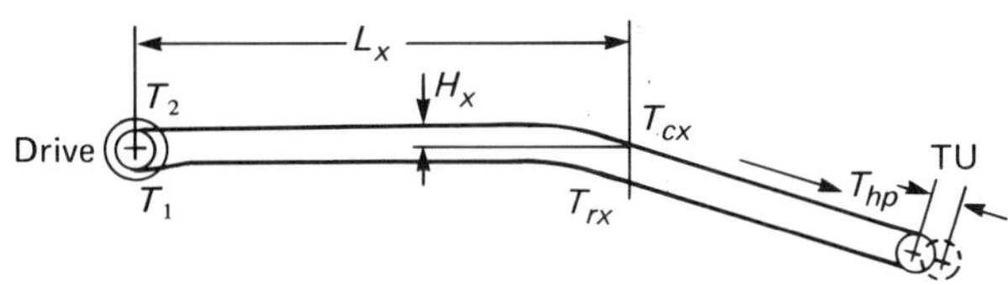

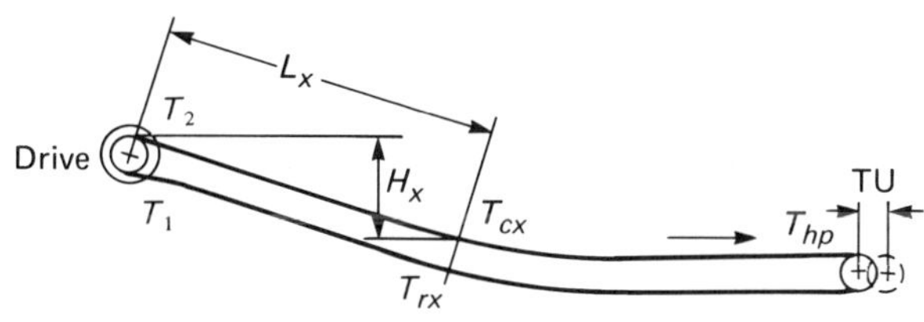

6.12C. *Conveyor with concave vertical curve, tail pulley drive. Lowering without regenerative load.*

FIGURE 6.13. ***Tail Pulley Drive—Lowering with Regenerative Load****

$T_e = T_1 - T_2$
$T_2 = C_w \times T_e$ or $T_2 = T_0 + T_b + T_{yr}$
Use the larger value of T_2

$T_{hp} = T_2 - T_b - T_{yr}$ or $T_{hp} = T_0$
$T_{hp} = T_{min}$
$T_1 = T_{max} = T_e + T_2$
$T_{cx} = T_1 - T_{wcx} + T_{fcx}$
$T_{rx} = T_2 - T_{wrx} - T_{frx}$
"Calculate belt tension required at takeup during acceleration and make takeup adequate to prevent lift-up. See page 126."

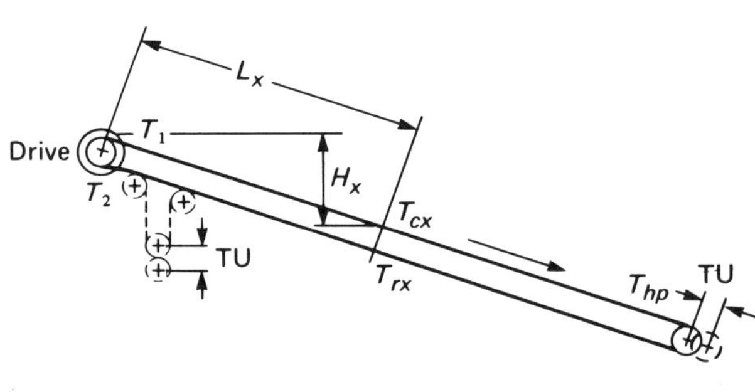

6.13A. *Declined belt conveyor with tail pulley drive. Lowering with regenerative load.*

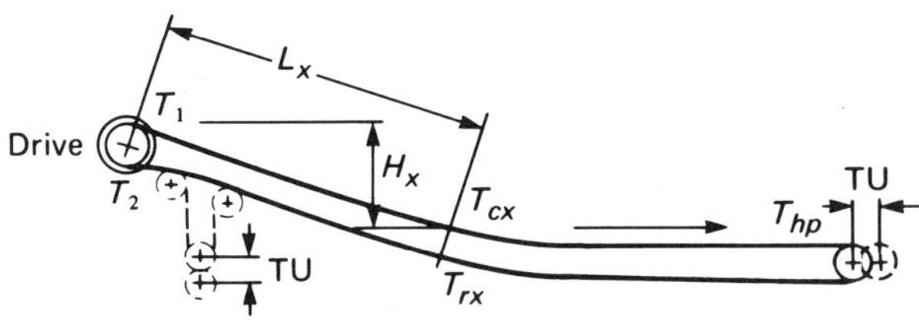

6.13B. *Conveyor with concave vertical curve, tail pulley drive. Lowering with regenerative load.*

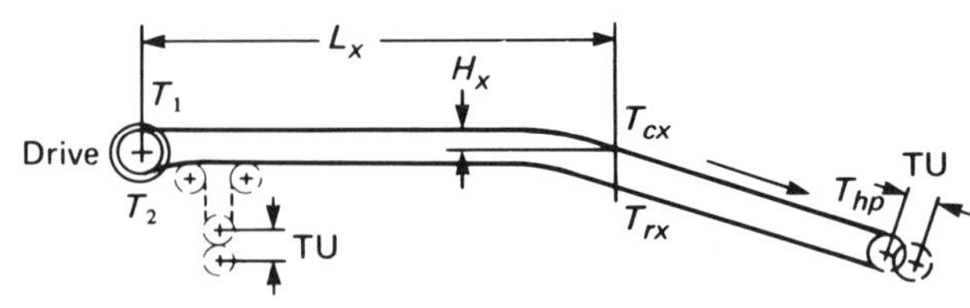

6.13C. *Conveyor with convex vertical curve, tail pulley drive. Lowering with regenerative load.*

*Note: Two takeups are shown only to illustrate alternatives. Two automatic takeups cannot function properly on the same conveyor.

FIGURE 6.14. *Drive on Return Run—Horizontal or Elevating**

$T_e = T_1 - T_2$
$T_2 = C_w \times T_e$ or
$T_2 = T_0 - 0.015\, W_b L_s + W_b H_d$
here
H_d = lift to the drive pulley
Use the larger value of T_2

$T_t = T_{min}$ and $T_t = T_0$
$T_t = T_2 + 0.015\, W_b L_s - W_b H_d$
Use the larger value of T_t

$T_{hp} = T_t + T_{fcx} + T_{wcx}$ where $L_x = L$
$T_{hp} = T_{max}$
$T_{cx} = T_t + T_{fcx} + T_{wcx}$
$T_{rx} = T_t - T_{frx} + T_{wrx}$

Takeups on return run or at tail pulley.

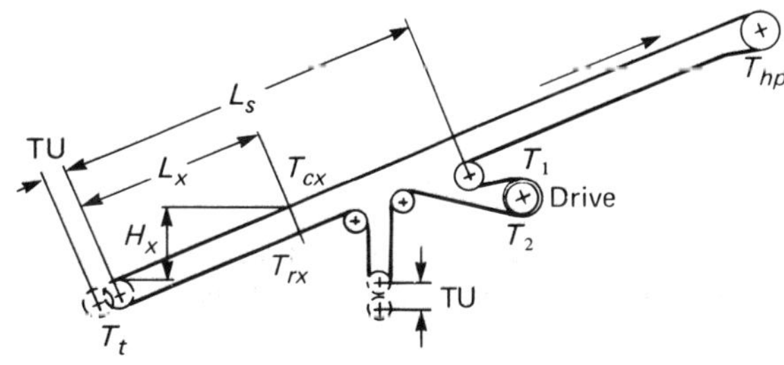

6.14A. *Inclined conveyor with drive on return run.*

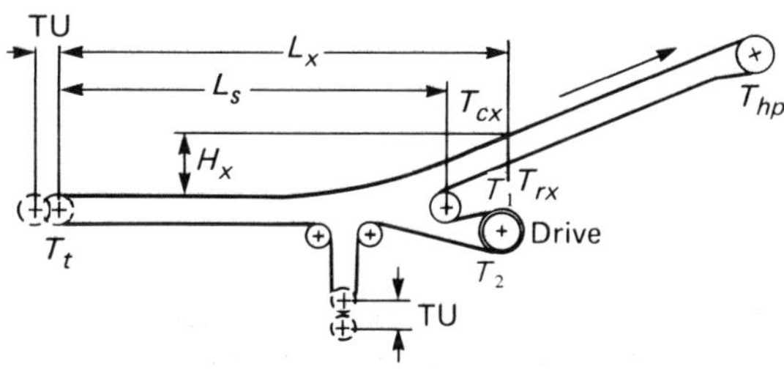

6.14B. *Horizontal belt conveyor with concave vertical curve, and drive on return run.*

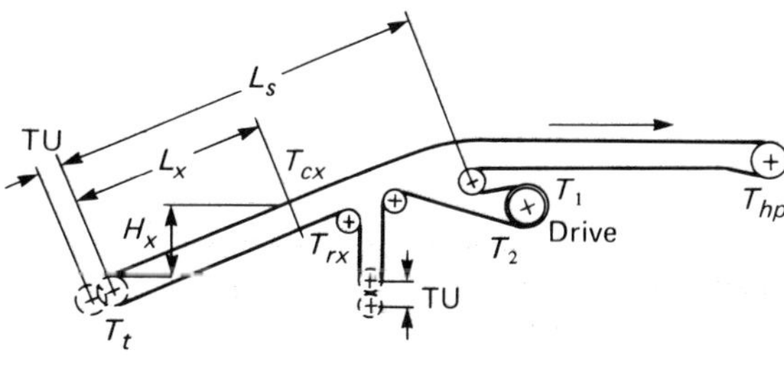

6.14C. *Horizontal belt conveyor with convex vertical curve, and drive on return run.*

*Note: Two takeups are shown only to illustrate alternatives. Two automatic takeups cannot function properly on the same conveyor.

Figure 6.15. ***Drive on Return Run—Lowering without Regenerative Load****

$T_e = T_1 - T_2$

$T_2 = C_w \times T_e$ or $T_2 = T_0 - T_e$

Use the larger value of T_2

$T_1 = T_e + T_2$

$T_t = T_2 + 0.015\, W_b L_s - W_b H_d$

here

H_d = lift to the drive pulley, or $T_t = T_0$

Use the larger value of T_t

$T_{max} = T_t$ or T_1

$T_{min} = T_t$ or T_1

$T_{hp} = T_t + T_{fcx} - T_{wcx}$

here

$L_x = L$

$T_{cx} = T_t + T_{fcx} - T_{wcx}$

$T_{rx} = T_t - T_{frx} - T_{wrx}$

Takeup on return run or at tail pulley

6.15A. *Declined conveyor, with drive on return run. Lowering without regenerative load.*

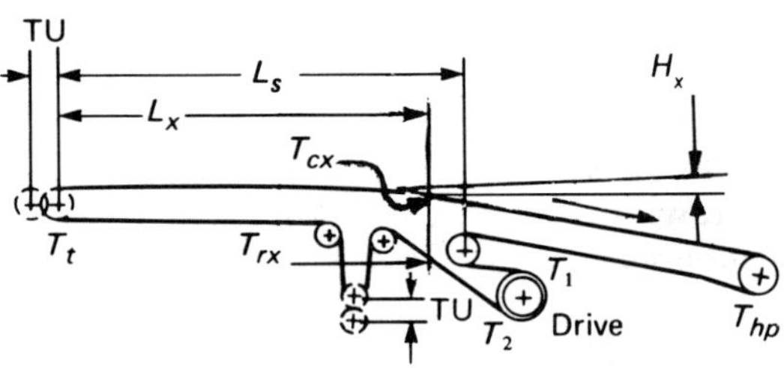

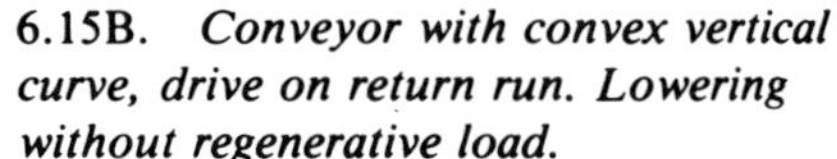
6.15B. *Conveyor with convex vertical curve, drive on return run. Lowering without regenerative load.*

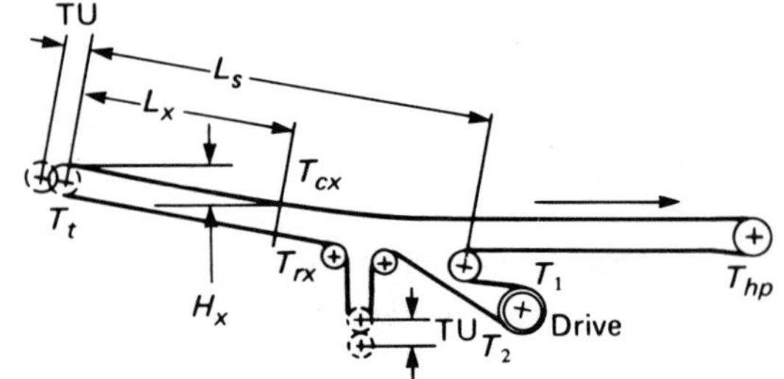

6.15C. *Conveyor with concave vertical curve, drive on return run. Lowering without regenerative load.*

**Note: Two takeups are shown only to illustrate alternatives. Two automatic takeups cannot function properly on the same conveyor.*

FIGURE 6.16. *Drive on Return Run—Lowering with Regenerative Load**

$T_e = T_1 - T_2$
$T_2 = C_w \times T_e$ or $T_2 = T_0$
Use the larger value of T_2

$T_1 = T_e + T_2$
$T_t = T_{max} = T_1 + 0.015\, W_b L_s + W_b H_d$
here
H_d = lift to the drive pulley
$T_{hp} = T_0$
$T_{hp} = T_2 - 0.015\, W_b\, (L - L_s) - W_b (H + H_d)$
Use the larger value of T_{hp}

$T_{cx} = T_t - T_{wcx} + T_{fcx}$
$T_{rx} = T_t - T_{wrx} - T_{frx}$

Takeup on return run or at head pulley

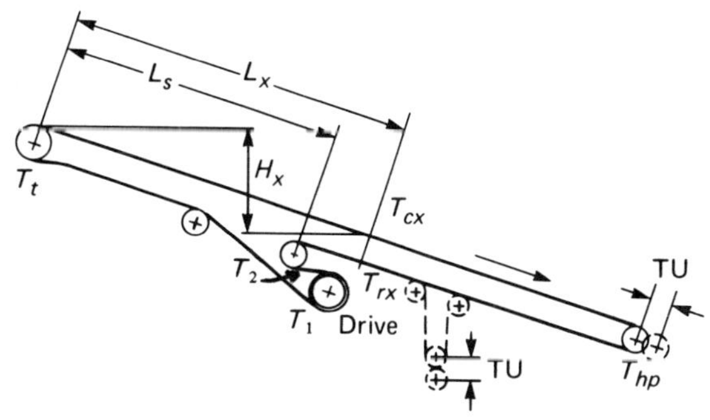

6.16A. *Declined conveyor with drive on return run. Lowering with regenerative load.*

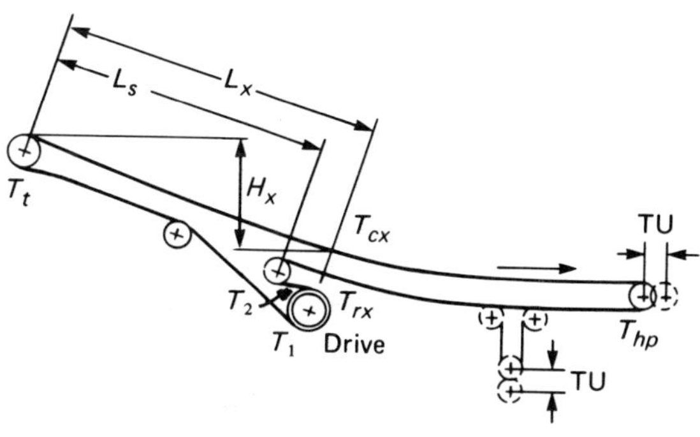

6.16B. *Conveyor with concave vertical curve, drive on return run. Lowering with regenerative load.*

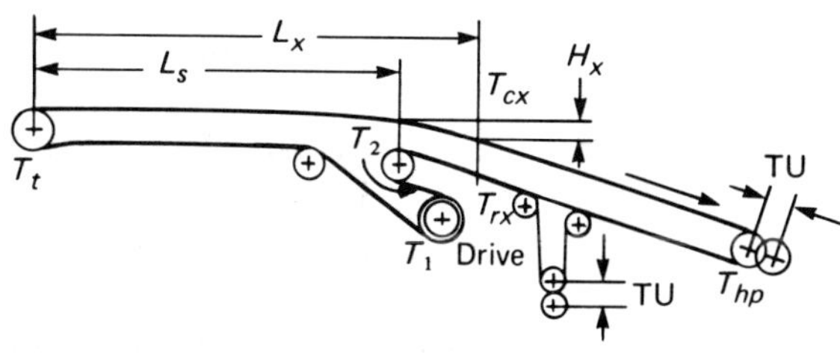

6.16C. *Conveyor with convex vertical curve, drive on return run. Lowering with regenerative load.*

**Note: Two takeups are shown only to illustrate alternatives. Two automatic takeups cannot function properly on the same conveyor.*

Belt Tension Examples

A typical calculation of the various tensions in a conveyor belt with single-pulley drive is given below.

Example 1

Calculate the various belt tensions for a 30-inch belt conveyor per Figure 6.8A, with 300-ft centers, and a lift of 50 ft. Capacity is 500 tph, of material weighing 100 lbs per cu ft, at a belt speed of 350 fpm. The belt is carried on 5-inch diam. Class C5 idlers (see Chapter 5) with ¾-inch shafts, 35° end roll angle. Idlers are spaced every 3½ ft. The material contains 50% lumps. Belt weight is 15.0 lbs per ft, W_b. Material weight is 47.5 lbs per ft, W_m. Temperature is 60°F. T_e has been calculated and is equal to 3,030 lbs.

Step 1: Determine C_w. Assume lagged pulley, gravity takeup, and 180° wrap. Table 6-8 gives $C_w = 0.50$.

Step 2: Determine belt tension, T_2. Minimum T_2 to drive $= T_e \times C_w$ (3,030)(0.50) = 1,515 lbs. T_0, the minimum allowable tension for a 2% belt sag, per page 104, is as follows:

$$T_0 = 6.25\ S_i\,(W_b + W_m) = (6.25)(3.5)(15.0 + 47.5) = 1,367 \text{ lbs}$$

Using the formula for determining return belt friction tension, page 79,

$$L = 300$$
$$W_b = 15$$
$$K_t = 1.0, \text{ at } 60°\text{F.}$$

$$T_{yr} = 0.015 L W_b K_t = (0.015)(300)(15)(1) = 68 \text{ lbs}$$

$$T_2 \text{ (considering } T_0) = T_0 + T_b - T_{yr}$$

$$T_b = HW_b = (50)(15) = 750 \text{ lbs}$$

therefore,

$$T_2 = 1,367 + 750 - 68 = 2,049 \text{ lbs}$$

Because this is larger than the 1515 lbs minimum T_2 to drive, use $T_2 = 2,049$ lbs.

Step 3: Calculate the T_1, T_{max}, and takeup tensions.

$$T_{max} = T_1 = T_e + T_2 = 3,030 + 2,049 = 5,079 \text{ lbs}$$

Takeup tension depends upon the location of the gravity takeup. If located near the head end, the belt tension at takeup is T_2 less the weight of nearly 3 ft of belt (45 lbs), or 2,049 − 45 = 2,004 lbs. If located near the tail pulley, the takeup tension will be approximately the same as T_0:

$$T_0 = T_2 + T_{yr} - T_b = 2,049 + 68 - 750 = 1,367 \text{ lbs}$$

(The nondriving pulley frictions have been omitted.)

A typical calculation of the belt tensions of a two-pulley (dual) drive is given below. The conveyor outline is per Figure 6.14A, but with a dual-pulley head drive, per Figure 6.7F.

Example 2

Conveyor length = 1,200 ft
Belt speed = 400 fpm
T_e at the drive pulleys = 20,625 lbs
Required horsepower at the drive pulleys = 250 hp

Total motor horsepower = 275 hp
Primary drive motor = 200 hp
Secondary drive motor = 75 hp
Belt weight, W_b, = 20 lbs per ft
C_w = 0.11, according to Table 6-8 (380° wrap, lagged pulleys)

Step 1: Calculate T_{ep} and T_{es}:

$$T_{ep} = \left(\frac{200}{275}\right)(250)\left(\frac{33{,}000}{400}\right) = 15{,}000 \text{ lbs}$$

$$T_{es} = \left(\frac{75}{275}\right)(250)\left(\frac{33{,}000}{400}\right) = 5{,}625 \text{ lbs}$$

Step 2: Calculate T_2, which is the minimum value that avoids slippage of the belt on the secondary pulley:

$$T_2 = T_e C_w = (20{,}625)(0.11) = 2{,}269 \text{ lbs}$$

Step 3: Calculate T_3: $T_3 = T_2 + T_{es} = 2{,}269 + 5{,}625 = 7{,}894$ lbs

Step 4: Calculate T_1: $T_1 = T_3 + T_{ep} = 7{,}894 + 15{,}000 = 22{,}894$ lbs

Step 5: Calculate C_{wp} and C_{ws}:

$$C_{wp} = \frac{T_3}{T_{ep}} = \frac{7{,}894}{15{,}000} = 0.53, \text{ requiring } 180° \text{ wrap angle.}$$

$$C_{ws} = \frac{T_3}{T_{es}} - 1 = \frac{7{,}894}{5{,}625} - 1 = 0.40,$$

requiring 205° wrap angle.

Step 6: Check T_2, using formula in Figure 6.14A. Assume conveyor is 1,200 ft long, the lift is 60 ft, $W_b = 20$, $W_m = 80$, idler spacing 3½ ft, drive at head of the conveyor.

$$T_0 = T_{min} = 6.25(W_b + W_m)S_i = 6.25(20 + 80)3.5$$
$$= 2{,}188 \text{ lbs}$$

(See "Minimum Belt Tension," page 103.)

$$T_{yr} = 0.015 L W_b K_t = (.015)(1{,}200)(20) = 360 \text{ lbs}$$

(See "Return Belt Friction Tension," page 105; assume temperature above 32°, $K_t = 1.0$)

$$T_b = HW_b = (60)(20) = 1{,}200 \text{ lbs}$$

(See "Tension, T_b" page 105.)

Then $$T_2 = T_{min} + T_b - T_{yr} = 2188 + 1200 - 360 = 3{,}028 \text{ lbs}$$

Therefore, as T_2 based on T_{min} is larger than T_2 minimum to prevent slippage (3,028 is greater than 2,269), use $T_2 = 3{,}028$ lbs.

Step 7: Calculate corrected values of T_3, T_1, C_{ws}, and C_{wp}:

$$T_3 = T_2 + T_{es} = 3{,}028 + 5{,}625 = 8{,}653 \text{ lbs}$$

$$T_{max} = T_1 = T_3 + T_{ep} = 8{,}653 + 15{,}000 = 23{,}653 \text{ lbs}$$

$$C_{ws} = \frac{T_3}{T_{es}} - 1 = \frac{8{,}653}{5{,}625} - 1 = 0.54$$

$$C_{wp} = \frac{T_3}{T_{ep}} = \frac{8{,}653}{15{,}000} = 0.58$$

Based on wrap factors from Step 5, which provide the minimum T_2 tension to drive without slippage, the secondary drive pulley would require a 205° angle of wrap, and the primary drive pulley would require a 180° angle of wrap.

The revised value of T_2 and the correspondingly revised values of C_{wp} and C_{ws}, per Step 7, indicate that both drives could have 180° angles of wrap. In order to have equal resistance to slip, both drives should have approximately the same wrap angle.

Belt Tension Calculations

Five illustrative examples are offered to make clear the use of the formulas in determining the belt tensions at point X on the belt conveyor.

Example 1 The basis for this example is the conveyor profile in Figure 6.8A from page 107, repeated below.

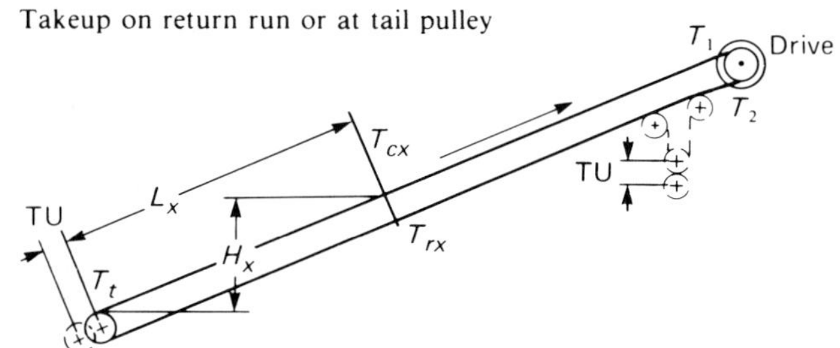

FIGURE 6.8A.

48-inch belt conveyor

W_b = weight of belt = 15 lbs per ft
W_m = weight of material = 106.6 lbs per ft

Troughing idlers, 20° angle, Class E6, 6-inch dia, spaced at 3½ ft, factor A_i = 2.8

Return idlers, Class C6, 6-inch dia, spaced at 10 ft

K_t = temperature correction factor = 1.0

$T_t = T_0$ = 1,788 lbs, as $T_0 = T_{min}$ here

L_x = 1,000 ft

H_x = 31.3 ft

To find the belt tension at point X on the carrying run:

T_{cx} = tension at point X on carrying run

T_{wcx} = tension resulting from weight of belt and material at point X

T_{fcx} = tension resulting from friction on carrying run at point X

$$T_{cx} = T_t + T_{wcx} + T_{fcx}$$

$$T_{wcx} = H_x(W_b + W_m) = (31.3)(121.6) = 3{,}806 \text{ lbs}$$

$T_{fcx} = L_x K_t(K_x + K_y W_b) + L_x K_y W_m$ when K_t = 1.0, then
$T_{fcx} = L_x [K_x + K_y(W_b + W_m)]$

$$K_x = .00068\ (W_b + W_m) + \frac{A_i}{S_i}$$
$$= 0.00068\ (15 + 106.6) + \frac{2.8}{3.5}$$
$= 0.883$ (for value of A_i, see tabulation on page 81)

$K_y = 0.025$ when conveyor length is 1,000 ft (see Table 6-2). At 3.13% slope, $W_b + W_m = 121.6$ (use 125 in tables); and 3½ ft standard idler spacing.

$$T_{fcx} = 1{,}000\ [0.883 + 0.025(121.6)] = 3{,}923 \text{ lbs}$$
$$T_{cx} = 1{,}788 + 3{,}806 + 3{,}923 = 9{,}517 \text{ lbs}$$

To find the belt tension at point X on the return run:

T_{rx} = tension at point X on the return run

T_{wrx} = tension at point X on the return run resulting from the weight of the belt

T_{frx} = tension at point X on the return run resulting from return run friction

$$T_{rx} = T_t + T_{wrx} - T_{frx}$$
$$T_{wrx} = H_x W_b - (31.3)(15) = 470 \text{ lbs}$$
$$T_{frx} = L_x .015\ W_b K_t = (1{,}000)(0.015)(15)(1) = 225 \text{ lbs}$$
$$T_{rx} = 1{,}788 + 470 - 225 = 2{,}033 \text{ lbs}$$

Example 2 The basis for this example is the conveyor profile shown in Figure 6.8B from page 107, repeated below; otherwise, the data is the same as for Example 1, above.

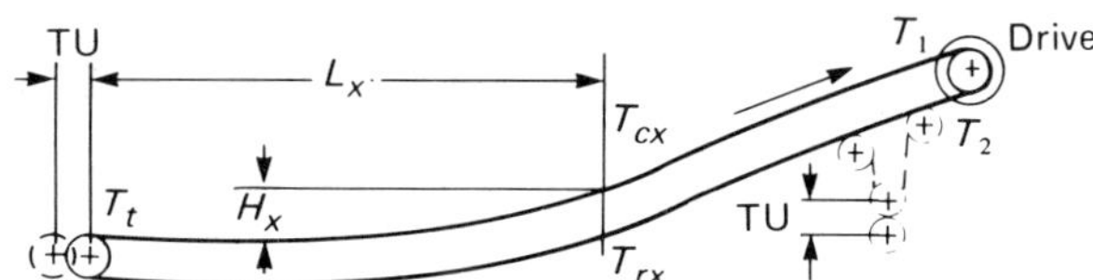

FIGURE 6.8B

$L_x = 1{,}915$ ft

$H_x = 31.3$ ft

$T_t = 1{,}788$ lbs

Horizontal portion 1,565 ft long, inclined portion 835 ft long, on 9% slope.

To find the T_{cx} tension in the carrying run at point X:

$$T_{cx} = T_t + T_{wcx} + T_{fcx}$$
$$T_{wcx} = H_x(W_b + W_m) = (31.3)(121.6) = 3{,}806 \text{ lbs}$$
$$T_{fcx} = L_x\ [K_x + K_y(W_b + W_m)] \text{ Since } K_t = 1.0$$

T_{fcx} is figured in two parts, first for the horizontal portion and then for the inclined portion.

For the horizontal portion of the carrying run:

$K_x = 0.883$

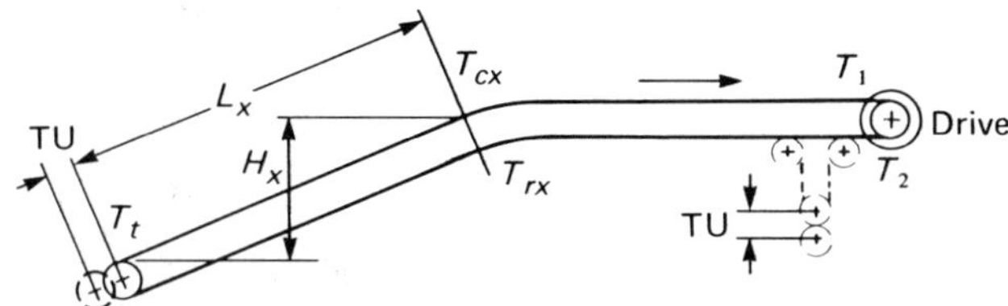

FIGURE 6.8C.

K_y = 0.0277 when conveyor length is 1,565 ft at 0° slope; $W_b + W_m$ = 121.6, use 125 in tables; 3½ ft standard idler spacing.

T_{fcx}, horizontal = 1,565 [(0.883) + (0.0277)(121.6)] = 6,653 lbs

For the inclined portion of the carrying run:

$K_x = 0.883$

K_y = 0.0217 when conveyor length is 1,915 ft at 1.63% average slope; $W_b + W_m$ = 121.6, use 125 in tables; 3½ ft standard idler spacing.

T_{fcx}, *incline,* = 352 [(0.883) + (0.0217)(121.6)] = 1,240 lbs, where 352 ft is the distance along the inclined portion to point X. Total T_{fcx}, horizontal plus incline = 6,653 + 1,240 = 7,893 lbs

T_{cx} = 1,788 + 3,806 + 7,893 = 13,487 lbs

To find the T_{rx} tension in the return belt at point X:

$T_{rx} = T_t + T_{wrx} - T_{frx}$

$T_{wrx} = (H_x)(W_b) = (31.3)(15) = 470$ lbs

$T_{frx} = L_x(0.015\ W_b)\ K_t = 1{,}915(0.015)(15)(1.0) = 431$ lbs

$T_{rx} = 1{,}788 + 470 - 431 = 1{,}827$ lbs

Example 3 The basis for this example is the conveyor profile shown in Figure 6.8C from page 107, repeated following. The data is the same as for Examples 1 and 2 above.

L_x = 350 ft, on 9% slope

H_x = 31.3 ft

T_t = 1,788

To find T_{cx} at point X on carrying run:

$$T_{cx} = T_t + T_{wcx} + T_{fcx}$$

$T_{wcx} = H_x(W_b + W_m) = (31.3)(121.6) = 3{,}806$ lbs

$T_{fcx} = L_x[K_x + K_y(W_b + W_m)]$

$K_x = 0.883$

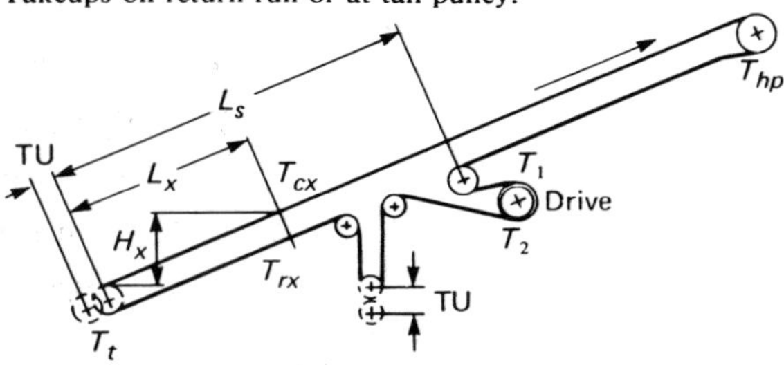

FIGURE 6.14A.

K_y = 0.0293 when conveyor length is 350 ft at 9% slope; $W_b + W_m$ = 121.6 (use 125 in tables); 3½ ft standard idler spacing.

$T_{fcx} = 350\ [0.883 + (0.0293)(121.6)] = 1{,}556$ lbs

$T_{cx} = 1{,}788 + 3{,}806 + 1{,}556 = 7{,}150$ lbs

To find T_{rx} at point X on return run:

$T_{rx} = T_t + T_{wrx} - T_{frx}$

$T_{wrx} = H_x W_b = (31.3)(15) = 470$ lbs

$T_{frx} = L_x(0.015\ W_b)\ K_t = 350(0.015)(15)(1.0) = 79$ lbs

$T_{rx} = 1{,}788 + 470 - 79 = 2{,}179$ lbs

Example 4 This example illustrates the problem of finding belt tensions when the drive is on the return run. The tension calculated at the head and tail pulleys is carried out as follows:

(1) Calculate T_e, T_1, T_2, and T_t, the same as for a conveyor driven at a terminal pulley, using the horsepower formula on page 77, and appropriate tension formulas indicated with the conveyor profiles, Figure 6.14. The T_e, T_1, and T_2 tensions so calculated apply at the drive pulley regardless of its location along the return run.

(2) Calculate the tension, T_{hp}, at the head pulley using the appropriate formula for T_{cx} as indicated for conveyor profile with drive on the return run, Figure 6.14. Calculate T_{wcx} and T_{fcx} from the formulae for determining belt tension at any point (page 106).

Conveyor data:

W_m = 120 lbs per ft

W_b = 15 lbs per ft

K_t = 1.0

K_x = 0.35

K_y = 0.0243

C_w = 0.35

S_i = 3.5 ft, idler spacing

36-inch belt conveyor, 600-ft centers, drive located midway of return run, lift 54 ft, slope 9%.

Calculate the head pulley tension, T_{hp}, and the tail pulley tension, T_t.

$$T_e = L\ K_t(K_x + K_y W_b + 0.015\ W_b) + W_m(LK_y \pm H)$$
$$= 600\ [0.35 + (0.0243)(15) + (0.015)(15)] + 120[(600)(0.0243) + 54] = 8,794 \text{ lbs}$$

For 3% belt sag,

$$T_0 = 4.2\ S_i(W_b + W_m) = (4.2)(3.5)(135) = 1{,}985 \text{ lbs}$$

T_2 minimum to drive = $T_e C_w$ = (8,794)(0.35) = 3,078 lbs

Corresponding $T_t = T_2 + (L/2)(0.015\ W_b) - (H/2)\ W_b$
= 3,078 + (300)(0.015)(15) − (27)(15) = 2,741 lbs.

This total tail pulley tension, 2,741 lbs, is greater than 1,985 lbs.

Therefore: $T_t = 2{,}741$ lbs, and $T_2 = 3{,}078$ lbs

$T_1 = T_e + T_2 = 8{,}794 + 3{,}078 = 11{,}872$ lbs

The tension at any point on the carrying run is:

$$T_{cx} = T_t + T_{wcx} + T_{fcx}$$

Now let $L_x = L$

$K_t = 1.0$

Then, tension at the head pulley, $T_{hp} = T_t + T_{wcx} + T_{fcx}$

$T_{wcx} = H_x(W_b + W_m) = (54)(135) = 7{,}290$ lbs

$T_{fcx} = L_x[K_tK_x + K_yW_b] + L_xK_yW_m$, and since $L = L_x$

$= 600\ [0.35 + (0.0243)(15)] + (600)(0.0243)(120)$
$= 2{,}178$ lbs

Therefore, $T_{hp} = 2{,}741 + 7{,}290 + 2{,}178 = 12{,}209$ lbs

This is the maximum belt tension. The T_1 tension at the drive pulley may be checked as follows:

$$T_1 = T_{hp} - 27W_b + L/2\ (0.015W_b) = 12{,}209 - (27)(15) + (300)(0.015)(15) = 11{,}872 \text{ lbs}$$

This checks with the 11,872 lbs calculated for T_1 from the formula, $T_1 = T_e T_2$.

Example 5 This example calculates the belt tension at any point in a declined regenerative conveyor. The calculation is substantially the same as that for a nonregenerative conveyor except that ⅔K_y is used in place of K_y, and the factor A_i is eliminated in the formula for K_x. The value of K_y is for the length L_x. Figure 6.10A from page 109 is repeated below.

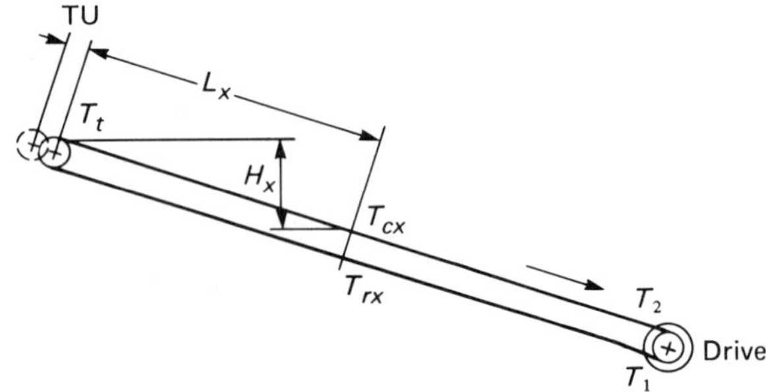

FIGURE 6.10A.

Conveyor data:

36-inch belt conveyor, 1000-ft centers, head pulley drive, drop 90 ft, slope 9%.

$W_b = 15$ lbs per ft

$W_m = 120$ lbs per ft

S_i = 3.5 ft idler spacing

$K_x = 0.00068\ (W_b + W_m) = 0.00068(135) = 0.0918$

$K_y = (0.0169)(0.666) = 0.01126$ for 1,000 ft, 135 for $(W_b + W_m)$ and 9% slope

$K_t = 1.0$

$C_w = 0.35$

$T_{cx} = T_t - T_{wcx} + T_{fcx}$

$T_{rx} = T_t - T_{wrx} - T_{frx}$

$T_e = LK_t(K_x + K_yW_b + 0.015W_b) + W_m(LK_y \pm H)$

$T_2 = T_eC_w$, or $T_2 = T_0$. If T_0 is the greater

$T_1 = T_e + T_2$

$T_t = T_1 + 0.015W_bL + W_bH$

$T_e = 1{,}000\ [0.0918 + (0.01126)(15) + (0.015)(15)] + (1{,}000)(0.01126)(120) - (90)(120) = 485.7 + 1{,}351.2 - 51.2 - 10{,}800 = -8{,}963$ lbs

The minus sign merely means that the belt drives the pulley (the conveyor is regenerative).

$T_2 = (8{,}963)(0.35) = 3{,}137$ lbs

T_0 for 3% sag = $(4.2)(3.5)(135) = 1{,}985$ lbs

Therefore, T_2 can be taken at 3,137 lbs

$T_1 = T_e + T_2 = 8{,}963 + 3{,}137 = 12{,}100$ lbs

$T_t = 12{,}100 + (1{,}000)(0.015)(15) + (90)(15) = 13{,}675$ lbs

To calculate T_{cx} at a point 500 ft from the tail shaft

$$T_{cx} = T_t - T_{wcx} + T_{fcx}$$

$T_{wcx} = H_x(W_b + W_m) = (45)(135) = 6{,}075$ lbs

$T_{fcx} = L_x(K_x + K_yW_b) + L_xK_yW_m$ when $K_t = 1.0$

$= 500\ [0.0918 + (0.0182)(15)] + (500)(0.0182)(120)$

$= 182 + 1{,}092 = 1{,}274$ lbs when K_y = ⅔ K_y for a 500-ft conveyor at 9% slope.

Then, $T_{cx} = 13{,}675 - 6{,}075 + 1{,}274 = 8{,}874$ lbs

To calculate T_{rx} at a point 500 ft from tail shaft

$$T_{rx} = T_t - T_{wrx} - T_{frx}$$

$T_{wrx} = H_xW_b = (45)(15) = 675$ lbs

$T_{frx} = L_x(0.015)\ W_b = (500)(0.015)(15) = 113$ lbs when $K_t = 1.0$

Then $T_{rx} = 13{,}675 - 675 - 113 = 12{,}887$ lbs

For conveyor profiles per Figures 6.10B and 6.10C, the portion of the conveyor on a given slope is calculated separately, as in Example 2, page 119, and Problems 5 and 6, pages 155–159.

Acceleration and Deceleration Forces

Investigation of the acceleration and deceleration forces is necessary for the following reasons.

Belt Stress

Economy of design dictates the selection of a belt having a carcass strength at or near the normal operating tensions. Consequently, the additional forces resulting from acceleration or deceleration may overstress the belt or its splices, particularly if mechanical splices are used. While this problem is most likely to exist with respect to the belt, there also is the possibility of overstressing the mechanical components such as pulleys, shafts, bearings, takeups, etc.

Vertical Curves

Two different problems may be encountered with vertical curves.

In the case of concave curves (where the center of curvature lies above the belt) if belt tensions are too high during starting, the belt will lift off the troughing idlers. This is discussed in detail in Chapter 9. It is necessary to analyze this problem in regard to full, partial, and no loads.

In the case of convex vertical curves, where the center of the curvature lies below the belt, there is the possibility of overloading certain idlers.

Loss of Tension Ratio

During both acceleration and deceleration there exists the distinct possibility of losing the required T_1/T_2 ratio necessary to maintain the desired control of the engagement of belt and drive pulley. This particularly is true if the takeup is located far from the drive.

If a screw takeup is used and improperly adjusted or the travel of a gravity takeup is too limited, the necessary ratio T_1/T_2 may be lost during the attempt to accelerate the belt conveyor.

During deceleration, the effect of the inertia load may cause a loss of the T_1/T_2 ratio necessary to transmit braking forces from the braking pulley to the belt. This would permit the continued motion of the belt and load, after the pulley had been stopped.

Load Conditions on the Belt

The belt conveyor may operate satisfactorily during stopping or starting if completely loaded or if empty. This, however, may not be so if only portions of the conveyor length are loaded. The conveyor, therefore, has to be analyzed under various conditions of loading.

For example, when a belt conveyor contains a concave curve, a critical condition of starting may be the lifting of the belt at the curve during acceleration because the portion of the belt ahead of the concave vertical curve is loaded, while the remainder of the belt is not. This may not be true if the conveyor is regenerative. Such conditions require careful analysis.

Coasting

Where there is a system of belt conveyors transferring from one to another, sequence starting or stopping is almost always a prerequisite of design. As an example, a belt with very long centers may transfer to a belt of short centers, in which case the time required to decelerate the two belts must obviously be synchronized, despite the differences in the braking forces required. During the acceleration period, the same synchronization is necessary. In either case, the consequences of not making a proper analysis and providing the necessary controls will result in a pile-up at the transfer point and possible destruction of the machinery and belt, plus an inoperative system.

Takeup Movement

During both the acceleration and deceleration cycles, where counterweighted takeups are used, the takeup travel may be insufficient unless these forces are considered. The engineer must consider not only the length of travel, but also the rate of travel, particularly where hydraulic, electric, or pneumatic controls are involved.

Effect on Material Carried

In certain instances, the rate of starting and stopping may exert influences on the material which result in intolerable conditions. Obviously, certain materials can be accelerated or decelerated more effectively by the belt than others. For example, if a declined belt conveyor handling pelletized iron ore is stopped too rapidly, the material may start to roll on the belt surface and result in a pile-up at the discharge point. Similarly, starting an inclined belt too rapidly may cause the material to roll backward.

Festooning

Without proper consideration of the starting and stopping forces, it is possible that belt tensions may drop to a point, at some spot in the line, where the belt will festoon (buckle). For example, a belt with a decline from the tail end, and an incline at the head, may be loaded at the tail end only. If braking is applied at the head pulley, the belt may have zero tension or even some slack on the carrying side. The obvious result is load spillage, entanglement, loss of alignment, etc.

Power Failure

In the event of power failure, the belt eventually will stop because of inherent friction forces. Depending upon the profile and conditions of loading, the time required for the friction forces to stop the belt may be intolerably long or short. In the case of a declined regenerative belt conveyor, it may completely unload itself. In a system of belt conveyors, a pile-up of material at transfer points is probable. Therefore, it is obvious that controlled stopping, in the event of a power failure, is very important.

Braking Tensions Taken by Return Run and Tail Pulley

When deceleration is accomplished by means of a brake, the belt tension resulting from the braking force is taken in a direction opposite to that for driving the belt.

For instance, if the drive is at the head end of a horizontal or lift conveyor, power is transmitted from the drive pulley to the carrying side of the belt when the motor is energized. When decelerating with a brake connected to the drive pulley and the motor de-energized, the braking force may be transmitted from the drive pulley to the return belt. The brake application, therefore, may be significant in determining the amount of the counterweight, the design of the takeup, and the shaft sizes.

These are some of the problems that result when acceleration and deceleration forces are ignored or are improperly evaluated. While other difficulties may also exist, those discussed above are sufficient to indicate the importance of proper consideration and analysis.

Analysis of Acceleration and Deceleration Forces

The accelerating and decelerating forces which act on a belt conveyor during the starting and stopping intervals are the same in either case. However, their magnitude and the algebraic signs governing these forces change, as do the means for dealing with them.

Acceleration

The acceleration of a belt conveyor is accomplished by some form of prime mover, usually by an electric motor. The resulting forces in a horizontal conveyor are determined by inertia plus friction; in an inclined conveyor, by inertia plus friction plus elevating of the load; in a declined conveyor, by inertia plus friction minus lowering of the load.

Deceleration

The deceleration of a belt conveyor is accomplished by some form of brake. The resulting forces in a horizontal conveyor are determined by inertia minus friction; in an inclined conveyor, by inertia minus friction minus elevating of the load; in a declined conveyor, by inertia minus friction plus lowering of the load.

If the conveyor contains several portions with different (positive or negative) slopes, a combination of these conditions may result.

Calculation of Acceleration and Deceleration Forces

The belt conveyor designer is then confronted with the necessity to compute for the conveyor in question the inertia of all its moving parts, the inertia of the load on the belt, total frictional forces, and forces caused by elevating or lowering the load and belt. To be useful, the first two quantities have to be converted to a pound force at the belt line.

Inasmuch as acceleration is defined as the second derivative of displacement with respect to time, and deceleration is simply negative acceleration, time is the basic variable in computing the force. To compute the time, Newton's second law is used. The basic approach is as follows:

$$F_a = Ma$$

where F_a = accelerating or decelerating force, lbs

M = mass, in slugs = $\frac{W_e}{g}$

W_e = equivalent weight of moving parts of the conveyor and load, lbs

g = acceleration by gravity = 32.2 ft/sec^2

a = acceleration, ft per second per second (ft/sec^2).

The force necessary to achieve the acceleration or deceleration is always directly proportional to the mass (or the weight) of the parts and material in motion.

For purposes of calculation, it can be assumed that the belt and the load on it move in a straight line. Other important parts of the system, however, rotate. This is true for all pulleys (including those on takeups and belt trippers), all idlers, and all the rotating parts of the drive.

It appears convenient to use the equation for linear motion as the basis for calculating the acceleration and deceleration forces. This makes it necessary to convert the physical properties of the rotating components of the system to a form in which they can be used in the basic linear relationship:

$$F = \left(\frac{W_e}{g}\right)a$$

In other words, one must find the "Equivalent Weight" of the rotating parts.

For rotating bodies, the mass actually distributed around the center of rotation is equivalent in its effect to the whole mass concentrated at a distance, K, (the polar radius of gyration, ft) from that center.

The WK^2 is the weight of the body multiplied by the square of the radius of gyration. If WK^2 is known for the rotating conveyor components, the Equivalent Weight of these components, at the belt line, can be found by solving the equation

$$\text{Equivalent Weight lbs} = WK^2 \left(\frac{2\pi \text{ rpm}}{V}\right)^2$$

where V = belt velocity, fpm.

Values of WK^2 (expressed in lb-ft²), which are difficult to compute, except for very simple shapes, must be obtained for each component from the manufacturers of the conveyor components, motors, transmission elements, etc.

So far considered have been the forces in the system caused by inertia of the moving parts of the conveyor, the moving parts of the drive, and the moving load. Two other forces also are involved, as mentioned on pages 126 and 127. These are: (1) Forces resulting from friction. (2) Forces resulting from the elevating or lowering of the load and belt. These simply represent the components of the weight of the material and belt, in the direction of motion of the belt, in the various portions of the conveyor.

Design Considerations

The belt conveyor designer is confronted with two problems: (1) The necessity to provide a prime mover powerful enough to start the conveyor, sometimes under adverse conditions. (2) To make sure, for the reasons outlined under "Acceleration and deceleration forces," pages 124 through 126, that the maximum force exerted on the conveyor is within safe limits.

In long, level, high-speed conveyors, a motor large enough for continuous full-load operation may be unable to start the fully loaded conveyor, particularly in cold weather. On the other hand, a motor capable of continuous full-load operation of an inclined conveyor may overstress the belt during starting, unless preventive measures are taken.

The maximum permissible accelerating forces are determined by the factors listed in pages 124 through 126 of this chapter. Minimum accelerating forces may be dictated by the time during which the prime mover, which usually is an electric motor, can exert its starting torque without being damaged. This limitation is also affected by the frequency of starting the conveyor system.

In the case of deceleration, maximums are governed by the same factors. A minimum deceleration may be dictated by safety or may be necessary because of the material flow at transfer points. In all deceleration calculations involving brakes, the energy-dissipating capacity of the brake will be an important factor to consider.

Necessary Assumptions

As in all engineering investigations of this type, the first question is, "To what degree of accuracy will the computations have to be carried out?" The answer is not simple. Important factors are the overall size, the importance of the installation, and the type and sensitivity of the equipment adjacent to it.

In any case, numerous simplifying assumptions will have to be made to keep the engineering work within reasonable limits. For examples of simplifying assumptions, refer to the problems connected with belt stretch (elastic elongation from accelerating or decelerating forces) and takeup reactions.

During both the acceleration and deceleration cycles, the transient forces imposed result in extra stretch not encountered during steady state operation. This may result in early splice failure, excessive takeup travel, and other difficulties. Because of the vast differences in carcass construction, from the standpoint of both materials used and methods of manufacture, no single numerical value can express belt stretch as a function of the applied force.

Most manufacturers have listed values of B_m (elastic constant) for their line of belts. These vary from 1.3×10^6 lbs per inch of belt width for steel-cable belt to 2.3×10^3 lbs per ply inch of belt width for cotton fabric belts. Other rubber manufacturers may list different values, but they also would vary over the same wide range.

For this reason, as well as many others, the calculations for acceleration and deceleration treat the system as a rigid body. This is a common practice in the solution of problems in dynamics. And while the results usually are quite satisfactory, there is more cause for concern over the accuracy of results in the case of belt conveyors.

No further attempt will be made to justify simplifying the assumption, because this usually is not of major significance. However, the belt conveyor designer should be aware that, for conveyor systems with very long center belts, stretch considerations should not be overlooked.

Calculations

While the calculations are relatively simple for a conveyor with only one slope, they become increasingly complex for belt conveyors which change slope several times, or which are loaded and unloaded at different points, or which have belt trippers operating on them.

All this results in a great number of possible combinations of load distribution, tripper position, etc. Although theoretically it suffices to investigate only the worst combination of conditions, without analysis, it is usually impossible for even the experienced designer to tell which combination of factors will lead to this extreme case.

In the more complicated cases, it will be necessary to divide the conveyor into portions or sections—within which neither the slope nor the conditions of loading change more than is permitted by the required accuracy of the calculations—and to determine the physical properties for each such portion discussed under "Analysis of acceleration and deceleration forces," page 126. Any really large rotating member of the belt conveyor, because of its very magnitude, may have to be considered a portion or section by itself.

A summation of these weights, forces, and stresses with proper consideration of their algebraic signs will indicate that portion of the system which will impose the most severe limitations to the allowable values for acceleration and deceleration. This, in turn, will permit the selection of the proper prime mover and the necessary control elements.

Conveyor Horsepower Determination—Graphical Method

The graphical method shown in Figures 6.17, 6.18, and 6.19 provides the means for estimating horsepower. Belt tensions can be calculated from the resulting horsepower. This method is suitable for conveyors of moderate capacity having relatively straight paths of travel. The results will be sufficiently accurate to establish horsepower requirements when actual weights of belt and revolving parts per foot of conveyor centers are used in Figure 6.17. However, for use in determining tentative or approximate horsepower, a convenient table of typical weights is superimposed on Figure 6.17.

The graphical method is not suitable for final calculations of horsepower for conveyors having decline portions, high capacity, or complex arrangements of terminals, nor for the extended use of rubber skirting and plows that substantially increase the frictional drag on the conveyor belt. On the other hand, it is useful for tentative estimates of horsepower under most of these conditions.

An example of the use of the graphical method follows.

Determining Required Horsepower—Graphical Method

The following example illustrates a method for determining the required horsepower for a belt conveyor. The example is the same as Problem 1, page 134, calculated by the analytical method.

In this graphical solution, only the horsepower requirements to move the belt horizontally, elevate the material, and convey the material horizontally are considered. Additional accessory factors such as pulley friction, skirtboard friction, material acceleration, and auxiliary device frictions are included as averages.

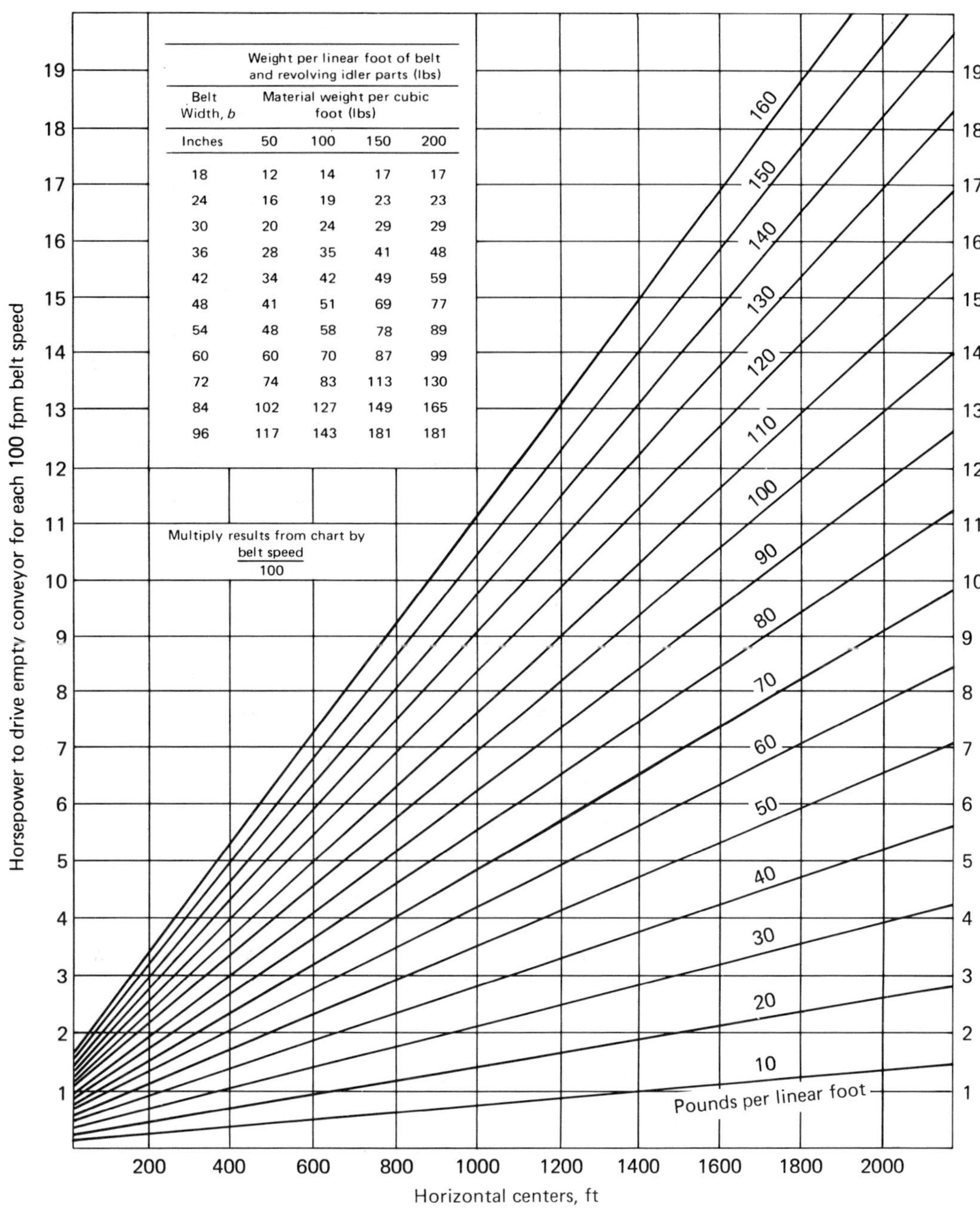

Belt Width, *b*	Weight per linear foot of belt and revolving idler parts (lbs): Material weight per cubic foot (lbs)			
Inches	50	100	150	200
18	12	14	17	17
24	16	19	23	23
30	20	24	29	29
36	28	35	41	48
42	34	42	49	59
48	41	51	69	77
54	48	58	78	89
60	60	70	87	99
72	74	83	113	130
84	102	127	149	165
96	117	143	181	181

FIGURE 6.17. *Horsepower required to drive empty conveyor.**

*Note: The table of weights is representative of average weights of revolving idler parts, as given in Chapter 5, and estimated belt weights, listed in Table 6-1, page 81. Where actual weights are known, these should be used in the graphical solution.

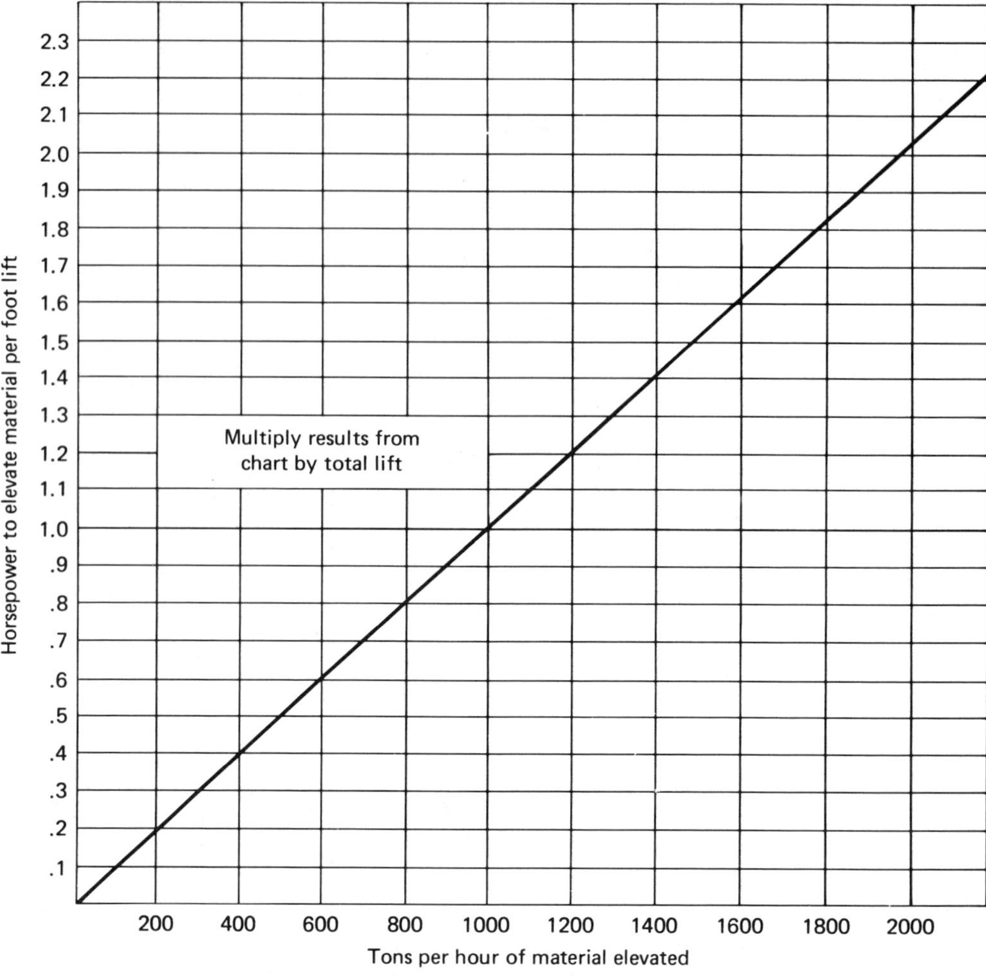

FIGURE 6.18. *Horsepower required to elevate material.*

Conveyor specifications.

Length, L = 2,000 ft

Lift, H = 75 ft

Capacity, Q = 1,600 tph

Belt speed, V = 500 fpm

Material density, d_m = 100 lbs/cu ft

Belt width, b = 48 inches

Graphical analysis. Referring to Figure 6.17, the weight/ft of belt and revolving idler parts for a 48-inch wide conveyor and 100 lbs/cu ft material is given as 51 lbs/ft. Using this value, the horsepower required to drive the empty conveyor at a speed of 100 fpm is 6.5.

Therefore, the horsepower to drive the empty conveyor at 500 fpm is equal to:

$$\frac{6.5 \times 500}{100} = 32.5 \text{ horsepower}$$

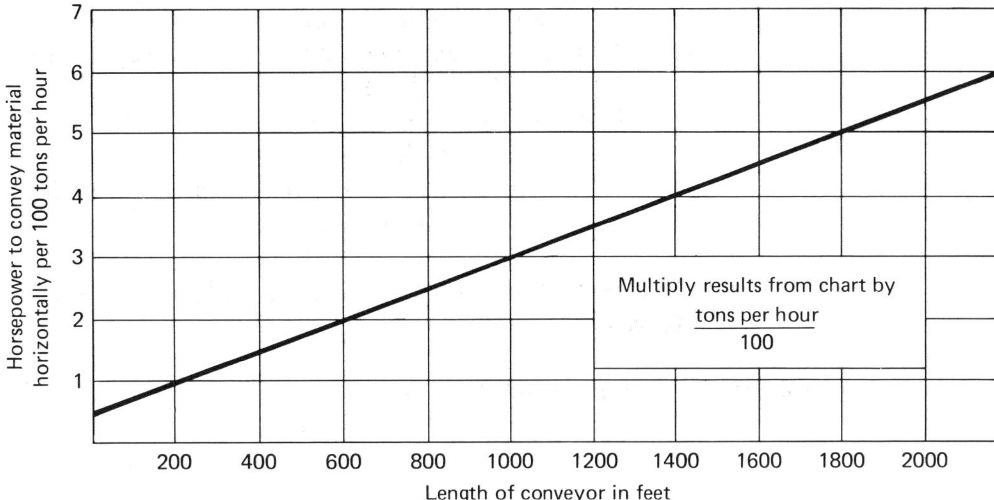

FIGURE 6.19. *Horsepower required to convey material horizontally.*

From using Figure 6.18, the horsepower required to elevate the material can be determined. The horsepower/ft of lift for the 1,600 tph capacity is 1.62. Therefore, the horsepower required to elevate the material 75 ft is given as:

$$1.62 \times 75 = 121.5 \text{ horsepower}$$

The horsepower needed to convey the material horizontally is determined through the use of Figure 6.19. Using the given conveyor specification of 2,000 ft of conveyor length, the horsepower required to convey 100 tph of material is equal to 5.5 horsepower. Therefore, the horsepower required for the 1,600 tph capacity is equal to:

$$\frac{5.5 \times 1{,}600}{100} = 88 \text{ horsepower}$$

The total required horsepower at the belt line is the sum of the above, and equals 32.5 + 121.5 + 88 = 242 horsepower.

Assuming a standard 5% loss of power through the drive components due to their inefficiencies, the required motor horsepower is:

$$\frac{242}{.95} = 254.7 \text{ horsepower}$$

A comparison of the horsepower derived by the analytical method, shown on pages 134 to 136 and the above illustrated graphical method shows that the results of the two methods are comparatively very close. This is coincidental inasmuch as the degree of accuracy of determinations made with the graphical solution is dependable only for estimating purposes. Final design should be made by the analytical method for greatest accuracy.

Examples of Belt Tension and Horsepower Calculations

Application of the CEMA horsepower formula and analysis of belt tensions and power requirements will be illustrated by the following six problems:

Problem 1—inclined conveyor;
Problem 2—declined conveyor with regenerative characteristics;
Problem 3—horizontal conveyor;
Problem 4—conveyor with a horizontal section, an inclined section, and vertical curves;
Problems 5,6—comparison of tension and horsepower values on two similar conveyors.

Problems 3 and 4 also include calculation of acceleration and deceleration forces.

Problem 1 *Inclined Belt Conveyor*

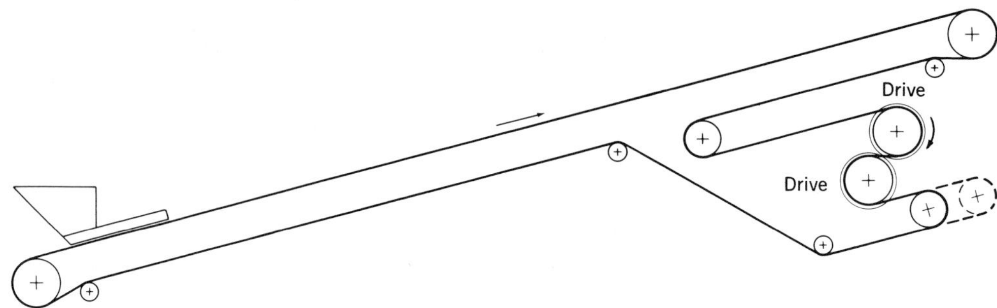

FIGURE 6.20. *Inclined belt conveyor.*

Problem:
Determine effective tension, T_e; slack-side tension, T_2; maximum tension, T_1; tail tension, T_t; belt and motor horsepower requirements; and type and location of drive.

In this problem, only two accessories are considered, pulley friction from nondriving pulleys and skirtboard friction. The belt speed is too low to involve any appreciable material acceleration force. The discharge is made freely over the head pulley and no cleaning devices are used.

Conveyor Specifications:
W_b = 15 lbs per ft, from Table 6-1
L = length = 2,000 ft
V = speed = 500 fpm
H = lift = 75 ft
Q = capacity = 1,600 tph
S_i = spacing = 3.5 ft
Ambient temperature = 60 °F
Belt width = 48 in
Material = phosphate rock at 80 lb/ft³, 15-inch maximum lump from a gyratory crusher
Drive = lagged head pulley or dual drive. Wrap is 240° or 380°, depending on which drive is to be used. See Figures 6.6B and 6.7A; also, Example 2, pages 116-118, and comments.
Troughing idlers = Class E6, 6-inch dia., 20° angle
Return idlers = Class C6, 6-inch dia., 10-ft spacing

Analysis:
Using Table 6-8, drive factor C_w = 0.30 or 0.11, depending on use of lagged head pulley or dual drive.

$$W_m = \frac{33.3\ Q}{V} = \frac{(33.3)(1{,}600)}{500} = 106.6 \text{ lbs per ft}$$

From Figure 6.1, for 60°F, $K_t = 1.0$
Formula:

$$T_e = LK_t(K_x + K_y W_b + 0.015 W_b) + W_m(LK_y + H) + T_{ac}$$

To find K_x and K_y, it is necessary to find

$$W_b + W_m = 15 + 106.6 = 121.6 \text{ lbs per ft}$$

Since 3½-ft idler spacing is given and the K_x value is calculated by using the formula:

$$K_x = 0.00068\ (W_b + W_m) + \frac{A_i}{S_i}$$

$$K_x = 0.00068\ (121.6) + \frac{2.8}{3.5} = 0.0826 + 0.800 = 0.8826$$

K_y for $L = 2{,}000$ ft, the slope is (75/2,000)(100%) = 3.75%, and $W_b + W_m$ = 121.6 lbs per ft. Table 6-2 gives $K_y = 0.018$.

Minimum tension for 3% sag, $T_o = 4.2\ S_i\ (W_b + W_m)$

$$T_o = (4.2)(3.5)(121.6) = 1{,}788 \text{ lbs}$$

Determine Accessories
In this case, the only accessories are nondriving pulley friction and skirtboards. Assume that skirtboards are 15 ft long and spaced apart two-thirds the width of the belt. Then the pull on the belt to overcome skirtboard friction is $T = C_s L_b h_s^2$. From the calculation of skirtboard friction, $h_s = (0.1)(48) = 4.8$ in. C_s is 0.1086, from Table 6-7, for phosphate rock at 80 lbs per cu ft. Thus, to solve the equation:

$$T = C_s L_b h_s^2 = (0.1086)(15)(4.8)^2 = 38 \text{ lbs}$$

For the 30 ft of rubber edging on the skirtboards, the additional resistance is (3)(30) = 90 lbs. The total skirtboard resistance is 38 + 90 = 128 lbs.

LK_tK_x	= (2,000)(1)(0.8826) =	1,765
$LK_tK_yW_b$	= (2,000)(1)(0.018)(15) =	540
$LK_t 0.015\ W_b$	= (2,000)(1)(0.015)(15) =	450
$K_y L W_m$	= (0.018)(2,000)(106.6) =	3,838
HW_m	= (75)(106.6) =	7,995
Nondriving pulley friction	= (2)(200) + (2)(150) + (4)(100) =	1,100
	Skirtboard resistance, T_{sb} =	128
	Effective tension, T_e =	15,816 lbs

Determine type of drive.
Analyze head pulley drive at 240° wrap, $C_w = 0.30$

$$T_2 = C_w T_e = (0.30)(15{,}816) = 4{,}745 \text{ lbs}$$

Analyze dual drive @ 380° wrap, $C_w = 0.11$

$$T_2 = C_w T_e = (0.11)(15{,}816) = 1{,}740 \text{ lbs}$$

However, the minimum tension, T_0 = 1,788 lbs. This minimum should exist close to the loading point on the carrying run of the belt, or at T_t, to avoid more than 3% sag between the troughing idlers spaced at 3.5 ft intervals.

If T_t = 1,788 lbs, the weight of the return belt is $HW_b = 75 \times 15$ = 1,125 lbs, and the resistance of the return belt is $0.015LW_b$, then T_2 = 1,788 + 1,125 − 450 = 2,463 lbs.

Using T_2 = 2,463 lbs, the saving in belt tension with the dual drive over a single-head pulley drive is 4,745 − 2,463 = 2,282 lbs, or 2,282/48 = 48 lbs per inch width of belt. This saving in belt cost may be enough to offset the cost of a drive.

T_2, by choice =		2,463
Return run friction (2,000)(0.015)(15) =	+	450
		2,913 lbs
Less weight of return belt (75)(15) =	−	1,125
Tail tension, T_t =		1,788 lbs

Final Tensions:

$$T_e = 15{,}816 \text{ lbs}$$

$$T_2 = 2{,}463 \text{ lbs}$$

$$T_1 = T_e + T_2 = 15{,}816 + 2{,}463 = 18{,}279 \text{ lbs}$$

$$T_t = 1{,}788 \text{ lbs}$$

Horsepower at Motor Shafts:

$$\text{Belt hp} = \frac{T_e V}{33{,}000} = \frac{(15{,}816)(500)}{33{,}000} = 239.64$$

$$\text{Drive pulley friction loss hp} = \frac{(2)(200)(500)}{33{,}000} = 6.06$$

$$\text{Add 5\% for speed reduction loss} = 0.05\,(239.64 + 6.06) = 12.29$$

$$\text{Total horsepower at motor shafts} = 257.99 \text{ hp.}$$

$$\text{Belt tension} = \frac{18{,}279}{48} = 381 \text{ lbs per inch of belt width.}$$

Problem 2 *Declined Belt Conveyor*

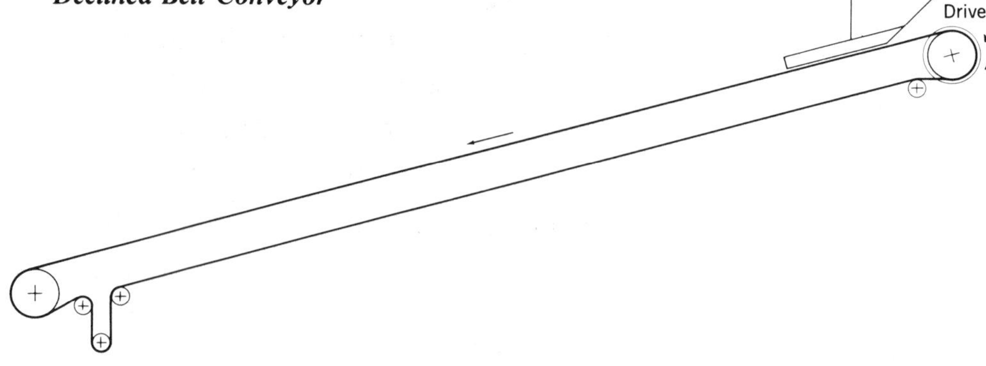

FIGURE 6.21. *Declined belt conveyor.*

Before attempting the solution of a declined conveyor, certain peculiar conditions must be considered.

A declined conveyor, which delivers material below the elevation at which it is received, will generate power if the net change in elevation is more than 2½% of the conveyor length. It may generate power at a lower slope, depending on conditions. An electric motor, acting as a generator, is used to retard the conveyor. A brake is used to stop the conveyor.

The motor size is determined by the maximum horsepower, either positive or negative, that it will be called on to produce, and it usually is set by the horsepower generated. The drive is usually located at the tail (feed) end of the conveyor, involving special design problems. One of these is that the motor must start the conveyor by driving through the gravity takeup without lifting the takeup pulley. Care must be taken to check the horsepower and belt tensions for an empty and partially loaded belt.

The brake must be large enough to absorb the torque generated and to decelerate the load. However, the retarding torque must be limited so that it does not overstress the belt. Frequently, on large conveyors, the limiting factor in brake selection will be its holding power within its ability to absorb and dissipate heat. Refer to "Brake Heat Absorption Capacity," page 175.

When a conveyor runs downhill, friction forces increase the belt tension in the direction of motion, while gravity forces decrease the belt tension, by the weight per foot of belt and load, for every foot that the belt and load are lowered.

Reduced friction. The belt, load, and idler friction absorb some of the power that the motor or brake would be compelled to absorb if these quantities did not exist. Therefore, it is important not to overestimate the friction forces or else the selected size of motor or brake might be too small. In order to avoid overestimating the friction forces, the effective tension, T_e, is calculated as follows:

$$T_e = LK_t(K_x + C_1K_yW_b + C_1 0.015W_b) + C_1W_mLK_y - HW_m + C_1T_{ac}$$

in which factor C_1 will vary from 0.5 to 0.7 and, for average conditions, will be 0.66.

For *declined conveyors only*, determine K_x by the formula $K_x = 0.00068(W_b + W_m)$. The additive term A_i/S_i is omitted because the allowance for grease and seal friction, represented by the factor A_i, is no longer on the safe side. It may under some conditions approach zero, so the safe course in declined conveyors is to make $A_i = 0$.

Problem:
Determine effective tension, T_e; slack-side tension, T_2; maximum tension, T_1; tail tension, T_t; and belt and motor horsepower requirements.

In this problem, only two accessories are considered, pulley friction of nondriving pulleys and skirtboard friction. The belt speed is too low to involve any appreciable material acceleration and no cleaning devices are employed.

Conveyor Specifications:

W_b = 10 lbs per ft, from Table 6-1
L = length = 1,200 ft
V = speed = 450 fpm
H = drop = 200 ft
Q = capacity = 1,000 tph
S_i = spacing = 4 ft
Ambient temperature = 32°F, minimum
Belt width = 36 inches
Material = limestone at 85 lbs per cu ft, 4-inch maximum lumps

Drive = lagged grooved tail pulley, wrap = 220°
Troughing idlers = Class C6, 6-inch dia., 20° angle, A_i = 1.5
Return idlers = Class C6, 6-inch dia., 10 ft spacing

Analysis:
From Table 6–8, drive factor, C_w = 0.35
From Figure 6.1, for 32 °F, K_t = 1.0

$$W_m = \frac{33.3\ Q}{V} = \frac{(33.3)(1{,}000)}{450} = 74 \text{ lbs per ft}$$

Formula (not considering C_1 factor):

$$T_e = LK_t(K_x + K_yW_b + 0.015W_b) + W_m(LK_y - H) + T_{ac}$$

To find K_x and K_y it is necessary to find

$$W_b + W_m = 10 + 74 = 84 \text{ lbs per ft}$$

K_x must be calculated for two cases. In the first calculation, K_x is taken at its normal value, so that the tension for the full friction can be determined. In the second calculation, K_x is taken at its reduced value, so that the tension for the reduced friction can be calculated.

$$\text{Normal } K_x = 0.00068\ (W_b + W_m) + \frac{A_i}{S_i}$$

$$K_x = (0.00068)(84) + \frac{1.5}{4} = 0.05712 + 0.375 = 0.4321$$

Reduced K_x = 0.00068(84) = 0.05712

K_y also must be determined for two cases. In the first instance, K_y will have its normal value as selected from Tables 6-2 and 6-3. This value is then employed for the full friction calculation of the tension. In the second instance, K_y is modified by the reduced friction factor C_1.

The slope is (200/1,200)(100%) = 16.6%. From Table 6-2, it can be seen that 4 ft is not a tabular spacing. For the 16.6% slope, L = 1,200 and $W_b + W_m$ = 84, the correct value of K_y is 0.01743, by double interpolation. For simplification in calculations, use K_y value of .018.

The minimum tension for 3% sag is:

$$T_0 = 4.2S_i(W_b + W_m) = (4.2)(4)(84) = 1{,}411 \text{ lbs}$$

Determine Accessories.
The accessories are nondriving pulley friction and skirtboard friction. Assume the skirtboards to be 10 ft long and spaced apart, two-thirds the width of the belt. Then the pull to overcome skirtboard friction is:

$$T = C_sL_bh_s^2$$

h_s = 10% of belt width or (0.1)(36) = 3.6 inches. From Table 6–7, C_s for limestone is 0.128. Thus, to solve the equation:

$$T = (0.128)(10)(3.6)^2 = 17 \text{ lbs.}$$

For the 20 ft of rubber skirtboard edging, the additional resistance is (3)(20) = 60 lbs. Total skirtboard resistance, T_{sb} = 17 + 60 = 77 lbs.

Full friction, T_e:

$LK_tK_x = (1{,}200)(1)(0.4321) =$	518.5
$LK_tK_yW_b = (1{,}200)(1)(0.018)(10) =$	216.0
$LK_t0.015W_b = (1{,}200)(1)(0.015)(10) =$	180.0
$K_yLW_m = (0.018)(1{,}200)(74) =$	1,598.4
$-HW_m = -(200)(74) =$	−14,800.0
Nondriving pulley friction = (2)(150) + (3)(100) =	600.0
Skirtboard resistance, T_{sb} =	77.0
Full friction, T_e =	−11,610.1 lbs

Reduced friction, T_e:

$$T_e = LK_t(K_x + C_1K_yW_b + C_10.015W_b) + C_1K_yLW_m - HW_m + C_1T_{ac}$$

and

$LK_tK_x = (1{,}200)(1)(0.05712) =$	68.5
$LK_tC_1K_yW_b = (1{,}200)(1)(0.66)(0.018)(10) =$	142.6
$LK_t0.015C_1W_b = (1{,}200)(1)(0.015)(0.66)(10) =$	118.8
$C_1K_yLW_m = (0.66)(0.018)(1{,}200)(74) =$	1,054.9
$-HW_m = -(200)(74) =$	−14,800.0
Nondriving pulley friction = [(2)(150) + (3)(100)]0.66 =	396.0
Skirtboard resistance = (77)(0.66) =	50.8
Reduced friction, T_e =	−12,968.4 lbs

Full friction, T_2:

$$T_2 = C_wT_e = (0.35)(11{,}610.1) = 4{,}064 \text{ lbs}$$

Reduced friction, T_2:

$$T_2 = C_wT_e = (0.35)(12{,}968.4) = 4{,}539 \text{ lbs}$$

Full friction, T_t:

$$T_t = T_2 - 0.015LW_b - HW_b = 4{,}064 - 180 - 2{,}000 = 1{,}884 \text{ lbs}$$

Reduced friction, T_t:

$$T_t = T_2 - 0.015C_1LW_b - HW_b = 4{,}539 - 120 - 2{,}000 = 2{,}419 \text{ lbs}$$

Full friction, T_1:

$$T_1 = T_e + T_2 = 11{,}610 + 4{,}064 = 15{,}674 \text{ lbs}$$

Reduced friction, T_1:

$$T_1 = T_e + T_2 = 12{,}968 + 4{,}539 = 17{,}507 \text{ lbs}$$

Final tensions	*Full friction, lbs*	*Reduced friction, lbs*
T_e	11,610	12,968
T_2	4,064	4,539
T_1	15,674	17,507
T_t	1,884	2,419

Horsepower at Motor Shaft:
The horsepower at the motor shaft should be based on the higher of the two values for T_e.

$$\text{Belt hp} = \frac{T_e V}{33{,}000} = \frac{(-12{,}968)(450)}{33{,}000} = -176.84$$

$$\text{Drive pulley friction hp} = \frac{(200)(450)}{33{,}000} = +\ \ 2.73$$

$$\text{Addition 5\% for speed reduction losses} = \underline{+\ \ 8.71}$$

$$\text{Horsepower at motor shaft} = -165.4\ \text{hp}$$

$$\text{Belt tension} = \frac{17{,}507}{36} = 486 \text{ lbs per inch of belt width.}$$

Problem 3 ***Horizontal Belt Conveyor***

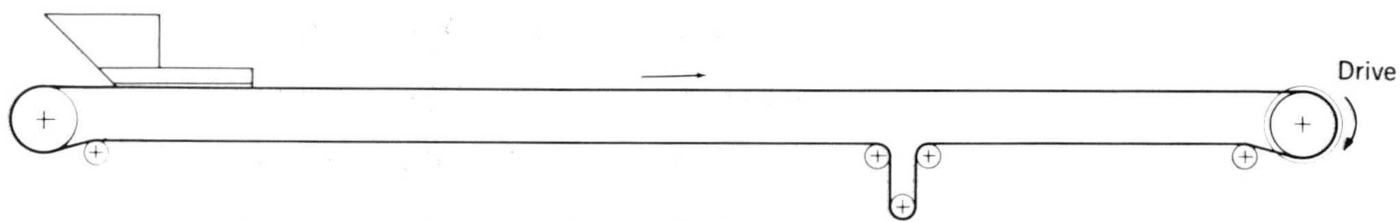

FIGURE 6.22. *Horizontal belt conveyor.*

Problem:
Determine effective tension, T_e; slack-side tension, T_2; maximum tension, T_1; tail tension, T_t; belt and motor horsepower requirements.

In this problem, only two accessories are considered, pulley friction from nondriving pulleys and skirtboard friction. Material acceleration force has been omitted in this example. The discharge is made freely over the head pulley. No belt-cleaning devices are employed.

Conveyor Specifications:

W_b = 17 lbs per ft, from Table 6-1
L = length = 2,400 ft
V = speed = 500 fpm
H = lift = 0
Q = capacity = 3,400 tph
S_i = spacing = 3 ft

Ambient temperature = 60 °F
Belt width = 48 inches
Material = iron ore at 150 lbs per cu ft, 10-inch maximum lumps from a gyratory crusher.
Drive = lagged and grooved head pulley, 220° wrap
Troughing idlers = Class E6, 6-inch dia., 20° angle.
Return idlers = rubber-disc type, Class C6, 6-inch dia., 10 ft spacing.

Analysis:
From Table 6-8, drive factor, C_w = 0.35

$$W_m = \frac{33.3\ Q}{V} = \frac{(33.3)(3{,}400)}{500} = 226.4 \text{ lbs per ft}$$

From Figure 6.1 for 60 °F, K_t = 1.0

Formula:

$$T_e = LK_t(K_x + K_yW_b + 0.015W_b) + W_m(LK_y + H) + T_{ac}$$

To find K_x and K_y it is necessary to find

$$W_b + W_m = 17 + 226.4 = 243.4 \text{ lbs per ft}$$

thus:

K_x = 1.099, for 3.0 spacing A_i = 2.8 and $W_b + W_m$ = 243.4 lbs from equation (3),

K_y = 0.021, for L = 2,400, slope 0° and $W_b + W_m$ = 243.4 lbs Refer to Table 6-2.

Minimum tension, T_0, for 3% sag = 4.2 S_i ($W_b + W_m$) = (4.2)(3)(243.4) = 3,067 lbs.

Determine Accessories:
In this case the only accessories are pulley friction and a loading chute plus skirtboards. Assume that skirtboards are 10 ft long and spaced apart, two-thirds the width of the belt. The pull on the belt to overcome skirtboard friction is $T = C_sL_bh_s^2$. From the calculation of skirtboard friction, from page 90, h_s = (0.1)(48) = 4.8 inches. C_s is (safely) .276, from Table 6-7, for iron ore @ 150 lbs per cu ft. Therefore, $T = (0.276)(10)(4.8)^2$ = 64 lbs. For the additional 20 ft of rubber edging on the skirtboards, additional resistance is (3)(20) = 60 lbs. Total skirtboard resistance, T_{sb} = 64 + 60 = 124 lbs.

LK_tK_x =	(2,400)(1)(1.099) =	2,638
$LK_tK_yW_b$ =	(2,400)(1)(0.021)(17) =	857
$LK_t0.015W_b$ =	(2,400)(1)(0.015)(17) =	612
K_yLW_m =	(0.021)(2,400)(226.4) =	11,411
HW_m =	(0)(226.4) =	0
Nondriving pulley friction =	(4)(100) + (2)(150) =	700
Skirtboard resistance, T_{sb} =		124
	Effective tension, T_e =	16,342 lbs
$T_2 = C_wT_e$ =	(0.35)(16,342) =	+ 5,720
Maximum tension, $T_1 = T_e + T_2$ =		22,062 lbs

Tail tension, $T_t = T_2 + .015LK_tW_b$ + pulley friction = 5,720 + 612 + 700 = 7,032 lbs

Final tensions:

T_e = 16,342 lbs
T_2 = 5,720 lbs
T_1 = 22,062 lbs
T_t = 7,032 lbs

Horsepower at Motor Shaft:

$$\text{Belt hp} = \frac{T_eV}{33{,}000} = \frac{(16{,}342)(500)}{33{,}000} = 247.61$$

$$\text{Drive pulley hp} = \frac{(200)(500)}{33{,}000} = 3.03$$

Add 5% for speed reduction loss = 0.05(247.61 + 3.03) = 12.53

Horsepower at motor shaft = 263.17 hp

$$\text{Belt tension} = \frac{T_1}{\text{width}} = \frac{22{,}062}{48} = 460 \text{ lbs per inch of belt width}$$

Hp at motor shaft = 263 (select 300 hp, 1,750 rpm, motor)

Acceleration Calculations:

The following calculations are used to determine the acceleration forces and times:

WK^2 of drive (all values are taken at motor speed and should be obtained from the equipment manufacturer):

WK^2 of motor = 101 lb-ft²

Equivalent WK^2 of reducer = 20 lb-ft² (common practice is to take 1/5 of WK^2 of motor)

WK^2 of coupling = 4 lb-ft²

Equivalent WK^2 of drive pulley = 5 lb-ft²

Total WK^2 of drive = 130 lb-ft², at motor speed

Converting this WK^2 value by using the equation for equivalent weight, page 128, 62,870 lbs is calculated as follows:

$$\text{Drive equivalent weight is } (130 \text{ lb. ft.}^2)\left(\frac{1{,}750 \text{ rpm}}{500 \text{ fpm}}\right)^2 (2\pi)^2 = 62{,}870 \text{ lbs.}$$

For purposes of calculating the equivalent weights, the pulley diameters must first be estimated. The diameters of the head and tail pulleys are assumed to be 42 inches; the rest of the pulleys are assumed to be 30 inches. The actual required pulley diameters are a function of the characteristics of the belt to be used and the belt tension at the pulley. The assumed pulley diameters should be reviewed after this information is known.

Conveyor equivalent weight is as follows:

Pulleys—From Table 8-1, one 42″ dia. × 51″ pulley with max. bore of 5″, weight = 1,275 lbs

From Table 8-1, five 30″ dia. × 51″ pulleys with max. bore of 4″, weight = 5 × 780 lbs = 3,900 lbs
6 (total) 1,275 + 3,900 = 5,175 lbs

Thus, 5,175 lbs is the approximate total weight for all of the nondriving pulleys on this conveyor. For more accurate calculations, the equipment manufacturer can supply actual weights. This weight is distributed among all of the elements that make up each pulley (rim, end and center discs, hubs, etc.). The belt must accelerate and decelerate these pulleys.

A generally accepted method for determining the equivalent weight of conveyor pulleys is to use ⅔ of the actual total weight. Therefore,

⅔(5,175 lbs) = 3,450 lbs

Belt, carrying run, from Table 6–1,
17 lbs per ft × 2,400 ft = 40,800 lbs

Belt, return run, 17 lbs per ft. × (2,400 ft + 30 ft) = 41,310 lbs

Idlers, troughing, from Table 5-13, for 48″ belt width and Class E6, the weight is 81.9 lbs

$$\text{thus, } 81.9 \text{ lbs} \times \frac{2{,}400 \text{ ft}}{3\text{-ft spacing}} = 65{,}520 \text{ lbs}$$

Idlers, return, from Table 5-14, for 48″ belt width and Class C6, the weight is 48.4 lbs

$$\text{thus, } 48.4 \text{ lbs} \times \frac{2{,}400 \text{ ft.}}{10\text{-ft spacing}} = \underline{11{,}616 \text{ lbs}}$$

Total conveyor equivalent weight = 162,696 lbs

Material load (226.4 lbs per ft)(2,400 ft) = 543,360 lbs

Total equivalent weight for system =
62,870 lbs + 162,696 lbs + 543,360 lbs = 768,926 lbs

Percent of total within the conveyor:

$$\frac{706{,}056}{768{,}926} \times 100\% = 91.8\%$$

Having selected a belt for T_1 = 22,062 lbs, as explained in Chapter 7, at an allowable rating of 90 lbs per inch per ply, 6 plies are required and the rated tension is 25,920 lbs.

If the starting tension is limited to 180% of the rated tension (see page 102), then the allowable extra belt tension for acceleration is

$$(1.80)(25{,}920) - (22{,}062) = 46{,}656 - 22{,}062 = 24{,}594 \text{ lbs}$$

The time for acceleration is found from the equation:

$$F_a t = M \frac{V_1 - V_0}{60}$$

where F_a = the allowable extra accelerating tension = 24,594 lbs
t = time, seconds
V_1 = final velocity = 500 fpm
V_0 = initial velocity = 0 fpm
M = mass of conveyor system = $\dfrac{706{,}056}{32.2}$ = 21,927 slugs.

Solving for t:

$$t = \frac{M}{F_a} \frac{(V_1 - V_0)}{60} = \frac{21{,}927}{24{,}594} \frac{(500 - 0)}{60} = 7.43 \text{ seconds}$$

This means that in order not to exceed the maximum permissible belt tension at 46,656 lbs, the time used for acceleration should not be less than 7.43 seconds.

Assume a 300 hp motor is used, with a maximum torque of 200% of the full-load torque. This corresponds to a force of 39,600 lbs acting at the belt line, if the friction losses of the drive are not considered and the belt speed is 500 fpm. This is not excessive when compared to the 46,656 lbs belt tension allowable at 180% of belt rating.

Another limiting factor may be the time which the motor needs to accelerate the system. The average torque—available during acceleration of the chosen motor—taken from its speed torque curve is 180% of full load torque. For a drive efficiency of 95%, it was found in Problem 3, that the horsepower at the motor shaft to operate the loaded conveyor is 263.17 hp.

$$\text{Horsepower} = \frac{(\text{pull in lbs})(\text{belt speed, fpm})}{33{,}000}$$

Therefore,

$$(\text{pull in lbs}) = \frac{(\text{hp})(33{,}000)}{(\text{belt speed, fpm})}$$

Also, the horsepower delivered by the motor is practically proportional to the torque, assuming no appreciable drop in speed from the full-load speed. Therefore, at 180% torque, the motor will deliver (1.8)(300) = 540 hp.

The force available for acceleration of the total equivalent mass of the loaded conveyor system, for a belt speed of 500 fpm, is:

$$F_a = \left[\frac{(300)(1.8)(33{,}000)}{500} - \frac{(263.17)(33{,}000)}{500}\right] 0.95$$

$$= 0.95[(300)(1.8) - 263.17] \frac{33{,}000}{500}$$

$$= \frac{(263)(33{,}000)}{500} = 17{,}358 \text{ lbs}$$

The total equivalent mass = 768,926/32.2 = 23,880 slugs.
From the equation,

$$F_a = Ma$$

$$\text{the acceleration, } a = \frac{F_a}{M} = \frac{17{,}358}{23{,}880} = 0.727 \text{ ft per sec}^2.$$

The time needed is:

$$t = \frac{V_1 - V_0}{60\ a}$$

Therefore,

$$t = \frac{500 - 0}{(60)(0.727)} = 11.46 \text{ seconds.}$$

The time required by the motor to accelerate the loaded conveyor, 11.46 seconds, is greater than the minimum acceleration time to stay within the maximum allowable belt tension, 7.43 seconds. Therefore, the conveyor is safe to start, fully loaded, with the equipment selected.

Had the starting belt stress been limited to 120% of the normal belt rating—instead of 180%—the allowable extra belt tension would have been (1.2)(25,920) − (22,062) = 9,042 lbs and the acceleration time, t = (21,927)(500 − 0)/(9,042)(60) = 20.21 sec minimum. This is more than the time calculated for the motor to accelerate the loaded system, 11.46 seconds. So, if such limitation had been placed on the starting belt stress, the system would not have been safe to start with the equipment selected. In fact, the belt stress during acceleration must be:

$$\text{extra belt tension} = \frac{(21{,}927)(500 - 0)}{(11.46)(60)} = 15{,}752 \text{ lbs}$$

$$\%\text{ of normal belt rating } \frac{(15{,}752) + (22{,}062)}{25{,}920}(100\%) = 146\,\%$$

The foregoing assumes that the mass between the slack side of the drive pulley and the takeup is negligible. If the takeup is far removed from the drive, this should be taken into account in the calculations.

In Chapter 12 it is indicated that the acceleration time for NEMA Type C motors, in general, be considered as 10 seconds or less. It, therefore, would be prudent to check with the motor manufacturer to make sure that the calculated acceleration time of 11.46 seconds would not cause the motor to overheat during starting. In the case of *this particular* problem, the motor manufacturer was asked what maximum safe acceleration time would be for his 300 hp NEMA Type C motor. The manufacturer stated that any time up to 20 seconds would be permissible.

Therefore, the conveyor in this problem could be safely started with the equipment selected, provided the allowable belt tension during starting was 146% of normal belt rating, or greater. The use of the 300 hp NEMA Type C motor is justified, provided that operating conditions of this particular conveyor are such that abnormal starting conditions (which would require forces considerably in excess of those calculated) are unlikely to occur. If abnormal starting conditions are likely to occur, even infrequently, consideration should be given to the use of a different means of starting which would satisfy all the requirements described earlier in the section, "Acceleration and deceleration forces."

Deceleration Calculations:

In the preceding example on acceleration, it was found that the total equivalent mass of the conveyor system under normal conditions of operation is equal to 23,880 slugs (see page 144). As these calculations are based on the belt speed of 500 fpm or 8.33 fps, the kinetic energy of the system is:

$$\frac{MV^2}{2} = \frac{(23{,}880)(8.33)^2}{2} = 828{,}503 \text{ ft lbs}$$

Earlier in this problem (3), it was found that 263.17 hp is required to operate this conveyor at its rated speed of 500 fpm. Because the conveyor is horizontal, this represents the product of the friction forces and the distance traveled in unit time. This means that the frictional retarding force is:

$$\frac{(263.17)(33{,}000)}{500} = 17{,}369 \text{ lbs}$$

The average velocity of the conveyor during the deceleration period would be $\frac{500 + 0}{2} = 250$ fpm.

Because the total work performed has to be equal to the kinetic energy of the total mass,

$$(t)(250 \text{ fpm})(17{,}369) = 828{,}503 \text{ ft lbs}$$

where, t = time in minutes.

Therefore,

$$t = \frac{828{,}503}{(250)(17{,}369)} = 0.191 \text{ minutes, or } 11.45 \text{ seconds}$$

and the belt will have moved (0.191)(250 fpm) = 47.75 ft in this time. As the belt is fully loaded (by assumption), the belt will discharge the following amount of material:

$$\left(\frac{3{,}400 \text{ tph}}{60}\right)\left(\frac{47.75}{500}\right) = 5.41 \text{ tons}$$

If 5.41 tons of material discharged is objectionable, the use of a brake has to be considered. Such a step, however, can be justified only if the reduced deceleration time is still greater than, or at least equal to, the deceleration cycle of whatever piece of equipment delivers to the conveyor in this example.

Also, another difficulty arises. Suppose it is desirable or necessary to reduce the deceleration time from 11.45 seconds to 7 seconds. Since the total retarding force is inversely proportional to the deceleration time, the additional braking force required must be:

$$17{,}369 \times \left(\frac{11.45 - 7}{7}\right) = 11{,}042 \text{ lbs}$$

If the brake is connected to the drive pulley shaft, the drive pulley is required to transmit to the belt a braking force equal to

$$11{,}042 \times \left(\frac{768{,}926 - 62{,}870}{768{,}926}\right) = 10{,}139 \text{ lbs}$$

The difference between the 11,042 lbs and the 10,139 lbs is the braking force required to decelerate the drive and drive pulley and is not transmitted to the belt.

However, under coasting conditions, the belt tension is principally governed by the gravity takeup which, if located adjacent to the head pulley, would provide a maximum tension equal to T_2, or 5,720 lbs. Obviously, it is impossible to secure a braking force of 10,139 lbs on the head pulley. Even a much smaller force than this would result in looseness of the belt around the head pulley.

The solution is to provide the braking action on the tail pulley, where it would increase rather than decrease the contact pressure between the belt and pulley. However, a further check on the tail pulley indicates that with 11,042 lbs braking tension, a plain bare tail pulley with 180° wrap angle could not produce a sufficient ratio of tight-side to slack-side tension.

Therefore, it would be necessary to do one or a combination of the following: increase the takeup tension weight, lag the tail pulley, or snub the tail pulley for a greater wrap angle. If the increasing takeup weight should result in a heavier and more costly belt carcass, the second and third remedies are preferable and more economical.

It should be noted that the above calculations are based on maximum friction losses and therefore will give a minimum coasting distance. Since most installations operate under variable conditions, braking and coasting problems should be investigated for a range of friction values. These lower friction values for K_x and K_y can be found by the methods outlined in Problem 2, and can result in a lower frictional retarding force approaching 60% of the original. This lower retarding force will show greater coasting distances or larger braking forces.

Problem 4

Complex Belt Line

In the previous examples, the application of the CEMA horsepower formula was limited to belt conveyors with a linear profile and an overall centers length not exceeding 3,000 ft. However, the CEMA horsepower formula can be applied to belt conveyors having more than one change in slope and a total centers length of more than 3,000 ft, provided certain procedures are followed. This problem entails a belt conveyor which has two changes in slope and a total centers length of 4,000 ft.

The K_y factor is dependent upon the average tension in that portion of the belt in which the tension is being analyzed. Tables 6-2 and 6-3 were

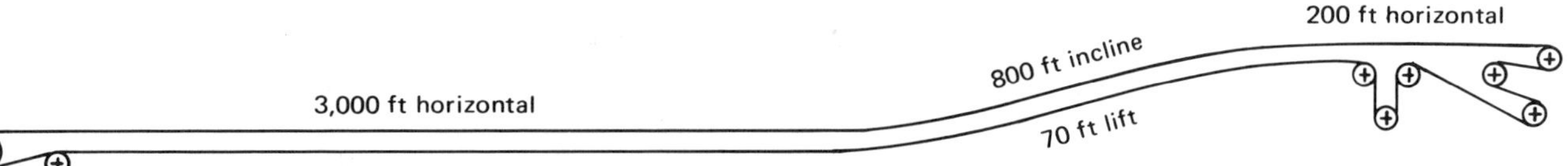

FIGURE 6.23. *Complex belt line.*

developed on the basis of the limitations and generalizations stated on page 82, and for normal average tensions in the belts within the limitations specified. For belt conveyors exceeding these limitations, it is necessary first to assume a tentative value for the average belt tension. The graphical method for conveyor horsepower determination, pages 130 to 133, may be of assistance in estimating this value. After estimating the average belt tension and idler spacing, reference to Table 6-4 will provide values for A and B, for use in equation (4), page 82. By using this equation, an initial value for K_y can be calculated. The comparison of this calculated average belt tension with the tentative value will determine the need to select another assumed belt tension. The process should be repeated until there is reasonable agreement between the estimated and calculated average belt tensions.

The following example of a belt tension analysis of a 4000-ft belt conveyor with two changes of slope demonstrates the method of calculation and the use of the table. See Figure 6.23.

Belt conveyors with different profiles can be analyzed in a similar manner, but the various problems, which increase in importance with long and complex belt conveyors, must be carefully analyzed. It is suggested that the designer of such complex conveyors check calculations with a CEMA member company before establishing the final conveyor design.

Problem:

Determine the effective tension, T_e; slack-side tension, T_2; maximum tension, T_1; tail tension, T_t; concave curve tension at bottom of incline run, convex curve tension at top of incline run; belt horsepower at drive terminal; belt stress; resulting drive factor.

Accessories are omitted in this example in order to clarify the procedure; however, they should be included in actual practice.

Conveyor Specifications:

Q = capacity = 800 tph
Material = crushed limestone, 85 lbs per cu ft, 8-in maximum lumps.
Ambient temperature, above freezing. Continuous operation.
L = length = 4,000 ft
H = lift = 70 ft (see Figure 6.23).
S_i = idler spacing = 4 ft
b = belt width = 36 inches
V = speed = 400 fpm
Drive = dual, 380° wrap, both pulleys lagged.
Troughing idlers = Class C6, 6 in dia, 20° angle, A_i = 1.5
Return idlers = rubber-disc type, Class C6, 6 in dia, 10 ft spacing
W_b = belt weight = 10 lbs per ft

Constants:

$$W_m = \frac{33.3Q}{V} = \frac{(33.3)(800)}{400} = 66.6 \text{ lbs per ft}$$

$$K_x = .00068\,(W_b + W_m) + \frac{A_i}{S_i} = .00068\,(10 + 66.6) + \frac{1.5}{4} = 0.427$$

Analysis:

Since each portion of this conveyor is analyzed separately, and each is less than or equal to 3,000 ft in length, Table 6–2, can be used to obtain a tentative K_y factor in order to calculate the average belt tension. This K_y value is then checked by using Table 6–4, equation (4), page 82, and the average belt tension. The final tensions in each portion of the conveyor are then accurately determined.

The profile is divided into three portions: (1) Initial horizontal portion, 3,000 ft long; (2) inclined portion, 800 ft long, 70-ft lift; (3) final horizontal portion, 200 ft long.

Initial horizontal section, 3000 ft long, 3% sag in belt.

where

$$K_t = 1.0$$
$$K_x = 0.427$$
$$W_b = 10 \text{ lbs per ft}$$
$$W_m = 66.6 \text{ lbs per ft}$$
$$(W_b + W_m) = 76.6 \text{ lbs per ft}$$

K_y from Table 6-2 would be 0.023.

The *average tension* is:

$$\frac{T_t + K_t[K_xL + K_yLW_b] + K_yLW_m + T_t}{2}$$

Here, T_t is at least equal to T_0. And T_0, for 3% sag, is

$$4.2\ S_i(W_b + W_m) = (4.2)(4)(76.6) = 1{,}287 \text{ lbs}$$

Thus, the *average tension* is:

$$\frac{1{,}287 + (0.427)(3{,}000) + (0.023)(3{,}000)(76.6) + 1{,}287}{2}$$

or,

$$\frac{1{,}287 + 1{,}281 + 5{,}285 + 1{,}287}{2} = \frac{9{,}140}{2} = 4{,}570 \text{ lbs}$$

Equation (4) indicates $K_y = 0.0255$, for 4,570 lbs average tension, and $(W_b + W_m) = 76.6$ lbs. Re-estimate using $K_y = 0.0255$. *Average* tension is:

$$\frac{1{,}287 + 1{,}281 + (0.0255)(3{,}000)(76.6) + 1{,}287}{2}$$

or,

$$\frac{1{,}287 + 1{,}281 + 5{,}860 + 1{,}287}{2} = \frac{9{,}715}{2} = 4{,}858 \text{ lbs}$$

Equation (4) checks $K_y = 0.0255$, for an average tension of 4,858 lbs, and $(W_b + W_m) = 76.6$ lbs.

The formula for the actual tension because of friction in the initial horizontal portion (see page 106) is:

$$T_{fcx} = L_x[K_t(K_x + K_yW_b)] + L_xK_yW_m$$

where,

$$L_x = 3{,}000 \text{ ft}$$
$$K_x = 0.427$$
$$K_t = 1.0$$
$$K_y = 0.0255$$
$$W_b = 10 \text{ lbs per ft}$$
$$W_m = 66.6 \text{ lbs per ft}$$

Therefore, $T_{fcx} = L_x[K_x + K_y(W_b + W_m)]$, since $K_t = 1.0$

$$T_{fcx} = 3{,}000[0.427 + 0.0255(76.6)]$$
$$= 3{,}000(2.38)$$
$$= 7{,}141 \text{ lbs}$$

The tension at the beginning of the vertical concave curve is calculated using the formula for belt tension at any point on the conveyor length, for point X on the carrying run (page 106), at the intersection of the initial horizontal run and the inclined run:

$$T_{cx} = T_t + T_{wcx} + T_{fcx}$$
$$T_{wcx} = H_x(W_b + W_m) = (0)(76.6) = 0$$

so

$$T_{cx} = 1{,}287 + 7{,}141 = 8{,}428 \text{ lbs}$$

The tension at the bottom of the incline, therefore, is 8,428 lbs. The estimated K_y is 0.024, for the first approximation of the calculation for the upper end of the incline from Table 6-2 for a value of $(W_b + W_m) = 76.6$, and a slope of $(70/800)(100\%) = 8.8\%$. Average tension is:

$$\frac{T_t + K_t(K_xL + K_yLW_b) + K_yLW_m + H(W_b + W_m) + T_t}{2}$$

in which T_t is the tension at the bottom of the incline, or 8,428 lbs, so,

$$\frac{8{,}428 + (0.427)(800) + (0.024)(800)(76.6) + (70)(76.6) + 8{,}428}{2}$$

or,

$$\frac{8{,}428 + 342 + 1{,}471 + 5{,}362 + 8{,}428}{2} = \frac{24{,}031}{2} = 12{,}016 \text{ lbs}$$

As stated on page 82, the minimum K_y value $= .016$ for 12,016 lbs tension and $W_b + W_m = 76.6$. Re-estimate using $K_y = .016$. *Average* tension then is:

$$\frac{8{,}428 + 342 + (0.016)(800)(76.6) + 5{,}362 + 8{,}428}{2}$$

or,

$$\frac{8{,}428 + 342 + 980 + 5{,}362 + 8{,}428}{2} = \frac{23{,}540}{2} = 11{,}770 \text{ lbs}$$

This checks $K_y = 0.016$ minimum value, for average tension of 11,770 lbs, and $W_b + W_m = 76.6$ lbs per ft.

From page 105, $T_{cx} = T_t + T_{wcx} + T_{fcx}$. Here, T_t is 8,428 lbs, the tension at the bottom of an incline.

$$T_{wcx} = H_x(W_b + W_m) = (70)(76.6) = 5{,}362 \text{ lbs}$$

$$T_{fcx} = L_xK_t[K_x + K_y(W_b + W_m)] \text{ where } L_x = 800 \text{ ft and } K_t = 1.0$$
$$= (800)(1.0)[0.427 + 0.016(76.6)]$$
$$= 800(1.653)$$
$$= 1{,}322 \text{ lbs}$$

$$T_{cx} = 8{,}428 + 5{,}362 + 1{,}322 = 15{,}112 \text{ lbs}$$

The tension at the top of the incline, then, is 15,112 lbs.

The final horizontal portion is 200 ft long:

$$K_t = 1.0$$
$$K_x = 0.427$$
$$W_b = 10 \text{ lbs per ft}$$
$$W_m = 66.6 \text{ lbs per ft}$$
$$W_b + W_m = 76.6 \text{ lbs per ft}$$

K_y will be at a minimum value because of the high tension which is obvious in this portion of the belt. From page 82, the minimum K_y of 0.016 is applicable at the indicated *average* tension (obviously more than 15,112 lbs) and with $W_b + W_m = 76.6$ lbs per ft.

From page 106, $T_{cx} = T_t + T_{wcx} + T_{fcx}$. Here, T_t, is the tension at the beginning of this horizontal section, or 15,112 lbs and $T_{wcx} = 0$, since $H_x = 0$

$$T_{fcx} = L_x K_t (K_x + K_y W_b) + L_x K_y W_m \text{ where } L_x = 200 \text{ ft and } K_t = 1.0$$
$$= (200)(1)(0.427 + (0.016)(10)) + (200)(0.016)(66.6)$$
$$= 117 + 213 = 330 \text{ lbs}$$

$$T_{cx} = 15{,}112 + 330 = 15{,}442 \text{ lbs}$$

In this case, $T_{cx} = T_1 = 15{,}442$ lbs
The final tension at the head pulley is 15,442 lbs. $T_e = T_1 - T_2$. To find T_2, refer to Figure 6.7A, where $T_t = T_2 - T_b + T_{yr}$. The tail tension, T_t , was taken at $T_0 = 1{,}287$ lbs to avoid more than 3% belt sag between idlers. Thus,

$$1{,}287 = T_2 - T_b + T_{yr}$$

$$T_b = HW_b = (70)(10) = 700 \text{ lbs}$$

$$T_{yr} = 0.015 L W_b K_t = (0.015)(3{,}000 + 800 + 200)(10)(1) = 600 \text{ lbs}$$

Thus,

$$1{,}287 = T_2 - 700 + 600, \text{ or } T_2 = 1{,}387 \text{ lbs}$$

$$T_e = T_1 - T_2 = 15{,}442 - 1{,}387 = 14{,}055 \text{ lbs}$$

With the T_e and T_2 tensions now known, it is necessary to check the wrap factor, C_w. A 380° wrap, dual-pulley drive with lagged pulleys, requires a $C_w = 0.11$. (See Table 6-8.) From the known tensions:

$$C_w = \frac{T_2}{T_e} = \frac{1{,}387}{14{,}055} = .099$$

Since this is less than the required 0.11, the belt may slip on the drive pulleys. This situation can be corrected in one of two ways: (1) The wrap on the drive pulleys can be increased from 380° to 405°, or (2) the takeup weight can be increased until $T_2/T_e = 0.11$. This requires an increase in all tensions of $(0.11 \times 14{,}055) - 1{,}387 = 160$ lbs.
Assume all tensions are increased by 160 lbs:

$$T_1 = 15{,}442 + 160 = 15{,}602$$

$$T_2 = 1{,}387 + 160 = 1{,}547$$

$$T_t = 1{,}287 + 160 = 1{,}447$$

$$\text{Belt stress} = \frac{T_1}{\text{Belt width}} = \frac{15{,}602}{36} = 433 \text{ lbs per inch of width (PIW)}$$

Horsepower at belt line, excluding all accessories, is as follows:

$$\text{Belt hp} = \frac{T_e V}{33{,}000} = \frac{(14{,}055)(400)}{33{,}000} = 170.36$$

If drive efficiency = .94, horsepower at motor shaft = 170.36/.94 = 181.23 hp. Acceleration and deceleration calculations for this example follow. For radii of concave and convex curves for this example, refer to Chapter 9.

Acceleration Calculations:

T_e = 14,055 lbs
T_2 = 1,547 lbs
T_1 = 15,602 lbs
T_t = 1,447 lbs
b = Belt width = 36 inches
L = Length = 4,000 ft
H = Lift = 70 ft
Q = Capacity = 800 tph
V = speed = 400 fpm
Material = crushed limestone at 85 lbs per cu ft
W_m = Material weight, = 66.6 lbs per ft
W_b = Belt weight, = 10 lbs per ft
Troughing idlers Class C6, 6-inch dia, 20° angle, at 4 ft spacing
Rubber-disc return idlers, Class C6 6-inch dia, at 10 ft spacing
Hp, at motor shaft = 181.23 (select one 75 hp and one 125 hp motor, each to be 1,750 rpm)
WK^2 of drive (all values are taken at motor speed and should be obtained from the equipment manufacturer)

WK^2 of motor =	58 lb-ft²
Equivalent WK^2 of reducer =	11.6 lb-ft²

(It is common practice to estimate the WK^2 of the reducer to be 20% of the WK^2 of the motor)

WK^2 of coupling =	2 lb-ft²
Equivalent WK^2 of drive pulley =	2 lb-ft²
Total WK^2 =	73.6 lb-ft², at motor speed

Converting this WK^2 value to the equivalent weight at the belt line,

$$\text{Drive equivalent weight (lbs)} = (WK^2)\left(\frac{2\pi \text{ rpm}}{V}\right)^2$$

$$\text{Drive equivalent weight is } (73.6 \text{ lb ft}^2)\left(\frac{1{,}750 \text{ rpm}}{400 \text{ fpm}}\right)^2(2\pi)^2 = 55{,}615 \text{ lbs}$$

Conveyor equivalent weight is as follows:

Pulleys—for reasons given on page 142, first assume the non-drive pulley diameter: two at 48 inches and four at 36 inches.

From Table 8-1, two (48 inches dia × 38 inches) pulleys with max. bore of 5 inches, weight = 2 × 1,270 = 2,540 lbs

From Table 8-1, four (36 inches dia × 38 inches) pulleys with max. bore of 4 inches, weight = 4 × 715 = 2,860 lbs

From page 142, ⅔(2,540 lbs + 2,860 lbs)	=	3,600 lbs
Belt, carrying run, from Table 6-1, 10 lbs per ft × 4,000 ft	=	40,000 lbs
Subtotal	=	43,600 lbs

Subtotal carried over = 43,600 lbs

Belt, return run, 10 lbs per ft × (4,000 ft + 50 ft) = 40,500 lbs

Idlers, troughing, from Table 5-13, for 36 inch belt width and class C6, the weight is 43.6 lbs

$$43.6\left(\frac{4{,}000 \text{ ft}}{4\text{-ft idler spacing}}\right) = 43{,}600 \text{ lbs}$$

Idlers, return, from Table 5-13, for 36 inch belt width and Class C6, the weight is 37.6 lbs

$$37.6\left(\frac{4{,}000 \text{ ft}}{10\text{-ft idler spacing}}\right) = 15{,}040 \text{ lbs}$$

Total conveyor equivalent weight = 142,740 lbs

Material load (66.6 lbs per ft)(4,000 ft) = 266,400 lbs

Total equivalent weight of system =
55,615 lbs + 142,740 lbs + 266,400 lbs = 464,755 lbs

Percent of total within the conveyor:

$$(142{,}740 + 266{,}400)/464{,}755 \times 100\% = 88\%$$

Having selected a belt for T_1 = 15,602 lbs, as explained in Chapter 7, at an allowable rating of 70 lbs per inch per ply, 7 plies are required and the rated tension is 17,640 lbs.

If the starting tension is limited to 180% of the rated tension (see page 102), then the allowable extra tension is

$$(1.80)(17{,}640) - (15{,}602) = 31{,}752 - 15{,}602 = 16{,}150 \text{ lbs}$$

The time for acceleration is found from the equation:

$$F_a t = M \frac{V_1 - V_0}{60}$$

where,

F_a = the allowable extra accelerating tension = 16,150 lbs
t = time, seconds
V_1 = final velocity = 400 fpm
V_0 = initial velocity = 0 fpm
M = mass of conveyor system = $\frac{409{,}140}{32.2}$ = 12,706 slugs.

Solving for t:

$$t = \frac{M}{F_a}\frac{(V_1 - V_0)}{60} = \frac{12{,}706}{16{,}150}\frac{(400 - 0)}{60} \doteq 5.24 \text{ seconds}$$

This means that in order not to exceed the maximum permissible belt tension at 31,752 lbs, the time used for acceleration should not be less than 5.24 seconds.

In determining the starting tension in the belt, the first step is to find the total horsepower available, in the form of tension. From this value, subtract the total tension to operate the loaded conveyor. The result will be the force available to accelerate the total system. The total horsepower, in the form of tension, available to accelerate the entire system comes from the 75 hp and 125 hp motors. The starting torque available from these NEMA Type C motors is a variable which should be confirmed by the motor manufacturer. For this example, the value is assumed to be 200% of the motor rating.

Then, the total tension available is:

$$\frac{2(75 + 125)(33{,}000)}{400} = 33{,}000 \text{ lbs}$$

From this value we subtract the tension required to operate the loaded conveyor:

$$33{,}000 - \frac{14{,}055}{.94(\text{drive efficiency})} = 18{,}048 \text{ lbs}$$

The resulting tension available to accelerate the loaded conveyor is 18,048 lbs. The acceleration of the total system consists of the acceleration of the drive (12% of total system) and of the conveyor (88% of the total system). However, in the process of acceleration, some amount of the available force (tension) is absorbed by the frictional losses (heat) in the drive machinery. Since this is a small amount compared to the total, a conservative approach is to ignore these losses because it is our aim to determine the effect of acceleration on the belt and its capacity to withstand tensile forces. Therefore, .88 × 18,048 = 15,882 lbs, which is the acceleration force (expressed in lbs of belt tension). The operating T_1 is then added to this value in order to obtain the actual starting tension in the belt, which is 15,602 + 15,882 = 31,484 lbs. This is not excessive when compared to the 31,752 lbs belt tension allowable at 180% of belt rating.

Another limiting factor may be the time which the motor needs to accelerate the system. The average torque available during acceleration of the chosen motor taken from its speed torque curve is 180% of full-load torque. For a drive efficiency of 94%, it was found in Problem 4, page 151, that the horsepower at the motor shaft to operate the loaded conveyor is 181.23 hp.

$$\text{Horsepower} = \frac{(\text{pull in lbs})(\text{belt speed, fpm})}{33{,}000}$$

Therefore,

$$(\text{pull in lbs}) = \frac{(\text{hp})(33{,}000)}{(\text{belt speed, fpm})}$$

The force available for acceleration of the total equivalent mass of the loaded conveyor system, for a belt speed of 400 fpm, is

$$F_a = \left[\frac{(200)(1.80)(33{,}000)}{400} - \frac{(181.23)(33{,}000)}{400}\right] 0.94$$

$$= \frac{0.94 \times 33{,}000}{400}(200 \times 1.80 - 181.23)$$

$$= 13{,}864 \text{ lbs}$$

The total equivalent mass = 464,755/32.2 = 14,433 slugs. From the equation,

$$F_a = Ma$$

$$\text{the acceleration, } a = \frac{F_a}{M} = \frac{13{,}864}{14{,}433} = 0.96 \text{ ft per sec}^2$$

The time needed is:

$$t = \frac{V_1 - V_0}{60a}$$

Therefore,

$$t = \frac{400 - 0}{(60)(0.96)} = 6.94 \text{ seconds}$$

The time required by the motor to accelerate the loaded conveyor, 6.94 seconds, is greater than the minimum acceleration time to stay within the maximum allowable belt tension, 5.24 seconds. Therefore, the conveyor is safe to start, fully loaded, with the equipment selected.

Had the starting belt stress been limited to 140% of the normal belt rating—instead of 180%—the allowable extra belt tension would have been (1.4)(17,640) − (15,602) = 9,094 lbs and the acceleration time t = (12,706)(400 − 0)/(9,094)(60) = 9.31 sec minimum. This is more than the time calculated for the motor to accelerate the loaded system, 6.94 seconds. If such a limitation had been placed on the starting belt stress, the system would not have been safe to start with the equipment selected. In fact, the belt stress during acceleration must be:

$$\text{extra belt tension} = \frac{(12{,}706)(400 - 0)}{(6.94)(60)} = 12{,}206 \text{ lbs}$$

$$\% \text{ of normal belt rating } \frac{12{,}206 + 15{,}602}{(17{,}640)} (100\%) = 158\%$$

The foregoing assumes that the mass between the slack side of the drive pulley and the takeup is negligible. If the takeup is far removed from the drive, this should be taken into account in the calculations.

In Chapter 12, "Accelerating time," it is indicated, in general, that the acceleration time for NEMA Type C motors be considered as 10 seconds or less. It is always prudent to check with the motor manufacturer to make sure that the calculated acceleration time will not cause the motor to overheat during starting.

Deceleration Calculations:

In the foregoing example of acceleration calculations, it was found that the total equivalent mass of the conveyor system under normal conditions of operation is equal to 14,433 slugs. As these calculations are based on the belt speed of 400 fpm or 6.67 fps, the kinetic energy of the system is:

$$\frac{MV^2}{2} = \frac{14{,}433 \times 6.67^2}{2} = 320{,}733 \text{ ft-lbs}$$

On page 151 it was calculated that 181.23 hp is required at the motor shaft to operate this conveyor at its rated speed of 400 fpm. This represents the product of the friction plus gravity forces and the distance traveled in unit time. This means that the frictional plus gravitational retarding force is:

$$\frac{(181.23)(33{,}000)}{400} = 14{,}951 \text{ lbs}$$

The average velocity of the conveyor during the deceleration period would be:

$$\frac{400 + 0}{2} = 200 \text{ fpm}$$

Because the total work performed has to be equal to the kinetic energy of the total mass:

$$(t)(200 \text{ fpm})(14{,}951) = 320{,}733 \text{ ft lbs}$$

where, t = time in minutes

Therefore,

$$t = \frac{320{,}733}{(200)(14{,}951)} = 0.1073 \text{ minutes, or } 6.44 \text{ seconds}$$

and the belt will have moved (0.1073)(200 fpm) = 21.46 ft in this time. As the belt is fully loaded (by assumption), it will discharge the following amount of material:

$$\left(\frac{800\text{ tph}}{60}\right)\left(\frac{21.46}{400}\right) = .72\text{ tons}$$

If .72 tons of material discharge is objectionable, the use of a brake has to be considered. Such a step, however, can be justified only if the reduced deceleration time is still greater than, or at least equal to, the deceleration cycle of whatever piece of equipment delivers to the conveyor in this example.

Also, another difficulty arises. Suppose it is desirable or necessary to reduce the deceleration time from 6.44 seconds to 5 seconds. Since the total retarding force is inversely proportional to the deceleration time, the additional braking force required must be:

$$14{,}951 \times \left(\frac{6.44 - 5}{5}\right) = 4{,}306\text{ lbs}$$

If the brake is connected to the drive pulley shaft, the drive pulley is required to transmit to the belt a braking force equal to

$$4{,}306\text{ lbs} \times .88 = 3{,}789\text{ lbs}$$

The difference between the 4,306 lbs and the 3,789 lbs is the braking force required to decelerate the drive and drive pulley and is not transmitted to the belt.

However, under coasting conditions, the belt tension is principally governed by the gravity takeup which, if located adjacent to the head pulley, would provide a maximum tension equal to T_2, or 1,547 lbs. Obviously, it is impossible to develop a braking force of 3,789 lbs on the head pulley. Even a much smaller force than this would result in looseness of the belt around the head pulley.

The solution is to provide the braking action on the tail pulley where it would increase rather than decrease the contact pressure between the belt and pulley. However, a further check on the tail pulley indicates that with 3,789 lbs braking tension, a plain, bare tail pulley with 180° wrap angle could not produce a sufficient ratio of tight-side to slack-side tension.

Therefore, it would be necessary to do one or a combination of the following: increase the takeup tension weight, lag the tail pulley, or snub the tail pulley for a greater wrap angle. If the increased takeup weight should result in a heavier and more costly belt carcass, the second and third remedies are preferable and more economical.

It should be noted that the above calculations are based on maximum friction losses and therefore will give a minimum coasting distance. Since most installations operate under variable conditions, braking and coasting problems should be investigated for a range of friction values. These lower friction values for K_x and K_y can be found by the methods outlined in Problem 2, and can result in a lower frictional retarding force approaching 60% of the original. This lower retarding force will show greater coasting distances or larger braking forces.

Problems 5 and 6

Comparison of Tension and Horsepower Values on Two Similar Conveyors

The two belt conveyors compared here in Problems 5 and 6 have the same load capacity, carry the same bulk material, have the same length, the same operating speed, and the same lift. The only difference is that one conveyor has a concave vertical curve while the other has a convex vertical curve.

The CEMA formula for power to operate belt conveyors determines the effective tension, T_e. The previous examples show how to obtain T_2, T_1, and T_t. *Frequently, the belt conveyor designer will require belt tensions elsewhere;* for instance, in the determination of the radius of a concave vertical curve. For a discussion of the belt tensions at any point on a belt conveyor, see "Belt tension at any point, *X,* on conveyor length," page 105. Formulas for the belt tensions in belt conveyors having concave and convex vertical curves are given in Figures 6.8 through 6.16, inclusive.

The comparison of the two belt conveyors in Problems 5 and 6 show how the factor K_y changes with increasing belt tension.

In Problem 5, Figure 6.24, the K_y factor for the tail half, L_1, of the conveyor is selected for a 300-ft horizontal conveyor. The K_y factor for the inclined drive half, L_2, is selected for the total conveyor length of 600 ft with an average slope of lift/total length = 36/600 = 6%, because the belt tension is higher than it would be for a 300-ft inclined conveyor, due to the belt pull at the end of the horizontal portion.

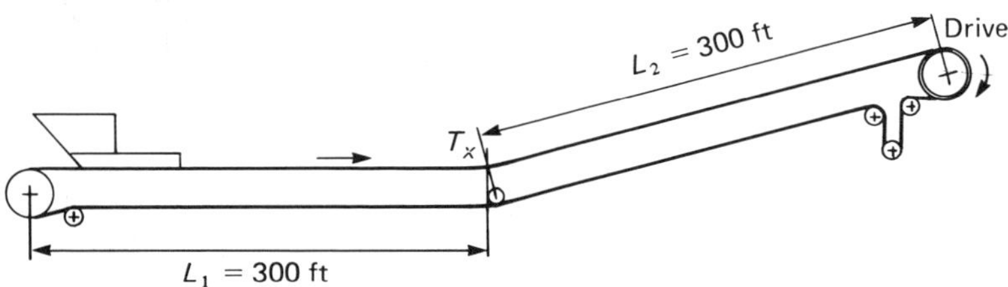

FIGURE 6.24. *Belt conveyor with concave vertical curve.*

In Problem 6, Figure 6.25, the K_y factor for the tail half, L_1, of the conveyor is selected for a 300-ft conveyor inclined at a slope of lift/300 ft = 36/300 = 12%. The K_y factor for the drive half, L_2, of the conveyor is less than it would be for a 300-ft horizontal conveyor, because of the high belt tension existing at the top of the inclined portion. The criterion for determining the K_y value to use for the horizontal drive half, L_2, of this conveyor is the K_y value of a 600-ft inclined conveyor at a 6% slope. The K_y value for the horizontal portion cannot be more, and probably is a little less than, this K_y value.

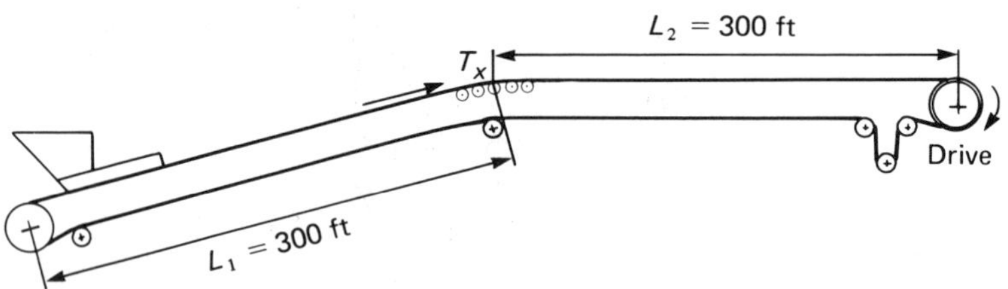

FIGURE 6.25. *Belt conveyor with convex vertical curve.*

The difference in the calculated effective tensions in Problems 5 and 6 is small. But larger and longer conveyors would entail more significant differences.

W_b = 10 lbs per ft, from Table 6-1
H = 36 ft
L = 600 ft

$L_1 = 300$ ft
$L_2 = 300$ ft
$V = 500$ fpm
$Q = 1{,}000$ tph
$S_i = 4.5$ ft
Ambient temperature = 60 °F
Belt width = 36 inches
Material = 100 lbs per cu ft
Drive = lagged head pulley, wrap = 220°
Troughing idlers = Class E6, 6-inch dia, 20° angle, $A_i = 2.8$
Return idlers = Class C6, 6-inch dia, 10 ft spacing

To simplify the calculations, all accessories are omitted.

Analysis (Problem 5, Figure 6.24):

From Table 6-8, wrap factor, $C_w = 0.35$. From Figure 6.1, for 60 °F, $K_t = 1.0$

$$W_m = \frac{33.3\ Q}{V} = \frac{(33.3)(1{,}000)}{500} = 66.6 \text{ lbs per ft}$$

$$W_b + W_m = 10 + 66.6 = 76.6 \text{ lbs per ft}$$

T_0, minimum tension for 3% sag = $4.2\ S_i\ (W_b + W_m) = (4.2)(4.5)(76.6)$ = 1,448 lbs

$$T_t \text{ is taken as } T_0 = 1{,}448 \text{ lbs.}$$

$$T_2 = T_t - 0.015LW_b + HW_b = 1{,}448 - (0.015)(600)(10) + (36)(10)$$
$$= 1{,}448 - 90 + 360 = 1{,}718 \text{ lbs}$$

$$K_x = (0.00068)(W_b + W_m) + \frac{A_i}{S_i} = (0.00068)(76.6) + \frac{2.8}{4.5} = 0.6743$$

The effective belt tension, T_e, is figured individually for each half of the conveyor.

Horizontal portion, 300 ft long, K_y from Table 6-2, for 0° slope, 300 ft and $(W_b + W_m) = 76.6$, is 0.0347. Corrected for 4.5-ft idler spacing, Table 6-3, gives $K_y = 0.0349$.

From "Belt tension at any point, X, on conveyor length," page 106, tension is:

$$T_{cx} = T_t + T_{wcx} + T_{fcx}$$

but, $T_{wcx} = H_x(W_b + W_m) = 0$, for a horizontal belt
and $T_{fcx} = L_x[K_t(K_x + K_yW_b)] + L_xK_yW_m$
and $K_t = 1.0$ for 60 °F

Therefore,

$$T_{fcx} = L_xK_x + L_xK_yW_b + L_xK_yW_m = L_xK_x + L_xK_y(W_b + W_m)$$

Thus,

$$T_{cx} = T_t + 0 + L_xK_x + L_xK_y(W_b + W_m)$$

Calling L_x equal to L_1 for the first (horizontal) half of the conveyor:

$$T_{cx} = T_t + L_1K_x + L_1K_y(W_b + W_m) = 1{,}448 + (300)(0.6743)$$
$$+ (300)(0.0349)(76.6) = 1{,}448 + 202 + 802$$
$$= 2{,}452 \text{ lbs}$$

For the inclined portion, drive half L_2.

$K_x = 0.6743$

K_y is 0.028, for a slope of (36/600)(100%) = 6%, and $W_b + W_m$ = 76.6, and a length of 600 ft, from Table 6-2 for the tabular idler spacing. The corrected value is 0.0298, from Table 6-3, for a 4½-ft spacing.

$T_{cx} = T_t + T_{wcx} + T_{fcx}$. However, T_t is the tension existing at the bottom of the incline, so:

$$T_{cx} = 2{,}452 + T_{wcx} + T_{fcx} = 2{,}452 + H_x(W_b + W_m) + L_xK_x + L_xK_y(W_b + W_m)$$

substituting L_2 for L_x, and 36 for H_x

$$\begin{aligned} T_{cx} &= 2{,}452 + 36(W_b + W_m) + L_2K_x + L_2K_y(W_b + W_m) \\ &= 2{,}452 + (36)(76.6) + (300)(.6743) + (300)(.0298)(76.6) \\ &= 2{,}452 + 2{,}757.6 + 202.3 + 684.8 = 6{,}097 \text{ lbs} \end{aligned}$$

Adding to T_{cx} the nondriving pulley friction, (2)(150) + (4)(100) = 700 lbs, the tension in the belt at the head pulley = $T_{cx} + 700 = T_1$ = 6,097 + 700 = 6,797 lbs

$$T_e = T_1 - T_2 = 6{,}797 - 1{,}718 = 5{,}079 \text{ lbs}$$

$$\text{Belt horsepower} = \frac{T_eV}{33{,}000} = \frac{(5{,}079)(500)}{33{,}000} = 77 \text{ hp}$$

Analysis (Problem 6, Figure 6.25):

T_0 has been calculated in Problem 5 as 1,448 lbs

Take $T_t = T_0$ = 1,448 lbs

$$T_t = T_2 + L(0.015W_b) - HW_b,$$

$$T_2 = T_t - L(0.015W_b) + HW_b = 1{,}448 - (600)(0.015)(10) + (36)(10) = 1{,}448 - 90 + 360 = 1{,}718 \text{ lbs}$$

Inclined portion, 300 ft long.

The slope of the incline is (36/300)(100%) = 12%. For this slope—a length of 300 ft, and $W_b + W_m$ = 76.6—the value of K_y, from Table 6-2, is 0.0293. Corrected for 4½-ft idler spacing, from Table 6-3, K_y = 0.0312. K_x already has been calculated in Problem 5 as 0.6743.

From "Belt tension at any point, X, on conveyor length," page 106, the tension at point X in the carrying run is:

$$T_{cx} = T_t + T_{wcx} + T_{fcx}$$

Since K_t = 1.0 for 60°F, the equation can be written:

$$T_{cx} = T_t + H_x(W_b + W_m) + L_xK_x + L_xK_y(W_b + W_m)$$

So, for the inclined tail half ($L_1 = L_x$) of the conveyor,

$$\begin{aligned} T_{cx} &= 1{,}448 + (36)(76.6) + (300)(0.6743) + (300)(0.0312)(76.6) \\ &= 1{,}448 + 2{,}757.6 + 202.3 + 716.9 = 5{,}125 \text{ lbs} \end{aligned}$$

T_{cx} = 5,125 lbs. This, then, is the tension in the belt at the top of the incline and at the beginning of the horizontal portion.

Horizontal portion = 300 ft long.

$$K_x = 0.6743$$

K_y is dependent on the average belt tension, which, as is seen from the preceding calculation, will be very high. Used as a criterion of K_y, the value of K_y is 0.028, calculated from a 600-ft-long inclined conveyor, at an average slope of 6%, from Table 6–2. Corrected for 4½-ft idler spacing, Table 6–3 gives $K_y = 0.0298$.

From "Belt tension at any point, X, on conveyor length," page 106, the tension at point X in the carrying run (at the head pulley, in this case) is $T_{cx} = T_t + T_{wcx} + T_{fcx}$.

However, T_t is the tension at the start of the horizontal run = 5,125 lbs. $T_{wcx} = H_x(W_b + W_m)$. And since $H_x = 0$, then $T_{wcx} = 0$. Also, $L_x = L_2 = 300$ ft.

$$\begin{aligned} T_{cx} &= 5{,}125 + T_{fcx} = 5{,}125 + L_2K_x + L_2K_y(W_b + W_m) \\ &= 5{,}125 + (300)(0.6743) + (300)(0.0298)(76.6) \\ &= 5{,}125 + 202 + 685 = 6{,}012 \text{ lbs} \end{aligned}$$

Add to T_{cx} the nondriving pulley friction (2)(150) + (4)(100) = 700 lbs

$$T_{cx} + 700 = T_1 = 6{,}012 + 700 = 6{,}712 \text{ lbs}$$

$$T_e = T_1 - T_2 = 6{,}712 - 1{,}718 = 4{,}994 \text{ lbs}$$

$$\text{Belt horsepower} = \frac{T_eV}{33{,}000} = \frac{(4{,}994)(500)}{33{,}000} = 75.7 \text{ hp}$$

The comparison of these two conveyors, each having the same length, lift, size, speed, and capacity, and handling the same material, shows that the conveyor with the concave curve will have a higher belt tension at the head pulley, will require a higher effective tension, and will require more horsepower than the conveyor with the convex curve.

Belt Conveyor Drive Equipment

The engineering of practically all belt conveyor installations involves a comprehensive knowledge of the proper application of conveyor drive equipment, including speed reduction mechanisms, electric motors and controls, and safety devices.

Belt Conveyor Drive Location

The best place for the drive of a belt conveyor is at the location which results in the lowest maximum belt tension. For horizontal or inclined conveyors, the drive usually is at the discharge end, while for declined conveyors the drive is usually at the loading end. For special conditions and requirements, it may be advisable to locate the drive elsewhere. See "Drive arrangements" and "Analysis of belt tensions."

Economics, accessibility, or maintenance requirements may make it preferable to locate the drive internally on the conveyor. Frequently, for the larger conveyors, a saving in supporting structures can be realized by doing so. Inclined boom conveyors sometimes are driven at the loading end for this reason.

Belt Conveyor Drive Arrangement

Belt conveyor drive equipment normally consists of a motor, speed reduction equipment, and drive shaft, together with the necessary machinery to transmit power from one unit to the next. The simplest drive, using the minimum number of units, usually is the best. However, economic reasons may dictate the inclusion of special-purpose units in the drive. These special-purpose units may be required to modify starting or stopping characteristics, to provide hold-back devices, or perhaps to vary the belt speed.

Speed-Reduction Mechanisms

The illustrations in Figures 6.26 through 6.33 show most of the belt conveyor drive equipment assemblies currently in common use.

The following comments apply to these figures:

- Figure 6.26—Gearmotor directly connected by flexible coupling to drive shaft—is a simple, reliable and economical drive.

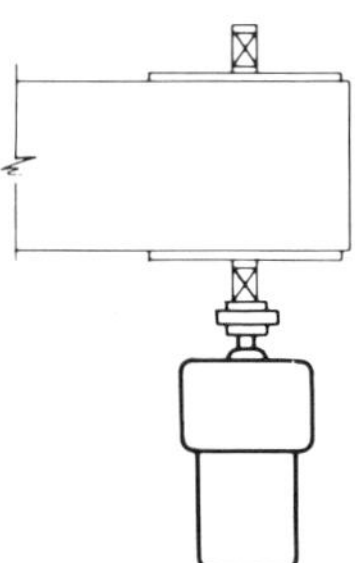

FIGURE 6.26. *Gearmotor is directly connected, by a flexible coupling, to the motor's drive shaft.*

- Figure 6.27—Gearmotor combined with chain drive to drive shaft—is one of the lowest cost flexible arrangements and is substantially reliable.

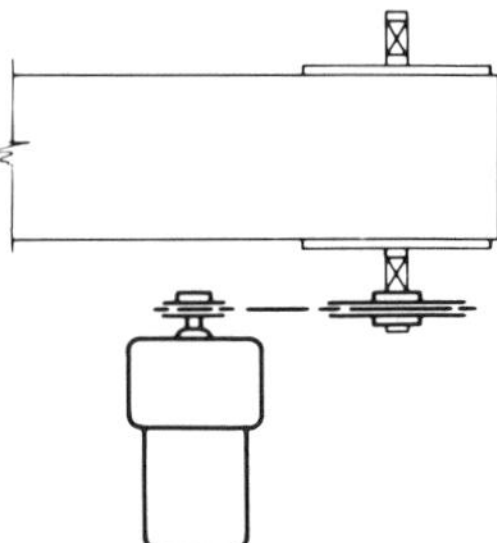

FIGURE 6.27. *Gearmotor connected, by chain drive, to drive shaft.*

- Figure 6.28—parallel-shaft speed reducer directly coupled to the motor and to drive shaft—is versatile, reliable, and generally heavier in construction and easy to maintain.

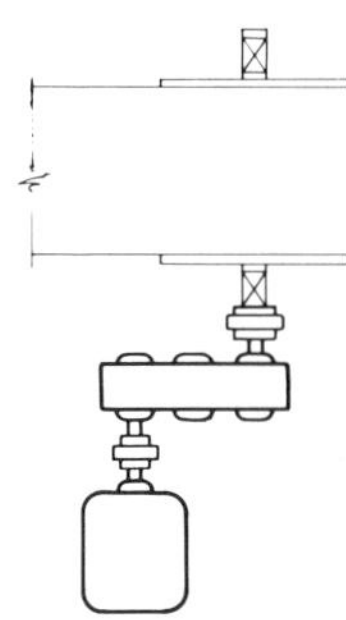

FIGURE 6.28. *Parallel-shaft speed reducer directly coupled to the motor and to the drive shaft.*

- Figure 6.29—parallel-shaft speed reducer coupled to motor, and with chain drive, to drive shaft—provides flexibility of location and also is suitable for the higher horsepower requirements.

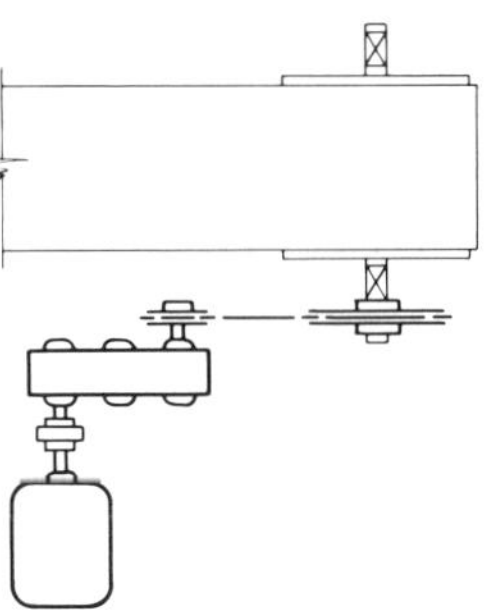

FIGURE 6.29. *Parallel shaft speed reducer coupled to the motor and, with a chain drive, to the drive shaft.*

- Figure 6.30—spiral-bevel helical speed reducer, or worm-gear speed reducer, directly coupled to motor and to drive shaft—is often desirable for space-saving reasons and simplicity of supports. The spiral-bevel speed reducer costs substantially more than the worm-gear speed reducer but is considerably more efficient.

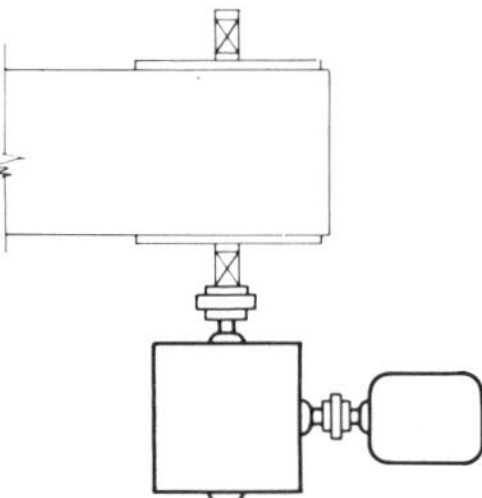

FIGURE 6.30. *Spiral-bevel helical speed reducer, or worm-gear speed reducer, directly coupled to the drive motor and to the drive shaft.*

- Figure 6.31—spiral-bevel helical speed reducer, or worm-gear speed reducer, coupled to motor and, with chain drive, to drive shaft—is a desirable selection for high reduction ratios in the lower horsepower requirements. This drive is slightly less efficient, but has lower initial costs and is most flexible in terms of location.

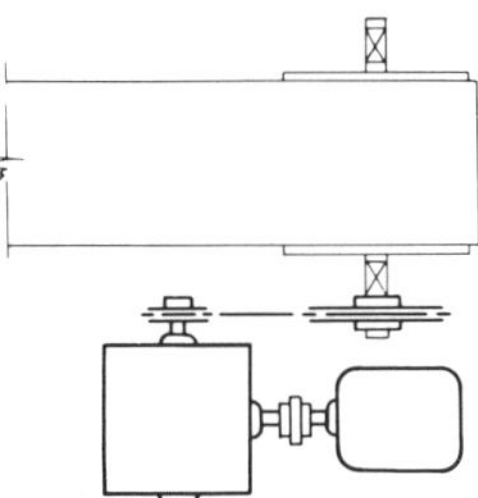

FIGURE 6.31. *Spiral-bevel helical speed reducer, or worm-gear speed reducer, directly coupled to motor and, with chain drive, to drive shaft.*

- Figure 6.32—drive-shaft-mounted speed reducer with V-belt reduction from motor—provides low initial cost, flexibility of location, and the possibility of some speed variation and space savings where large speed reduction ratios are not required and where horsepower requirements are not too large.

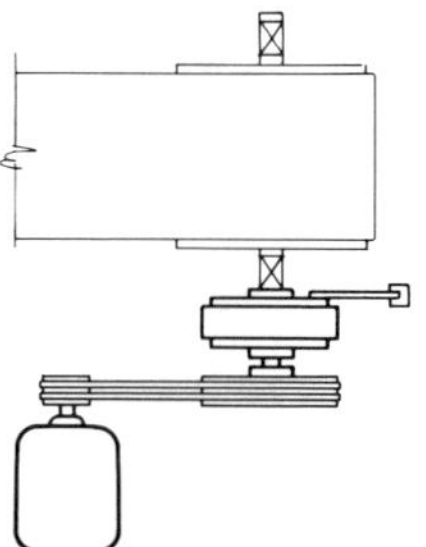

FIGURE 6.32. *Drive-shaft-mounted speed reducer with V belt reduction from motor.*

Dual-pulley drive, shown in Figure 6.33, is used where power requirements are very large, and use of heavy drive equipment may be economical by reducing belt tensions.

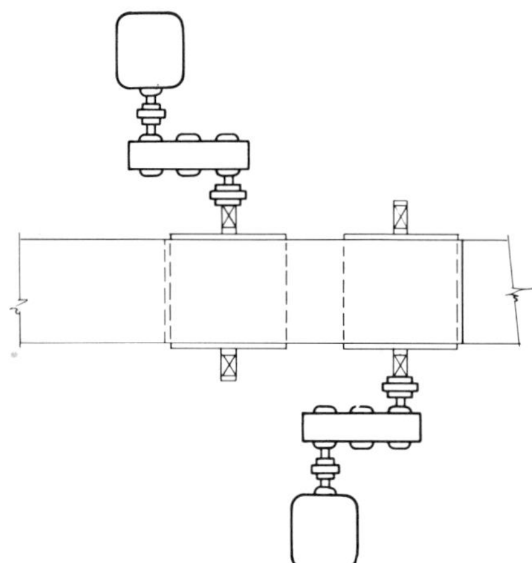

FIGURE 6.33. *Two motors (dual-pulley drive) coupled to helical or herringbone gear speed reducers, directly coupled to drive shafts.*

Selection of the type of speed-reduction mechanism can be determined by preference, cost, power limitations, limitations of the speed-reduction mechanism, limitations of available space, or desirability of drive location. The use of speed reducers in the drives for belt conveyors is almost universal today. However, space-saving considerations and low initial cost sometimes may dictate the use of countershaft drives with guarded gear or chain speed reductions.

All of the drives shown can be assembled in either left- or right-hand arrangement.

Drive Efficiencies

To determine the minimum horsepower at the motor, it is necessary to divide the horsepower at the drive shaft by the overall efficiency of the speed-reduction machinery.

To determine the overall efficiency, the efficiencies of each unit of the drive train are multiplied together. The final product is the overall efficiency.

The efficiencies of various speed-reduction mechanisms are listed in Table 6-11. These efficiencies represent conservative figures for the various types of drive equipment as they apply to belt conveyor usage. They do not necessarily represent the specific efficiencies of the drive units by themselves. Rather, they take into account the possible unforeseen adverse field conditions involving misalignment, uncertain maintenance, and the effects of temperature changes. While there are some variations in efficiency among different manufacturers' products, the data in Table 6–11 generally cover the efficiencies of the various speed-reduction mechanisms.

Table 6-11. *Mechanical Efficiencies of Speed–Reduction Mechanisms*

Type of speed reduction mechanism	*Approximate mechanical efficiency*
V-belts and sheaves	0.94
Roller chain and cut sprockets, open guard	0.93
Roller chain and cut sprockets, oil-tight enclosure	0.95
Single reduction helical or herringbone gear speed reducer or gearmotor	0.95
Double reduction helical or herringbone gear speed reducer or gearmotor	0.94
Triple reduction helical or herringbone gear speed reducer or gearmotor	0.93
Double reduction helical gear, shaft-mounted speed reducers	0.94
Low ratio (up to 20:1 range) worm-gear speed reducers	0.90
Medium ratio (20:1 to 60:1 range) worm-gear speed reducers	0.70
High ratio (60:1 to 100:1 range) worm-gear speed reducers	0.50
Cut spur gears	0.90
Cast spur gears	0.85

As an example of the application of the overall drive efficiency—the result of combining equipment unit efficiencies—consider a belt conveyor drive consisting of a double-helical-gear speed reducer and an open-guarded roller chain on cut sprockets. The approximate overall efficiency, according to Table 6-11, is (0.94)(0.93) = 0.874. If the calculated minimum horsepower at the drive shaft is 13.92 hp, then the required motor horsepower is 13.92/0.874 = 15.9 hp. Therefore, it is necessary to use at least a 20-hp motor.

Mechanical Variable Speed Devices

The most common mechanical methods of obtaining variable speeds of belt conveyors are: V-belt drives on variable-pitch diameter sheaves or pulleys, variable-speed transmissions, and variable-speed hydraulic couplings.

The choice of these devices depends upon the power and torque to be transmitted, the speed range and accuracy of control required, how well the chosen control works into the system, and the relative initial and maintenance costs.

Creeper Drives

In a climate with low temperatures that cause ice to form on the conveyor belt, with resulting loss in conveyor effectiveness, it is good practice to consider the installation of a creeper drive in connection with the drive equipment. The creeper drive consists of an auxiliary small motor and drive machinery, which, through a clutch arrangement, takes over the driving of the empty conveyor at a very slow speed. This creeper drive is arranged to be operative at all times when the conveyor is not handling any load, thus preventing the formation of harmful ice deposits on the conveyor belt.

Backstops

A loaded inclined belt conveyor of sufficient slope tends to move backwards, when forward motion is stopped by a cessation or interruption of power or a mechanical failure in the driving machinery. Should the loaded belt move backward, the material would pile up at the tail end of the conveyor. This could seriously damage the belt, impose a safety hazard, and result in the need to clean up and dispose of the spilled material. To prevent this reversal of motion, a backstop is used.

A backstop is a mechanical device that allows the conveyor to operate only in the desired direction. It permits free rotation of the drive pulley in the forward direction but automatically prevents rotation of the drive pulley in the opposite direction.

There are three general backstop designs: ratchet and pawl, differential band brake, and over-running clutch. See Figures 6.34, 6.35, and 6.36.

Determine Need and Capacity of Backstop, Inclined Conveyors

When the force required to lift the load vertically is greater than one-half the force required to move the belt and load horizontally, a backstop is required. That is, when:

$$HW_m > \frac{L(K_x + K_y W_b + 0.015 W_b) + W_m L K_y}{2}$$

See the formula for T_e, pages 78 and 91-92, omitting T_p, T_{am}, and T_{ac}.

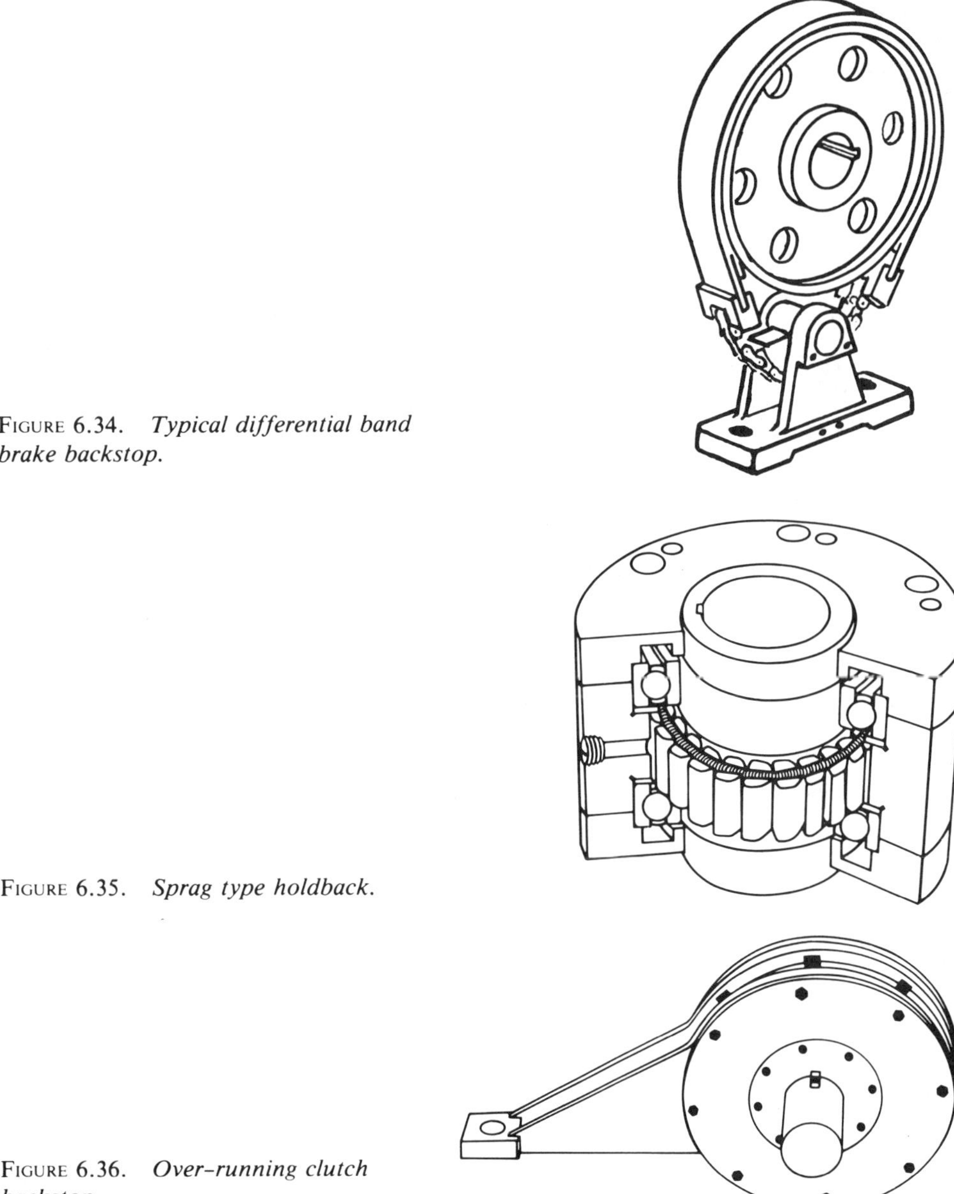

FIGURE 6.34. *Typical differential band brake backstop.*

FIGURE 6.35. *Sprag type holdback.*

FIGURE 6.36. *Over-running clutch backstop.*

Because a backstop is a safety device, it is important that the friction forces which retard the reverse motion of the conveyor are not overestimated. The above formula reduces these friction forces by 50% and **eliminates both the temperature correction factor and the friction introduced by the conveyor accessories.**

Backstops are rated on the basis of the pound-feet (lb–ft) of torque they can safely develop. To determine the approximate amount of torque required of a backstop mounted on the drive pulley shaft, the following analysis applies:

r = radius of drive pulley, ft

rpm = revolutions per minute of drive pulley shaft

Torque required of backstop:

$$\text{Torque} = r\left[HW_m - \frac{L\ (K_x + K_y\ W_b + 0.015\ W_b) + W_m\ L\ K_y}{2}\right]$$

$$\text{Torque} = \frac{\text{hp} \times 5{,}250}{\text{rpm}} \quad \text{and hp} = \frac{\text{rpm} \times \text{torque}}{5{,}250}$$

Hp of brake:

$$\text{hp}_b = \left(\frac{\text{rpm} \times \text{r}}{5{,}250}\right) \left[HW_m - \frac{L\ (K_x + K_y\ W_b + 0.015\ W_b) + W_m\ L\ K_y}{2}\right]$$

This analysis applies to straight inclined conveyors. For conveyors with irregular profiles, a special analysis must be made.

Most manufacturers of backstops recommend size selection based on maximum breakdown or stalled torque of motor. Refer to specific manufacturer for his selection procedure.

Brakes

A loaded declined regenerative conveyor, when operating, is restrained from running away by the power source. Any interruption of power or mechanical failure of the drive will permit the belt and load to run out of control. To prevent this, a properly located brake is required.

A horizontal conveyor, or a declined conveyor that is not regenerative, may coast to a degree that is not tolerable. In such cases, a brake is used to regulate the stopping time and distance.

A brake is a friction device for bringing a conveyor belt to a controlled stop. While brakes are used to bring a conveyor to rest in the event of power failure or mechanical drive failure, they also are used to control the coasting distance of a conveyor as it is being decelerated, in order to limit the amount of material which will discharge over the head pulley during the stopping interval. Brakes are used instead of backstops on inclined reversible conveyors, because backstops are unidirectional.

The brakes used in belt conveyor control operate on the principle that the braking surfaces are engaged by springs and disengaged either by a magnet or by hydraulic pressure induced by an electric motor-hydraulic pump combination. These two types generally are classified by the method of disengaging the braking surfaces. Eddy-current brakes are also used for deceleration.

Practically all conveyors involving lift or lowering need, in addition to the braking force, a holding action after the conveyor has come to a standstill, if for no other reason than safety. In the case of an inclined conveyor, this holding action could be provided by a backstop. However, for any declined conveyor, there is an obvious need for some device which permits application of a controlled torque to decelerate the load at a reasonable reduction in the rate of speed, yet allows sufficient holding power to keep the conveyor belt securely at a standstill when fully loaded but not in operation.

Any conveyor which, under some condition of loading, is regenerative must, for purposes of deceleration analysis and holding power of the brake, be considered as a declined conveyor.

Mechanical Friction Brakes

The mechanical friction brakes are commonly operated electrically. For safety reasons (power failure) such brakes should be spring-set, and power-released.

These mechanical friction brakes provide both the necessary decelerating torque and final holding action. They are interconnected electrically with the motor such that when the power to the motor is off, the holding coil, on the brake also is de-energized, thus allowing a spring to set the brake. For this reason, these brakes are "fail-safe."

The designer should bear in mind, though, that a friction brake is not a precision device, because of the inherently disadvantageous properties of brake linings. The coefficient of friction of brake lining, and with it the actual braking torque, is affected by temperature, humidity, and the degree to which the lining has become worn.

Eddy-Current Brakes

Eddy-current brakes produce a dynamic braking torque by means of a smooth drum which rotates in a magnetic field produced by a stationary field coil. Eddy currents are generated in the surface of the drum as it rotates. A magnetic attraction between these eddy currents and the poles of the field assembly produces a braking torque on the drum. This torque varies directly with the field current and the speed of the drum. It can be adjusted in a stepless manner by a control system.

For holding action, because the eddy-current brake is not effective in case of power failure, it should be combined with an auxiliary mechanical friction brake. As an eddy-current brake drum slows down, the torque that it is capable of exerting diminishes and is zero when the drum ceases to rotate. Thus, an eddy-current brake cannot be expected to hold a conveyor belt in a standstill position. The auxiliary friction brake also serves to decelerate the conveyor in case of power failure.

Deceleration can also be achieved within the drive motor and its control. There are three basic ways of achieving this braking action, none of which provides holding power after the conveyor belt has come to rest. For this reason, some type of auxiliary external brake is always needed to hold the conveyor belt at a standstill.

Plugging the Motor

Here, the current is reversed and counter torque is developed. This force attempts to rotate the motor in a direction opposite to the existing motion. The energy is dissipated as heat. The motor must be de-energized when zero speed is reached, otherwise the motor will attempt to accelerate in the reverse direction. Among others, squirrel-cage motors are most suitable for this application. There is no holding effect at zero speed and the electrical power losses during plugging are high.

Dynamic Braking

Dynamic braking is a system of electric braking in which motors are used as generators and the kinetic energy of the load is employed as the actuating means of exerting a retarding force. To dynamic brake a.c. motors, it is necessary to provide a source of d.c. excitation during the braking period. The control is so arranged that when the stop button is depressed and the a.c. line contactor is opened, another contactor closes to connect the d.c. excitation to one phase of the motor primary. The motor now acts as a generator and is loaded by the induced current flowing through its squirrel-cage winding. The braking torque, which varies in proportion to the exciting current, rapidly increases as the motor slows down but then decreases at near zero speed. The braking torque disappears near zero speed and there is no holding effect at zero speed.

Regenerative Braking

Squirrel-cage motors operating above synchronous speed have inherent retarding torque characteristics. This retarding condition, known as regenerative braking, is applicable above the synchronous speed of the motor (or for multi-speed motors above their synchronous speeds). The energy generated by the motor flows back into the electric power line. Care must be taken to insure that the electric power system is capable of absorbing the power generated by the motor.

This fundamental type of braking is found to be especially useful for declined conveyors operating at a speed which drives the motor at its synchronous speed, plus slip.

Brakes and Backstops in Combination

Often a brake is used to control the stopping interval on an inclined conveyor. If the conveyor is a large and important one, which may reverse and run backward in the event of a mechanical failure, prudence dictates the use of a mechanical backstop as a safety precaution, in addition to the electrically operated brake.

Friction surfaces on brakes, and brakes used as backstops, do not develop the design friction factors until the braking surfaces have worn in to effect full contact. Therefore, friction brakes used as such or as backstops must be adjusted to compensate for this "wearing in" process.

Restraint of Declined Conveyors

Declined conveyors of the regenerative type, are restrained in normal operation by the drive motor which acts as a generator, when the belt and its load force the motor to rotate faster than its synchronous speed. The motor may fail to restrain the belt and load when it is forced to a speed where its current

Table 6-12. ***Backstop and Brake Recommendation***

Type of conveyor	*Backstop*	*Brake*	*Forces to be controlled*
Level or horizontal conveyor	Not required	Required when coasting of belt and load is not allowable or needs to be controlled	Decelerating force minus resisting friction forces
Inclined conveyor	Required if hp of lift equals or exceeds hp of friction	Not usually required unless preferred over backstop	Incline load tension minus resisting friction forces
Declined conveyor	Not required	Required	Decelerating force plus incline load tension minus resisting friction

output is excessive and the overload protective device breaks the circuit. Proper selection of the motor and controls will avoid this contingency. Nevertheless, a brake must be supplied, one which will set when the power circuit is broken.

A centrifugal switch is often used on declined belt conveyors to open the electrical control circuit at a predetermined overspeed, and thus to set the brake. This will act as a safety against mechanical failure in the drive machinery.

A brake is usually located at the tail end of a declined conveyor.

Backstop and Brake Recommendations

Table 6–12 lists recommendations for the use of backstops and brakes on horizontal, inclined, and declined conveyors.

Deceleration by Brakes

Brakes are a necessity on declined conveyors so that the loaded belt may be stopped without excessive or runaway coasting. Brakes are also applied to horizontal and inclined belt conveyors for the same reason. Excessive coasting may discharge far more material than the succeeding conveyor or other units can handle. Mathematical calculation and the careful selection of a properly sized brake will eliminate such difficulties.

Devices for Acceleration, Deceleration, and Torque Control

Starting the Conveyor

Smooth starting of a conveyor belt is important. It can be accomplished by the use of torque-control equipment, either mechanical or electrical, or a combination of the two. The belt conveyor designer should investigate acceleration stresses of conveyor components to insure that the overall stresses remain within safe limits.

Smooth starting can be an important consideration, where excess horsepower may have been installed to provide for future increased capacity or for future extensions of the conveyor. In cases of conveyors having vertical curves or trippers, too rapid a start may cause excessive lifting of the belt from the idlers. This would necessitate a provision for gradual acceleration of the conveyor belt.

Controlled Acceleration

Acceleration can be controlled by several types of electrical devices.

Wound–Rotor Motors with Step Starting. By the addition of external resistance in the secondary winding, electrically accessible through slip rings, starting torque can be controlled by planned steps. This allows a program designed to suit the particular conveyor, and overcome the problems of excessive belt tension, shape of the vertical curves, and other problems which are solved by starting time control.

This type of electrical control device has been widely used for many years on large belt conveyor systems.

Squirrel–Cage Induction Motor with Autotransformer. Another method of controlling the torque, and with it the acceleration time, is the use of an induction motor (normal or high–torque) with autotransformer starting. Its use must be checked because the low–starting torque caused by the reduced voltage may not be enough to overcome the breakaway static friction in level or inclined conveyors.

Eddy-Current Couplings. These are electromagnetic devices composed of three basic parts: a rotor made up of multiple pole pieces (and secured to one shaft), a hollow iron cylinder or drum which surrounds the rotor (and is secured to the other shaft), and a stationary electromagnetic coil which surrounds both rotor and drum and provides the magnetic field in which they operate.

The electromagnetic coil is energized by a low-power, direct current supply. When either the rotor or the drum is rotated, eddy currents are induced. These eddy currents set up a secondary field and thus create a torque between the rotor and drum. The driven or output member never attains the same speed as the driving or input member. This inherent difference in speed is called "slip." The slip loss appears as heat, which must be dissipated by air or water cooling.

In a conveyor drive, the eddy-current coupling is placed between the squirrel cage motor and speed reducer, on the motor shaft, and on the speed reducer input shaft. Because the degree of excitation of the coil determines the slip between the driving and driven members, it is obvious that eddy-current couplings provide an ideal means of controlled acceleration. Excitation of the coil can be increased over a definite time period, or it can be made responsive to tachometer feedback speed-regulating control. Sophisticated electronic control can be employed to regulate the coil excitation to produce virtually any desired result.

There are several advantages of eddy-current couplings. (1) They require low-power coil excitation. (2) They permit smooth, controlled starting. (3) The motor can be started and accelerated without connecting the load. On frequent start and stop applications the motor can run continuously. (4) Variable speed can be obtained. However, in variable speed applications the additional slip creates more heat which must, of course, be dissipated. (5) They make possible the use of squirrel-cage motors and across-the-line starters. (6) A modified eddy-current coupling can be used as a decelerating brake (not as a "holding brake," however).

The disadvantages of eddy-current couplings are: (1) They require additional drive space. (2) Water cooling must be provided for the larger sizes. (3) Generally, they are more expensive than a wound-rotor motor and reduced-voltage starting.

Fluid Couplings. These are mechanical two-piece mechanisms consisting of an impeller and a runner, both within a housing filled with oil. The impeller is connected to the driving shaft; the runner is connected to the driven shaft. In conveyor drives, the fluid coupling usually is placed between the motor and speed reducer. When the impeller is revolved, the oil is thrown to the periphery and impinges on the blades of the runner, producing a torque on the runner proportional to the weight and rate of fluid flow. The fluid coupling basically is a slip clutch. And, as with an eddy-current coupling, the slip loss appears as heat. Unlike its electrical counterpart, a fluid coupling is not used as a variable-speed device.

When properly applied, a fluid coupling can produce reasonably smooth acceleration of high-inertia loads. The motor speed rises rapidly to a point near the maximum torque condition before the load is engaged. This makes the standard squirrel-cage motor an ideal driver, as its peak torque is about 200% of full-load torque. The fluid coupling permits the use of squirrel-cage motors with across-the-line starters. The fluid coupling also requires less space than an eddy-current coupling, but lacks the controlling features of the latter.

Variable-Speed Hydraulic Couplings. These devices have been used very successfully, especially in Europe. The variable-speed hydraulic coupling consists of a fluid coupling with input and output shafts, heat exchanger, charging oil pump, and associated control. The amount of oil in the coupling is variable. The control can be manual or completely automatic. Speed variation over a 4:1 range is possible. This hydraulic device has most of the features of the eddy-current coupling.

Dry Fluid Couplings. These are similar to oil-filled fluid couplings, except that they consist of a housing which is keyed to the motor shaft and a rotor which is connected to the load. The housing contains a charge of steel shot instead of a fluid. When the motor is started, centrifugal force throws the charge of steel shot to the inner periphery of the housing, where it packs around the rotor. Some slippage takes place before the housing and rotor are finally locked together. Power thus is transmitted from the motor to the load.

The amount of the charge of steel shot determines the torque during acceleration. It can also determine the torque–limiting feature of this coupling.

Miscellaneous Fluid Couplings. These are similar to dry fluid couplings except, in lieu of a charge of steel shot, *these* fluid couplings employ silicone fluid, mercury, etc. Characteristics of these couplings are such that manufacturers should be consulted for specific performance details before an application is made.

Centrifugal Clutch Couplings. These consist of a driving hub, a driven sleeve or drum, and a series of shoes connected to or driven by the hub. The periphery of each shoe is provided with a brake–lining material. The hub is carried by the driving shaft, the drum by the driven shaft. When the hub rotates, centrifugal force impels the shoes outward against the inside of the drum, to transmit power to the load. Slippage occurs, which produces the effect of smooth starting.

Flywheel. Mechanical control of starting and stopping can be accomplished by means of a flywheel, which adds to the WK^2 of the prime mover, thus increasing the starting time and limiting the torque input to the belt conveyor system, as well as increasing stopping time and distance.

Mechanical Clutches. This device can effectively control starting torque; it allows adjustment of the amount of torque, as well as the rate at which it is applied. The mechanical clutch can be preset for both rate of application and maximum limit.

Brake Requirement Determination (Deceleration Calculations)

In order to determine whether any braking action, other than the friction forces inherent in the system, is required, several different circumstances under which the conveyor might be stopped have to be considered. For instance, is the stopping intentional, or is it the result of power failure? Also, if another conveyor feeds onto it, or if the conveyor in question delivers its load to an additional belt, it is necessary to consider their respective motions and deceleration cycles.

It is obvious that the drives of a conveyor system which consists of more than one belt—and in which at least one belt feeds onto another—have to be interconnected electrically in such a way that if one conveyor is stopped for any reason, the one feeding onto it is also stopped. This precaution alone does not suffice, however, if the physical properties of the first conveyor are such that it would coast longer than the second one. If this were to occur, it would result in a pile-up of material on the second belt and could cause a hazardous situation.

Generally speaking, in any system with more than one conveyor, the length of the deceleration cycle of any successive conveyor should be equal to or more than that of the preceding one.

If the inherent properties of the various units do not result in deceleration cycles which agree with this basic rule, two remedies are possible. (1) A brake can be applied to those conveyors which coast too long. This is a straightforward solution, and relatively easy to accomplish. (2) The stored energy of those conveyors which come to a stop too quickly can be augmented, for instance, by a flywheel. Although a flywheel will lengthen the stopping distance of a conveyor, it will also increase its acceleration time. This must be taken into consideration by the belt conveyor designer.

However, in most cases, the application of a brake will be found more convenient, unless its use overstresses any member of the unit to which it is applied.

Material Discharged During Braking Interval

To determine the amount of material discharged during the braking interval, it must be assumed that the conveyor decelerates at a constant rate. Therefore, the distance travelled, while stopping from full speed, is the average velocity multiplied by the time of braking interval.

$$\text{Distance, ft, conveyor travels} = \left(\frac{V+0}{2}\right)\left(\frac{t_d}{60}\right) = \frac{Vt_d}{120}$$

where V = belt speed, fpm
t_d = actual stopping time, seconds

If the amount of material which can be safely discharged to the succeeding conveyor (or other) unit is known, the maximum length of time of the braking of decelerating interval can be determined as follows:

$$W_d = \frac{Vt_m}{120}\,(W_m)$$

Therefore,

$$t_m = \frac{120\,W_d}{W_m V}$$

where t_m = maximum permissible stopping time, seconds (braking or deceleration interval)
W_d = weight, lbs, which can be discharged
W_m = weight of material, lbs per ft of belt

Forces Acting During Braking or Deceleration

The forces which act on the conveyor during a braked stop (deceleration) include inertia; frictional resistance; gravity material load force; inclines or declines; and braking force.

The frictional resistance forces and the gravity material load forces, if any, are equal to T_e. The braking force is equal to the algebraic sum of the other forces.

Therefore, for horizontal, inclined, and nonregenerative declined belt conveyors, the braking force = inertial forces − T_e, or:

$$F_d = \frac{M_e V}{60t_m} - T_e = \frac{W_e V}{60gt_m} - T_e$$

For regenerative declined belt conveyors, braking force = inertial forces + T_e, or:

$$F_d = \frac{M_e V}{60t_m} + T_e = \frac{W_e V}{60gt_m} + T_e$$

where F_d = braking force, lbs, at belt line
M_e = equivalent moving mass, slugs
g = acceleration of gravity, 32.2 ft per sec^2
W_e = equivalent weight of moving parts of the conveyor and its load, lbs. See Problems 3 and 4, pages 142–146 and 151–155, respectively.
V = speed of belt, fpm
t_m = maximum permissible stopping time, seconds (braking or deceleration interval)
T_e = effective or driving horsepower tension, lbs

Brake Location

An analysis of the belt tension diagram during deceleration should be made to determine the appropriate pulley on which to apply the brake. The braking force will be additional to the friction and positive lift forces.

If the brake is installed on a head-end drive pulley, the automatic takeup force must be sufficient to transmit the braking force through the takeup. The wrap factor at the braking pulley must be checked for adequacy during braking. Also, the minimum belt tension in the carrying run of the conveyor must be maintained during braking. The maximum permissible belt tension must not be exceeded during deceleration.

For inclined or short horizontal conveyors, it may be possible to brake through the head or drive pulley, providing the takeup has sufficient force to absorb the braking force and still maintain a slack-side tension to meet the wrap factor requirements. If this is not practical, as in the case of a long horizontal or declined conveyor, then the braking force must be applied to the tail pulley.

The maximum belt tension during deceleration should be calculated to insure that it does not exceed the recommended allowable starting (or braking) tension. (See page 102, and Chapter 12, "Controlled deceleration.") If it is found that the belt tension does exceed the allowable amount, a heavier belt may be required. Or the belt conveyor can be re-analyzed to provide for a smaller braking force acting over a longer time period. If the conveyor is subjected to frequent stops, the pulleys and shafts must be selected for the higher tensions introduced during deceleration.

Braking Torque

The braking force (lbs) determined above and acting at the belt, multiplied by the radius (ft) of the braked conveyor pulley, gives the required torque rating of the brake (lb-ft), provided the brake is installed on the same shaft that carries the braking pulley.

$$\text{Torque} = F_d r$$

where F_d = braking force at the belt
r = radius of conveyor pulley, ft, on the same shaft as the brake

If the brake is to be installed on some shaft other than the pulley shaft, the torque requirement is converted by multiplying the above torque by the revolutions per minute of the shaft for which the torque was determined. This product is then divided by the revolutions per minute of the shaft on which the brake will be mounted. Select the brake with the next higher torque rating.

Brake Heat Absorption Capacity

The discussions above relate to the selection of a brake on the basis of torque only. Stopping a moving mass involves the absorption of the kinetic energy of the belt, the load, and the moving machinery. This energy only can be dissipated in the form of heat at the brake. The resulting temperature rise of the brake elements must not damage the brake. For this reason, a discussion of brake design and brake heat absorption follows.

Industrial brake linings usually are made of either woven or molded asbestos, plus various fillers and adhesives. The coefficient of friction of these linings against a brake wheel varies considerably with different ambient conditions. Because of the nature of these variations, definite values of the friction coefficient cannot be given. Nevertheless, some variations that can be expected are approximated below.

Coefficients, and consequently torque values, may vary widely for new linings and/or new wheels, until both lining and wheel surfaces are worn in. This requires approximately 4,000 to 6,000 full-torque brake-setting operations. During this period static torque may drop as much as 30% below the initial setting, and dynamic torque as much as 50%. For this reason, the discussion will relate only to well-worn-in linings and wheels.

Both static and dynamic torque vary with wheel surface temperatures. At 50 °C to 75 °C, the static torque may be as much as 30% to 35% high. But it then drops off rather rapidly with increase in wheel temperature. At 115 °C to 135 °C, the static torque is about normal. At 150 °C, it may be 5% to 7%

below normal. The dynamic torque may be 10% to 15% high at 40 °C to 60 °C, and then rise rapidly, until at 115 °C to 150 °C it may be as high as 140%. It then drops off rather rapidly with further temperature rise.

Because of these variations, brake wheels are rated at 120 °C rise for normal energy dissipations. The ratings, which are expressed in "hp seconds," are based on a maximum temperature rise of 120 °C at the brake wheel, when the brake is applied at the listed time intervals. The brakes have lower ratings for more frequent stops because they will not cool sufficiently between stops to absorb the heat of rapidly repeated stopping.

Humidity will also have an adverse effect on braking torque because industrial brake linings absorb moisture. If a brake is allowed to stand inoperative for some time in high ambient humidity, the braking torque may be reduced as much as 30% when the brake is first set. This condition is self-correcting, because the heat generated in braking rapidly drives off the moisture. Usually the torque will be restored to almost normal at the end of the first braking cycle. In this case, the only effect is a longer time than usual to make the first stop.

Variations in any given lining material, and in surface conditions of the lining and wheel, may result in a 10% plus or minus variation in torque during successive stops.

From the above factors, it is evident that industrial brakes are not precision devices. The normal method of setting brake torque by measuring either spring length or adjusting-bolt length is at best an approximation. Where braking effects are important to a conveyor operation, the brake should be readjusted for optimum braking by actually stopping and holding the load after the brake first is installed. For critical conveyor applications, it may be necessary to readjust the brake more than one time during the break-in period for new linings.

Brake Calculations

To check the brake wheel heat absorption for a single stop of a loaded conveyor, first determine the actual stopping time for the brake selected.

$$t_d = \frac{\dfrac{W_e V}{(32.2)(60)}}{\left(\dfrac{Z_b}{r}\right)\left(\dfrac{\text{rpm}_b}{\text{rpm}_p}\right) + T_e}$$

where
t_d = actual stopping time, seconds
W_e = equivalent weight of the moving mass, lbs
V = velocity or speed of belt, fpm
Z_b = torque rating or setting of brake, lb-ft
rpm_b = revolutions per minute of brake shaft
rpm_p = revolutions per minute of drive pulley shaft

Note: For regenerative belt conveyors, T_e may be negative.

The energy which must be absorbed by the brake when making a single stop of a loaded conveyor is expressed as follows:

$$\text{Energy, in horsepower seconds} = \frac{(Z_b)(\text{rpm}_b)(t_d)}{10{,}500}$$

The symbols are the same as above, but with the brake on the drive pulley shaft, $\text{rpm}_p = \text{rpm}_b$.

The heat absorption should be approved by the brake manufacturer for the anticipated duty cycle.

If the brake selected does not have the heat-absorption capacity required, either a modified or larger brake with the necessary heat-absorption capacity should be used. The spring should be adjusted to the torque desired.

Example As a numerical example of brake selection, the belt conveyor specifications for Problem 3, page 140, will be used. Since the WK^2 and total equivalent weight for the conveyor have been calculated, only the essential portions of these specifications are repeated here.

Conveyor Specifications:

V = belt speed = 500 fpm		
W_m = weight of material per ft of belt =	226	lbs
T_e = effective tension =	16,405	lbs
T_2 = slack-side tension =	5,720	lbs
T_0 = minimum tension =	3,067	lbs
T_t = tail tension =	7,054	lbs
Equivalent weight of conveyor moving parts =	162,696	lbs
Weight of material load =	543,360	lbs
Total equivalent weight for belt tension determination =	706,056	lbs
Drive equivalent at belt =	62,870	lbs
W_e = total equivalent weight for brake determination =	768,926	lbs

Assuming that the conveyor discharges into a hopper which holds 9,000 lbs of material, the maximum permissible stopping time is as follows:

$$t_m = \frac{120W_d}{W_mV} = \frac{(120)(9{,}000)}{(226)(500)} = 9.54 \text{ seconds}$$

Brake force at belt line is:

$$F_d = \frac{W_eV}{60gt_m} - T_e = \frac{(768{,}926)(500)}{(60)(32.2)(9.54)} - 16{,}342 = 4{,}517 \text{ lbs}$$

Analysis:

If the total equivalent retarding force of 4,517 lbs is applied to the head-end drive shaft, a proportion equal to:

$$\frac{(62{,}870)(500)}{(60)(32.2)(9.54)} = 1{,}706 \text{ lbs}$$

of the equivalent force would be absorbed in retarding the drive components. The remainder, 4,517 − 1,706 = 2,811 lbs of the equivalent force would be transmitted to the belt by the pulley to retard the conveyor moving parts and the load. This force is T_{eb}.

During braking, the highest tension in the belt will be T_{1b}, on the return run just past the drive pulley. If the automatic takeup is to be on the verge of yielding to the braking force, T_{1b} can be assumed to be equal to T_2, the slack-side tension during normal operation of the belt. Because:

$$T_{1b} - T_{2b} = T_{eb}$$

and substituting T_2 for T_{1b}:

$$T_2 - T_{2b} = T_{eb} = 5{,}720 - T_{2b} = 2{,}811$$

Therefore,

$$T_{2b} = 5{,}720 - 2{,}811 = 2{,}909 \text{ lbs}$$

This is the tension in the carrying run at the head-end drive pulley during braking. It is insufficient, for the minimum tension, T_0 = 3,067 lbs. Also, $T_{2b} = (C_{wb})(T_{eb})$, or:

$$C_{wb} = \frac{T_{2b}}{T_{eb}} = \frac{2{,}909}{2{,}811} = 1.03 \text{ wrap factor during braking.}$$

This is sufficient, as the wrap factor for the drive is 0.35 to prevent slip between the pulley and belt.

When it is shown that braking at the head-end drive produces too low a tension in the carrying run, or too small a wrap factor, it is necessary to increase the belt tensions by increasing the automatic takeup force.

The alternative to braking at the head-end drive pulley is to apply the braking force to the tail pulley. In this case, the entire braking force of 4,517 lbs must be transmitted to the belt.

When the power of the drive is cut off, and just as the brake takes effect, the tension in the return run at the tail pulley will be:

$$T_{2b} = T_2 + \text{pulley friction} + \text{return-belt idler friction}$$
$$- \text{interia of moving parts of the return run}$$

Pulley friction = (4)(100) + (1)(150) = 550 lbs
Return idler friction = $L\,(0.015\,W_b)$
= (2,400)(0.015)(17) = 612 lbs

Equivalent weights of the moving parts of the return run are:

Return belt, LW_b = (2,400)(17) = 40,800 lbs

Return idler weight of rotating parts is 48.4 lbs, from Table 5-14, for a 48-inch-wide belt and Class C6 idlers.

Total return idler rotating weight = $\left(\frac{2{,}400}{10}\right)(48.4)$ = 11,616 lbs

Rotating weight of pulleys, from Problem 3, page 142, is 3,450 lbs

Total equivalent moving parts of return run = 40,800 + 11,616 + 3,450 = 55,866 lbs

Equivalent force at the belt line

$$= \frac{W_e V}{60 g t_m}$$

$$= \frac{(55{,}866)(500)}{(60)(32.2)(9.54)}$$

$$= 1{,}516 \text{ lbs}$$

Therefore, $T_{2b} = 5{,}720 + 550 + 612 - 1{,}516 = 5{,}366$ lbs

And because $T_{eb} = 4{,}517$, and $C_{wb} = T_{2b}/T_{eb}$:

$$C_{wb} = \frac{5{,}366}{4{,}517} = 1.19$$

This is very satisfactory, since a 180° wrap, bare pulley requires only that the wrap factor, C_{wb}, be 0.84 or larger. See Table 6-8. The maximum belt tension when braking = 4,517 + 5,366 = 9,883 lbs. This is well within the maximum of 1.80(25,920) − 9,883 = 36,773 lbs (see page 143). It therefore is appropriate to place the brake on the tail pulley shaft of this conveyor.

Assuming the tail pulley radius is 1.5 ft and the pulley is revolving at 53 rpm, the torque at the tail pulley shaft is:

$$F_d r = (4{,}517)(1.5) = 6{,}776 \text{ lb-ft}$$

As the brake will be mounted directly on the tail pulley shaft, the required brake torque will be 6,776 lb-ft.

For this problem, assuming alternating current electric power is available, the brake selected from the brake manufacturers' catalog is an a.c. magnetic brake with a 10,000 lb-ft rating. This is the next larger size than that calculated at 6,776 lb-ft.

Actual stopping time, using this 10,000 lb-ft brake is:

$$t_d = \frac{\dfrac{W_e V}{(32.2)(60)}}{\left(\dfrac{Z_b}{r}\right)\left(\dfrac{\text{rpm}_b}{\text{rpm}_p}\right) + T_e}$$

$$= \frac{\dfrac{(768{,}926)(500)}{(32.2)(60)}}{\left(\dfrac{10{,}000}{1.5}\right)\left(\dfrac{53}{53}\right) + 16{,}342}$$

$$= 8.65 \text{ seconds}$$

This is less than the maximum of 9.54 seconds that is permissible.

Energy absorbed is:

$$p = \frac{(Z_b)(\text{rpm}_p)(t_d)}{10{,}500} = \frac{(10{,}000)(53)(8.65)}{10{,}500} = 436.62 \text{ hp seconds}$$

The brake selected is capable of heat absorption of 3,400 hp–seconds every 15 minutes. This indicates that a loaded stop can be made safely without overheating the brake.

Chapter 7

Belt Selection

Contents

The purpose of this chapter is to familiarize the user with the general requirements and method of proper selection of conveyor belting. Because of the multitude of conveyor belt constructions presently available and the impossibility of treating them all in this chapter, only those basic types and grades of conveyor belting that apply to a majority of conveyor applications will be covered.

Anyone using the data in this chapter should recognize that a belt selection determined by the data will be conservative. And while the selected belt will meet the specified conditions, it may not always be the most economical construction available. This is particularly true because of the continuing developments in the fields of elastomers and synthetic fibers for use in conveyor belts.

For major conveyor belt applications, the complete duty requirements of the conveyor must be analyzed to develop final specifications.

A complete analysis of a conveyor system for determining the conveyor belt specifications requires consideration of the design details which follow.

1. Material conveyed:
   general description;
   density, pounds per cubic foot, lbs/ft^3;
   lump size;
   presence of oils or chemicals, if any;
   maximum temperature of load, if hot;
   requirements for fire resistance.
2. Maximum loading rate or required maximum capacity, tons (2000 lbs) per hour (tph).
3. Belt width, inches.
4. Belt speed, feet per minute (fpm).
5. Profile of conveyor:
   profile distance along conveyor path, tail to head, feet;
   lift or drop, ± feet, *or* elevations of top and bottom of any inclines or declines;
   angles of slope of all inclines or declines;
   locations and radii of all vertical curves.
6. Drive:
   single-pulley or two-pulley;
   if dual drive, distribution of total motor horsepower at primary and secondary drive pulleys;
   angle of belt wrap on drive pulley(s);
   location of drive;
   pulley surface, bare or lagged; type of lagging;
   type of starting to be employed.

7. Pulley diameters. These may require confirmation according to the belt requirement.
8. Takeup:
   type;
   location.
9. Idlers:
   type, roll diameter, angle of trough;
   spacing, including transition distance at head and tail.
10. Type of loading arrangement:
    chutes;
    freefall distance, lumps to belt;
    skirtboard length.
11. Lowest cold weather operating temperature anticipated, if applicable.
12. Type of belt splice to be used.

Factors in the Composition of Conveyor Belting

While a belt conveyor system is composed of many important parts, none is more economically important than the conveyor belt itself, which, in most cases, will represent a substantial part of the initial cost. Therefore, the selection of the conveyor belt must be made with great care.

In general, a conveyor belt consists of three elements: top cover, carcass, and bottom cover. Figure 7.1 illustrates a cross section of a typical belt. The primary purpose of the covers is to protect the belt carcass against damage and any special deteriorating factors that may be present in the operating environment. The belt carcass carries the tension forces necessary in starting and moving the loaded belt, absorbs the impact energy of material loading, and provides the necessary stability for proper alignment and load support over idlers under all conditions of loading.

FIGURE 7.1. *Cross-section of a fabric-reinforced belt.*

Although these elements are treated here as separate components, successful operation depends upon their working together to provide the necessary overall belt characteristics. It will be shown later in this chapter that the distinction between covers and carcass as individual components is not as pronounced in some types of conveyor belting as it is in others.

Covers

Rubber or rubber-like compounds are used for the top and bottom covers of conveyor belting and for bonding together various components of the belt carcass. These compounds are produced by mixing rubbers or elastomers with various chemicals in order to obtain reinforcement and to develop the physical properties necessary for service conditions.

By definition, an elastomer is an elastic, rubber-like substance. In the case of conveyor belting, the term is extended to refer to all the thermosetting materials which require definite times and temperatures for cure, such as natural and synthetic rubbers, as well as such thermoplastic materials as polyvinyl chloride plastic (PVC).

Many years ago, a range of tensile strengths and elongations was adopted by the rubber industry for establishing the quality of the grades of covers. At that time, only natural rubber was available; thus, tensile strength and elongation were the criteria for evaluating rubber compound quality. At the present time, however, there is a wide choice of rubbers or elastomers available, and each can be utilized alone or blended with others to obtain a great number of combinations with intermediate properties particularly suitable for a wide variety of service conditions. The ranges of tensile strength and elongation specified previously are no longer necessarily valid as measures of the quality of a cover and, specifically, of its abrasion resistance.

It is possible to classify covers by the basic elastomer used, but evaluation of the quality of the cover should be based on its suitability for particular service, rather than on the kind of elastomer it contains. Each cover has characteristics which, when properly utilized, will provide a conveyor belt for the lowest cost per unit of material carried under specified conditions of service.

Because the primary function of the cover is to protect the carcass, it must resist the wearing effects of abrasion and gouging, which vary according to the type of material conveyed. The top cover will generally be greater in thickness than the bottom cover because the concentration of wear is usually on the top, or carrying, side. However, specific characteristics of the material to be conveyed and the operating conditions may require a belt with equal cover thickness on top and bottom.

Adhesion Values

At the same time that the criteria for belt cover quality were adopted, values were also established for the adhesion of the various elements making up the conveyor belt. These values were indicators of the general flex life of the belt at a time when only natural rubber and all-cotton carcass fabrics were available. However, since the introduction of synthetic rubbers, plastics, synthetic fibers, and steel cable into conveyor belt design, and with the variety of methods for using these materials today, these earlier adhesion values no longer represent a realistic measurement of the flexing capabilities of the many available types of modern conveyor belting.

Belt Carcass

The belt carcass is the tension element of a conveyor belt. It is the primary reinforcement for belt tear resistance, impact resistance, load support, and mechanical fastener-holding ability.

Most conveyor belt carcasses are made of one or more plies of woven fabric. Some high-tension carcasses employ a single layer of parallel steel cables.

Conveyor belt fabric is made of warp yarns, which run lengthwise, and weft yarns, or filling, which run crosswise or transversely. Four types of weave patterns are commonly used: plain weave, straight-warp weave, solid-woven weave, and woven-cord weave.

Plain weave. This is the oldest and most common type of belt fabric, in which the warp and filling yarns cross each other alternately. In some cases, warp or filling yarns appear in pairs rather than as single yarns. See Figure 7.2.

Straight-warp weave. This type of fabric has straight-laid, noninterwoven, or crimped warp yarns, and straight-laid, noninterwoven, or crimped filling yarns, plus binder warp yarns which are interwoven with the filling yarns to hold the structure together. The straight-laid warp yarns are the primary tension element in this type of fabric. See Figure 7.3.

Solid-woven weave. This type of fabric is a multiple weave, with at least two warp systems and two or more planes of filling yarns. Solid-woven fabric is primarily used in single-ply belts. See Figure 7.4.

FIGURE 7.2. *Plain weave.*

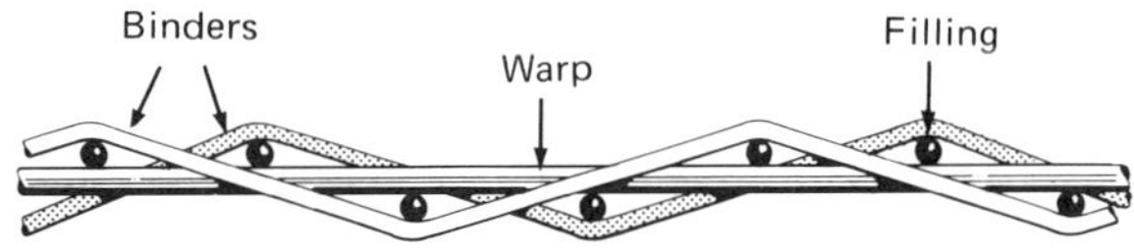

FIGURE 7.3. *Straight-warp weave.*

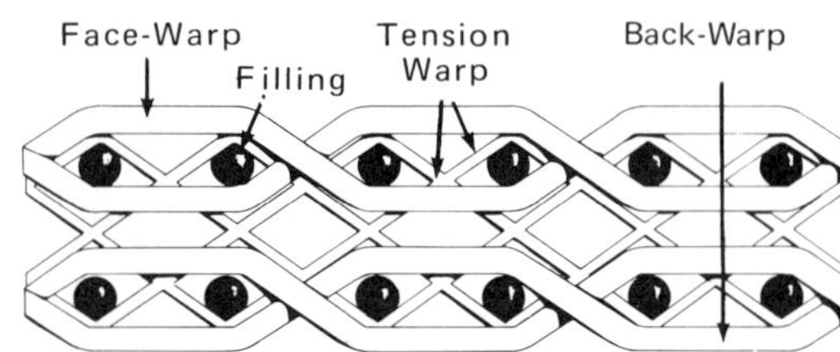

FIGURE 7.4. *Solid-woven weave.*

Woven-cord weave. This type of fabric has strong warp yarns and very light weight, interwoven filler yarns, which serve only to hold the structure together during belt manufacture. Woven-cord fabric is used in combination with plies of plain weave fabric as a conveyor belt carcass.

The types of textiles used by belt manufacturers vary greatly. Cotton, viscose rayon, nylon, and polyester are widely used, either in pure form or in various blends and combinations. Typical combinations are cotton warp-nylon filling, rayon warp-nylon filling, and polyester warp-nylon filling.

The fabric in textile carcasses is impregnated with elastomeric compound; if more than one ply is used, a skim coat of compound is usually provided between plies. For the PVC type of solid-woven belting, the carcass is impregnated with liquid PVC plastisol.

The major types of conveyor belt carcass are multiple-ply, reduced-ply, steel-cable, and solid-woven.

Multiple-ply belt carcasses. The carcass of the multiple-ply belt is usually made up of three or more plies, or layers, of woven belt fabric which are bonded together by an elastomeric compound. Belt strength and load support characteristics vary according to the number of plies and the fabric used, but practical considerations usually limit the number of plies in a carcass to a maximum of eight.

Multiple-ply belts have been standardized on the basis of tension rating into a range of "MP" designations. However, these designations do not limit the fabric plies to the use of any particular textile fiber; a variety of belt fabrics can be used. The warp and filling yarns can be made from the same textile fiber, or the warp yarns can be one type and the filling yarns another. In some cases, blends of fibers in the yarns are also used.

The multiple-ply conveyor belt was the most widely used type through the mid-1960s, but today it is often supplanted by reduced-ply belting.

Reduced-ply belts. These belts consist of carcasses with either fewer plies than comparable multiple-ply belts, or special weaves representing a total departure from the ply concept. Figure 7.5 shows a reduced-ply belt. The textile carcasses of such belts comprise high-strength synthetic fibers—usually nylon, polyester, or combinations—in plain-weave or special fabric designs. There is little or no standardization, and wide variations exist in the types of belts offered by manufacturers. In most cases, the reduced-ply belt depends

FIGURE 7.5. *Reduced-ply belt.*

upon the exclusive or extensive use of high-strength synthetic textile fibers concentrated in a carcass of higher unit strength and fewer plies than in a comparable multiple-ply belt.

The belt is made with plies of plain-weave fabric or with one or more plies of straight-warp fabric, depending on the manufacturer. Load support is provided by extra thick rubber skim layers between plies and/or in the weave design of the fabric itself.

Because of the differences in carcass design, the MP designations do not apply to reduced-ply belts. However, the technical data available from belt manufacturers generally indicate that reduced-ply belting can be used for the full range of applications specified for multiple-ply belting, and in some cases beyond it.

Steel-cable belts. Steel-cable conveyor belts are made with a single layer of parallel steel cables, completely imbedded in rubber, as the tension element. The carcass of steel-cable belting is available in two types of construction, depending on the manufacturer and the service conditions. The all-gum construction uses only cables and cable rubber, as shown in Figure 7.6. Fabric-reinforced construction has one or more plies of fabric above and below the cables, but separated from the cables by the cable rubber. See Figure 7.7. Breakers may be included in the fabric-reinforced type. Both types have appropriate top and bottom covers.

Steel-cable belting is produced using a broad range of cable diameters and spacing, depending primarily on the desired belt strength. This type of belting is often used in applications requiring operating tensions beyond the range of fabric belts and/or in installations where takeup travel limitations are such that changes in the length of a fabric belt cannot be accommodated.

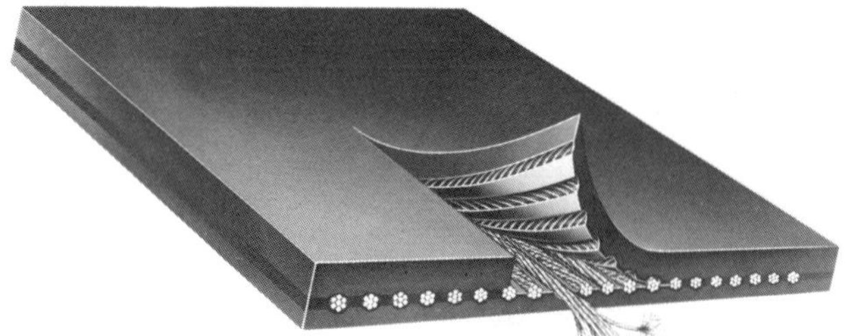

FIGURE 7.6. *Steel-cable belt—all-gum construction.*

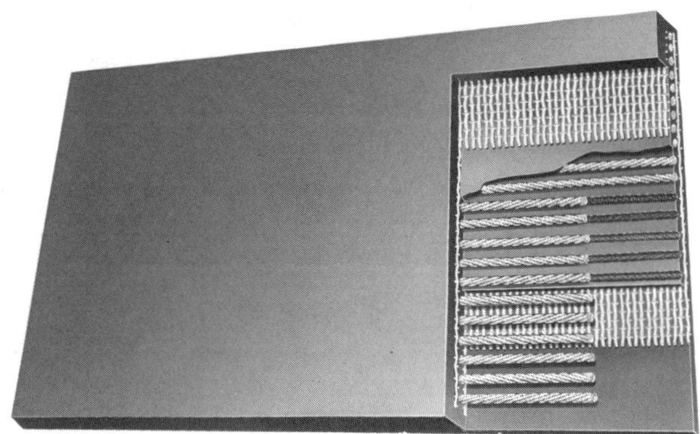

FIGURE 7.7. *Steel-cable belt—fabric-reinforced construction.*

Solid-woven belts. This type of belting consists of a single ply of solid-woven fabric, usually impregnated and covered with PVC, with relatively thin top and bottom covers. Abrasion resistance is provided by the combination of PVC and the surface yarns of the fabric. Some belting is produced with heavier covers, and thus is not dependent on the fabric for abrasion resistance.

Tension ratings and other criteria for solid-woven belting vary among manufacturers and are not related to other types of conveyor belting. Individual manufacturers should be consulted for proper belt selection.

Breakers

Breakers are fabric layers used primarily to increase the adhesion between the cover and the carcass under conditions of impact, and also to aid in distributing the shock of lump impacts. Breakers can be made from the same cotton and synthetic fibers used in carcass fabrics.

Two general types of fabrics are used as breakers: woven cord and leno weave. As described above, the woven-cord fabric is made with strong warp yarns held together by relatively fine filling yarns. Leno-weave fabric is an open-mesh weave in which a warp yarn passing over a filling yarn and a companion warp yarn passing under the same filling yarn are twisted to cross each other in the space between the filling yarns.

Although these fabrics are commonly used as breakers in the covers of multiple-ply belts, they are sometimes used as carcass plies in combination with plies of plain-weave fabric to produce variations of the multiple-ply carcass.

Breakers are generally recommended for use under the top cover of multiple-ply belts when lumps of 2 inches or more are handled. They are not usually required with reduced-ply belting.

Grades of Conveyor Belting and Their Uses

Several grades of conveyor belting are covered in this section and their service applications defined. Table 7-1 provides information to assist in cover quality selection.

TABLE 7-1. *Conveyor Belt Cover Quality Selection*

Cover grade	*Major advantages*			*General applications*
	Cut & tear resistance	*Abrasion resistance*	*Oil resistance*	
General Service				
Grade 1	Excellent	Excellent	Not recommended	Large size ore, sharp cutting materials. For extremely rugged service.
Grade 2	Good	Excellent	Not recommended	Sized materials with limited cutting action—primarily abrasion. For heavy duty service.
Oil & chemical service				
Chloroprene oil resistant (commonly called Neoprene)	Good	Very good	Very good for petroleum oils fair for vegetable & animal oils	Heavily oil sprayed coal (petroleum oil up to 20% aromatics, No. 2 Diesel fuel). Any material treated with or containing large amounts of petroleum oil.
Buna N oil resistant	Good	Good	Very good for petroleum vegetable & animal oils	Oily grain or seed service (soybeans, crushed corn, etc.) Food handling. Greasy, oil sprayed coal (petroleum oil up to 40% aromatics, No. 2 heating oil).
Medium oil resistant	Good	Good	Limited for petroleum, vegetable & animal oils	Lightly sprayed coal, mildly oily grains and feeds, wood chips, phosphates.

Grade 1 Belting

Grade 1 conveyor belting has covers made from natural rubber, synthetic rubber, combinations of natural and synthetic rubbers, or combinations of synthetic rubbers. These are selected to provide the optimum combination of cut, gouge, and abrasion resistance.

Grade 1 conveyor belting has a skim coat of rubber compound between the plies. The type of rubber compound used with the particular fabric insures the highest degree of flex life.

Grade 2 Belting

Grade 2 conveyor belting has covers made from natural rubber, synthetic rubber, combinations of natural and synthetic rubbers, or combinations of synthetic rubbers. Grade 2 covers are designed to provide good abrasion resistance, but not as high a degree of cut and gouge resistance as that of Grade 1.

Grade 2 conveyor belting also has a skim coat of rubber compound between the plies. The type of rubber compound used with the particular fabric provides excellent flex life for normal service conditions where recommended pulley diameters are used, and where overall operating conditions are less severe than those requiring a Grade 1 conveyor belt.

Hot Materials Service

For hot material belting, covers are made from synthetic rubber or combinations of synthetic rubbers. These are selected to provide the best resistance to deterioration from the effects of elevated temperatures, as well as adequate resistance to abrasion, cuts, and gouges for the specific material and material size handled. Special compounds may be necessary where a combination of heat and oil resistance is required.

Hot material covers, which can withstand temperatures from a bed of hot fines up to 325 °F and hot lumps up to 400 °F, should be considered where constant material temperatures over 150 °F will be encountered. Individual belting manufacturers should be consulted for selection of the best covers for specific hot material conditions.

Fire-Resistant Service

Fire-resistant belting is manufactured to conform to the standards of the Mine Safety and Health Administration (MSHA). An acceptance designation number is issued by MSHA to identify belting which meets its qualifications. Special fire-resistant cover compounds may be required where resistance to oil, as well as abrasion, cuts, and gouges, is needed. Individual belting manufacturers should be consulted for selection of the best covers for specific fire-resistant conditions.

Conveyor Belt Splices

Conveyor belting is made endless, usually at the job site, by the use of either mechanical fasteners or vulcanized splices. Figure 7.8 illustrates a vulcanized fabric belt splice and Figure 7.9 illustrates a steel-cable belt splice. The vulcanized-splice method provides the stronger connection and longer service life. However, in many cases, a mechanical fastener splice is acceptable, and in certain cases it actually may be preferred. Some of the advantages and disadvantages of vulcanized versus mechanically-fastened splices are described below.

Vulcanized Splice

Advantages. (1) Strength. It has the highest practical strength. (2) Long service life. Correctly applied on the appropriate conveyor equipment and properly cared for, a vulcanized splice can last for years. However, a vulcanized splice normally will not last for the life of the belt. (3) Cleanliness. A vulcanized splice is smooth and continuous. Thus, conveyed material cannot seep through it. Also, a vulcanized splice does not damage or interfere with belt wipers, as is the case with mechanical-fastener splices.

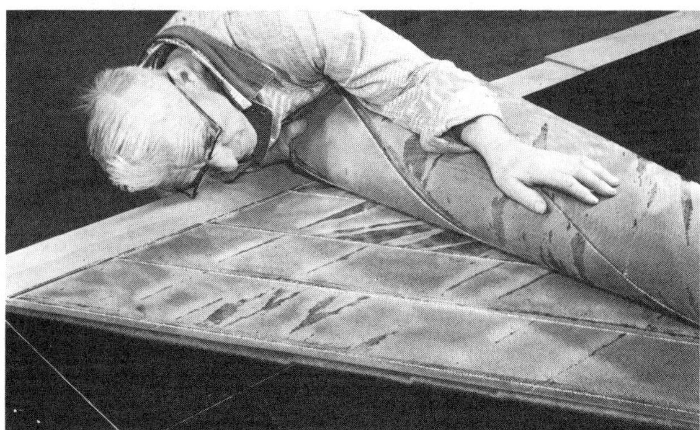

FIGURE 7.8. *Vulcanized-fabric belt splice.*

FIGURE 7.9. *Steel-cable belt splice.*

Disadvantages. (1) Greater initial expense. The initial cost of a vulcanized splice is many times greater than that of a mechanically fastened splice. The vulcanizer is heavy and difficult to move, and supports for it must be provided. (2) More costly takeup provisions. To insure sufficient takeup travel for accommodating both elastic and permanent variation in belt length, longer takeup travel must be provided. Renewing a vulcanized splice is time consuming and costly.

Mechanically Fastened Splice

Figure 7.10 illustrates a hinged-plate type of mechanical splice.

Advantages. (1) Quick to make. A mechanically fastened splice can be installed by experienced personnel in a very short time, whereas it takes hours to complete a vulcanized splice. (2) Low initial expense. The cost of labor and fasteners for a mechanically fastened splice will be a fraction of the cost of a vulcanized splice. Usually, only hand tools are required. (3) Takeup travel problems are minimized. If belt length variations exceed the amount which the takeup is capable of accommodating, the belt can be shortened and respliced quickly at relatively small cost.

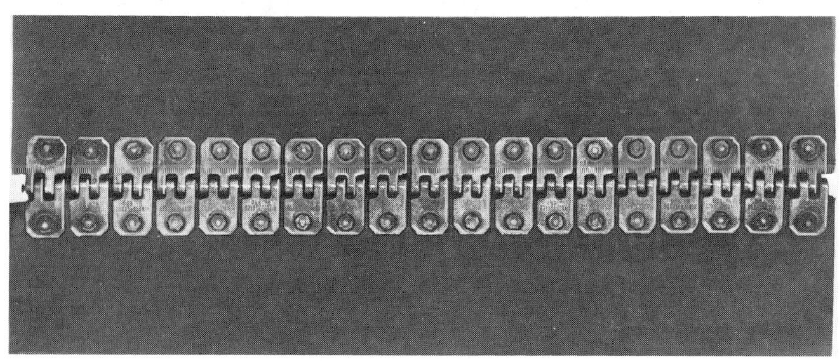

FIGURE 7.10. *Hinged-plate type of mechanical splice.*

Disadvantages. (1) Lower strength. Full belt tension rating cannot be utilized in most cases. (2) Exposure of cut belt ends to the effects of moisture and materials may have a deleterious effect on the belt carcass fabric. (3) Rough surface. Mechanical fasteners cannot be applied to produce as smooth a surface at the splice as that of a vulcanized splice. Belt wipers sometimes catch on the fasteners and suffer damage. (4) It is very difficult to produce a mechanically fastened splice that can be considered leak-proof in conveying fine materials. (5) In hot service, fasteners retain heat and transmit it directly into the belt carcass. This may cause local carcass degradation and early splice failure.

Applicable Service Conditions:

*Normal Mechanical-Fastener Splice, Industrial Use.**

- Pulley diameter as recommended in Table 7-6.
- No abnormal conditions that quickly reduce fabric strength such as heat or acids.

*Maximum Mechanical-Fastener Splice, Normal Plus 15%.**

- Heavy duty mechanical fasteners of the size and type recommended by the manufacturer.
- Service such as portable equipment.
- Where greater than normal fastener replacement is acceptable.
- No abnormal conditions that quickly reduce fabric strength, such as heat or acids.
- Pulley diameters as recommended in Table 7-6.

*Normal Vulcanized Splice.**

- Pulley diameters as recommended in Table 7-6.
- Automatic takeup with adequate takeup travel.
- Good maintenance of equipment.
- Starting tension limited to 150% of normal vulcanized splice tension rating.

Maximum Vulcanized Splice, Normal Plus 8%.* All requirements under "Normal Vulcanized Splice," above, apply, plus the additional requirements listed below:

- Conveyor manufacturer to approve the engineering of the belt conveyor.
- Pulley diameter to be chosen for a belt one ply thicker.
- Skim coated carcass in the belt.
- Provision for prompt vulcanized repair of belt damage.

*See Table 7-2 for Normal Tension Ratings.

High temperatures. For hot material service, a good rule is to limit the stress on the carcass fabric to 75% of its normal service rating. See section on high temperature.

Conveyor Belt Selection

After the conveyor belt tension requirement in pounds per inch of width has been established by the procedures outlined in Chapter 6, the conveyor belting must be selected. This is done on the basis of the following factors:

Tension

The belt carcass must, of course, be sufficient to carry the required tension. Table 7-2 shows the maximum allowable working tension for multiple-ply belt constructions with mechanical or vulcanized splices.

Note that more than one combination of fabric strength and plies may satisfy a given tension requirement. For example, 5 plies of MP 43 fabric and 6 plies of MP 35 fabric have the same rating (approximately 165 pounds per inch of width) when a mechanical splice is used. In cases such as this, other listed factors or comparative prices will determine what carcass to use.

Table 7-2. *Tension Rating of Multiple-Ply Belts*

Fabric Identification	*Tension ratings, lbs per inch per ply*	
	Normal mechanical-fastener splice	*Normal vulcanized splice*
Multiple-ply 35	27	35
Multiple-ply 43	33	43
Multiple-ply 50	40	50
Multiple-ply 60	45	60
Multiple-ply 70	55	70
Multiple-ply 90	—	90
Multiple-ply 120	—	120
Multiple-ply 155	—	155
Multiple-ply 195	—	195
Multiple-ply 240	—	240

Tension ratings for reduced-ply belts are established on the basis of the whole belt rating, rather than per ply. Thus, the ratings are expressed in pounds per inch of belt width (PIW) and will vary according to the belt manufacturer and belt type. As with the multiple-ply belts, the other listed factors which follow must also be considered in making a belt selection. Because of the wide variety of fabrics, strengths, constructions, and other characteristics offered in this type of belt, no list of standard ratings can be published, and it is necessary to consult the various manufacturers for specific data. Most manufacturers offer belts at several levels of rated tension, up to approximately 700 PIW, and some constructions are available for up to 1,000 PIW or more.

Troughability

A conveyor belt must be chosen with sufficient transverse flexibility so that it will conform to the general shape of the troughing idlers when running empty. Figure 7.11 illustrates incorrect and correct troughing. Table 7-4 shows the maximum number of plies allowable for troughing empty multiple-ply belts on 20°, 35°, and 45° idlers. For reduced-ply belts, this characteristic is shown in Table 7-3 as the minimum width that will trough properly when empty. These tables are intended only as guides for typical empty troughing requirements. Manufacturers should be consulted for accurate data on specific belts.

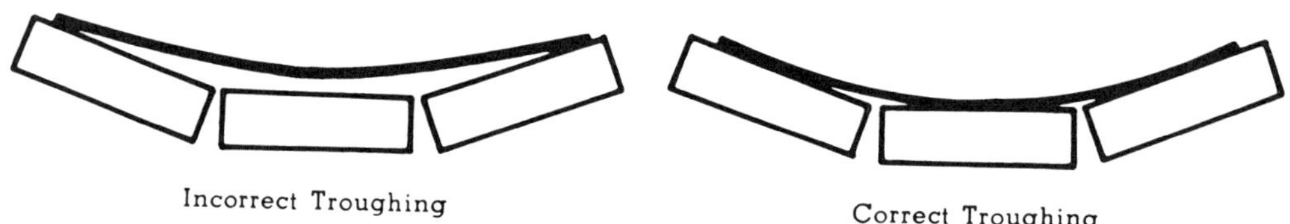

FIGURE 7.11. *Incorrect and correct troughing.*

TABLE 7-3. *Troughability of Reduced-Ply Type Belts*

Minimum Widths for Empty Troughing			
Rated Belt Tension	20° *Idlers*	35° *Idlers*	45° *Idlers*
To 150 PIW	14	18	18
To 200 PIW	16	24	24
To 250 PIW	24	24	30
To 300 PIW	30	30	30
To 500 PIW	36	36	36
To 700 PIW	42	42	48

Note: The above table is *only* a guide. Because of the wide variety of belt constructions offered by manufacturers, it is important that they be consulted for more accurate values.

Pulley Considerations

The diameter and face width of belt conveyor pulleys affect the selection of a conveyor belt. The relationship of these factors to the belt selection is as follows:

Table 7-4. *Maximum Plies for Troughing of Empty Multiple-Ply Type Belts*

Belt width (inches)	*Idler angle*	*MP* 35 *MP* 43	*MP* 50 *MP* 60 *MP* 70	*MP* 90 *MP* 120	*MP* 155	*MP* 195 *MP* 240
	20°	5	—	—	—	—
18	35°	4	—	—	—	
	45°	—	—	—	—	—
	20°	6	5	4	4	—
24	35°	5	4	4	—	—
	45°	4	—	—	—	—
	20°	7	6	5	5	4
30	35°	6	5	5	4	4
	45°	5	4	4	—	—
	20°	8	7	6	6	5
36	35°	7	6	6	5	5
	45°	6	5	5	4	4
	20°	8	8	7	6	6
42	35°	8	7	7	6	5
	45°	7	6	6	5	5
	20°	8	8	8	7	6
48	35°	8	8	7	6	6
	45°	8	7	6	5	5
	20°	8	8	8	8	7
54	35°	8	8	8	7	6
	45°	8	8	7	6	6
	20°	8	8	8	8	8
60	35°	8	8	8	8	7
	45°	8	8	8	7	6
	20°	8	8	8	8	8
72-84-	35°	8	8	8	8	8
96	45°	8	8	8	8	7

Note: The maximum number of plies is limited to 8 by practical manufacturing and belt-handling considerations.

Pulley Diameter. The primary object in selecting adequate pulley diameters is to make sure that, as the belt wraps around the pulley under tension, the stress in the belt carcass is below the fatigue limit of the bond between the belt components. Overstressing of the belt carcass will result in ply separation and premature belt failure, particularly at the belt splices. In some applications where available space is limited, smaller than recommended pulleys may be used. This, however, will result in more frequent splice replacement and will affect the service life of the belt. Tables 7-5, 7-6, and 7-7 indicate the recommended minimum diameters for drive and other pulleys. Reverse bending on a dual-pulley drive or high-tension reverse bending before a single-pulley drive on the return run require pulley diameters 6 inches larger than listed.

TABLE 7-5. *Minimum Pulley Diameters for Reduced-Ply Type Belts, Inches*

Maximum belt Tension	80-100% *Tension*	60-80% *Tension*	40-60% *Tension*
To 100 PIW	14	12	12
To 150 PIW	16	14	12
To 200 PIW	18	16	14
To 300 PIW	24	20	18
To 400 PIW	30	24	20
To 500 PIW	36	30	24
To 700 PIW	42	36	30

TABLE 7-6. *Minimum Pulley Diameters for Multiple-Ply Type Belts, Inches*

	MP 35			*MP* 43, 50			*MP* 60, 70, 90, 120			*MP* 155			*MP* 195, 240		
	% *Tension*			% *Tension*			% *Tension*			% *Tension*			% *Tension*		
Number of Plies	80 *to* 100	60 *to* 80	40 *to* 60	80 *to* 100	60 *to* 80	40 *to* 60	80 *to* 100	60 *to* 80	40 *to* 60	80 *to* 100	60 *to* 80	40 *to* 60	80 *to* 100	60 *to* 80	40 *to* 60
3	18	14	12	20	18	14	24	20	16	30	24	20	36	30	24
4	20	18	16	24	20	18	30	24	20	36	30	24	42	36	30
5	24	20	18	30	24	20	36	30	24	42	36	30	48	42	36
6	30	24	20	36	30	24	42	36	30	48	42	36	54	48	42
7	36	30	24	42	36	30	48	42	36	54	48	42	60	54	48
8	42	36	30	48	42	36	54	48	42	60	54	48	66	60	54

TABLE 7-7. ***Minimum Pulley Diameters for Steel-Cable Belts, Inches***

Rated Tension	80-100% *Tension*	60-80% *Tension*	40-60% *Tension*
To 1,000 PIW	30	30	24
To 1,800 PIW	42	36	30
To 2,400 PIW	48	36	30
To 2,800 PIW	54	42	36
To 3,500 PIW	54	48	36

Note: Regarding Tables 7-5, 7-6, and 7-7, reverse bending encountered on dual-pulley drives or on a high-tension reverse bend before a single-drive pulley on the return run requires pulley diameters 6 inches larger than listed.

Pulley Face. A crowned face is effective in centering a belt on the pulley if the approach to the pulley is an unsupported span that is unaffected by the steering action of idlers. Consequently, for troughed belts the crown on the head pulley of a conveyor is of little value in training the belt. Some benefit may be realized in terms of centering the belt through a tripper by having a

TABLE 7-8. *Recommended Pulley Face Width and Belt Clearances*

Conveyor belt widths b (inches)	*Pulley face width Pf (inches)*	*Distance between discharge chute plates (inches)*	*Return belt clearance* minimum each side (inches)*
42 and under	$b + 2$	$p_f + 3$	2½
over 42	$b + 3$	$p_f + 4$	3

* It may be desirable on conveyors with 500-foot centers and longer, for greater belt edge protection, that the next wider standard pulley face be used over that shown in the above table. For these conveyors, the stringers to carry the idlers should be spaced wider, allowing a return belt clearance of 5 or 6 inches or more, each side. This will result in increased cost of stringers, idler frames, chutes, but often is considered worthwhile.

crowned face on the tripper discharge pulley. Crowned face on low-tension bend pulleys in the return run and on a tail pulley where there is a relatively long unsupported span between return idler and the pulley may be slightly beneficial in centering the belt, but the contribution to overall training is minor.

Crowned-face pulleys should never be used for any pulleys on conveyors using steel-cable belt.

Multiple-ply belts should not be used on a crowned-face pulley where the tension will exceed 76 pounds per inch per ply. However, for all belts with textile carcasses, the best recommendation is that crowned pulleys be limited to locations where the belt will only be subjected to less than 40% of its rated tension.

Only straight-face pulleys should be used for all two-pulley drives and for drive snub pulleys.

The recommendations for the face width of belt conveyor pulleys are shown in Table 7-8. Return belt clearances are also indicated.

Load Support

While a belt must be chosen with sufficient transverse flexibility for empty troughing, it must also be able to support the load properly on idlers of a given end-roll angle. Tables 7-9 and 7-10 can be used as guides to determine constructions that will properly support the load. Load support characteristics of reduced-ply belts are usually expressed as the maximum width allowable in a given type of belt for conveying the load. Because of the wide variety of fabrics, strengths, constructions, and other characteristics offered in this type of belt, no standard load support table can be published, and it is necessary to consult the various manufacturers for specific data.

Impact Resistance

The belt carcass must be selected to take the expected impact of the material being loaded on the belt conveyor. Tables 7-11 and 7-12 can be used as a general guide for determining the range of belt constructions appropriate for

Table 7-9. ***Minimum Plies for Load Support, 20° Idlers, Multiple-Ply Belts***

Belt width (inches)	*25-49 PCF Material*							*50-74 PCF Material*							*75-99 PCF Material*							*100-150 PCF Material*						
	MP fabric I.D.							*MP fabric I.D.*							*MP fabric I.D.*							*MP fabric I.D.*						
	35 43	50	60 70	90 120	155	195	240	35 43	50	60 70	90 120	155	195	240	35 43	50	60 70	90 120	155	195	240	35 43	50	60 70	90 120	155	195	240
18	3	3	3					3	3	3	3	3			4	3	3	3	3			4	4	3	3	3		
24	3	3	3					4	3	3	3	3			5	4	3	3	3			5	4	4	4	4		
30	4	3	3	3	3	3	3	4	4	3	3	3	3	3	5	4	4	4	4	4	4	6	5	4	4	4	4	4
36	4	4	3	3	3	3	3	5	4	4	4	4	4	4	6	5	4	4	4	4	4	6	6	5	5	5	5	5
42	4	4	4	4	4	3	3	5	5	4	4	4	4	4	6	5	5	5	5	5	4	7	6	6	6	6	5	5
48	5	4	4	4	4	4	3	6	5	5	5	5	4	4	7	6	6	6	5	5	5	7	7	6	6	6	6	5
54	5	5	4	4	4	4	4	6	6	5	5	5	5	4	7	7	6	6	6	5	5	8	8	7	7	7	6	6
60	6	5	5	5	5	4	4	7	6	6	6	5	5	4	8	7	7	7	6	6	5			8	7	7	7	6
72	6	6	5	5	5	5	4	8	7	6	6	6	6	5		8	7	7	7	7	6				8	7	7	7
84	7	7	6	6	6	6	5		8	7	7	6	6	6			8	8	7	7	6					8	8	7
96	8	8	7	7	7	7	6			8	8	7	7	6					8	8	7							8

use under various loading conditions. It should be noted that more than one belt construction may satisfy the impact requirements. Manufacturers should always be consulted when the maximum impact energy (measured in foot-pounds) exceeds the values listed in Table 7-11.

Reduced-ply belting often provides greater impact resistance than multiple-ply (MP) fabric belting with the same tension rating. Individual belting manufacturers should be consulted for recommendation of proper reduced-ply belting for the impact conditions involved.

Table 7-10. ***Minimum Plies for Load Support, 35°-45° Idlers, Multiple-Ply Belts***

Belt width (inches)	*25-49 PCF Material*							*50-74 PCF Material*							*75-99 PCF Material*							*100-150 PCF Material*						
	MP fabric I.D.							*MP fabric I.D.*							*MP fabric I.D.*							*MP fabric I.D.*						
	35 43	50	60 70	90 120	155	195	240	35 43	50	60 70	90 120	155	195	240	35 43	50	60 70	90 120	155	195	240	35 43	50	60 70	90 120	155	195	240
24	4	3	3	3				4	4	4	4	4			5	4	4	4	4	4		5	5	4	4			
30	4	4	3	3				5	4	4	4	4	4	4	5	5	4	4	4	4	4	6	5	5	5	5	5	4
36	4	4	4	4	4	3	3	5	5	5	5	4	4	4	6	5	5	5	5	5	4	6	6	6	6	6	6	5
42	5	4	4	4	4	4	3	6	5	5	5	5	5	4	6	6	6	6	6	5	5	7	7	6	6	6	6	5
48	5	4	4	4	4	4	4	6	6	6	6	5	5	5	7	6	6	6	6	6	5	8	7	7	7	7	7	6
54	6	5	4	4	4	4	4	7	6	6	6	6	5	5	8	7	7	7	7	6	6	8	8	7	7	7	7	6
60	6	5	5	5	5	4	4	7	7	6	6	6	6	5	8	7	7	7	7	7	6			8	8	7	7	7
72	6	6	5	5	5	5	5	8	7	7	7	7	6	6		8	7	7	7	7	6						8	7
84	7	7	6	6	6	6	6		8	8	8	8	7	7			8	8	8	8	7							8
96	8	8	7	7	7	7	7						8	8							8							

Table 7-11. *Carcass Impact Rating MP Fabrics**

Number of plies	Maximum impact rating, foot-pounds				
	MP fabric identification				
	35	43	50	60 & 70	90-240
3	8	16	20	38	48
4	16	28	38	62	80
5	40	60	75	175	320
6	120	160	210	475	700
7	240	320	410	775	1060
8	—	520	660	1050	1440

* Carcass impact rating in foot-pounds is the maximum impact energy which the carcass can safely absorb based on the use of rubber impact idlers and good design of loading and transfer conditions.

Actual impact energy in foot-pounds = lump weight factor (see Table 7-12) times the equivalent free-fall in feet.

Where the actual impact energy in foot-pounds is greater than shown above in Table 7-11, consult the conveyor belting manufacturer.

The above impact ratings are based on 10% lumps, 90% fines. If there are more than 10% lumps, add one more ply to the belt than above table indicates. For sized material 4 inches and below, use tabular values; for sized material above 4 inches, add one more ply.

$$\text{Equivalent free-fall} = H_f + [H_v(\sin^2\Delta)]$$

where H_f = total free-fall, ft
H_v = vertical height on loading chute slope, ft
Δ = angle, degrees, that chute slope makes with horizontal

See Figure 7.12 below for illustration of values of H_f and H_r.

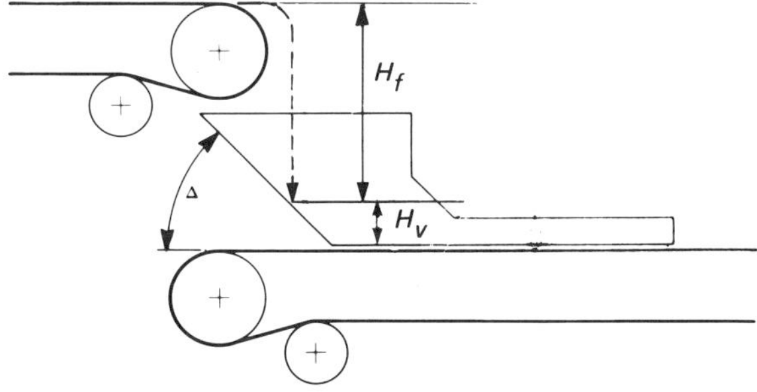

Figure 7.12. *Equivalent free fall and location of values H_f and H_v.*

Table 7-12. *Lump Weight Factor*

Weight of material (lbs per cu ft)	Lump weight factor (weight lump, lbs)												
	Lump size, inches												
	2	3	4	5	6	7	8	9	10	12	14	16	18
50	0.4	1.3	3.0	5.8	10	14	21	30	40	70	100	148	211
75	0.6	1.9	4.5	8.6	15	21	31	44	61	105	149	222	316
100	0.7	2.6	5.9	12.0	20	28	41	59	81	140	199	296	421
125	0.9	3.2	7.4	14.0	25	35	52	74	101	175	248	371	527
150	1.1	3.8	9.0	17.0	30	42	62	89	121	210	298	444	632
175	1.3	4.5	10.4	20.2	35	49	73	104	142	245	348	518	737

The above table is a close approximation of the weight of a lump, based on cubic lump and slab breakage characteristics.

Cover Considerations

The covers should be of sufficient thickness and quality to protect the carcass. Covers for general service applications are listed in Tables 7-13 and 7-14, which list suggested minimum thickness for carrying and pulley side covers, respectively.

The cover gauge required for a specific belt is a function of the material conveyed and the handling methods used. Increased cover thickness is required as the following conditions become more severe: material abrasiveness, maximum lump size of material, material weight, height of material drop onto the belt, loading angle, belt speed, and frequency of loading as determined by the frequency factor.

Table 7-13. ***Suggested Minimum Carry Cover Thickness for Normal Conditions RMA—Grade 2 Belting***

Class of material	*Examples*	*Thickness (inches)**
Package handling	Cartons, food products	Fric. to 1/32
Light or fine, nonabrasive	Wood chips, pulp, grain, bituminous coal, potash ore	1/16 to 1/8
Fine and abrasive	Sharp sand, clinker	1/8 to 3/16
Heavy, crushed to 3″ (76 mm)	Sand and gravel, crushed stone	1/8 to 3/16
Heavy, crushed to 8″ (203 mm)	R.O.M. coal, rock, ores	3/16 to 1/4
Heavy, large lumps	Hard ores, slag	1/4 to 5/16

*Note: Cover thicknesses are nominal values subject to manufacturers' tolerances.

Table 7-14. ***Suggested Minimum Pulley Cover Thickness RMA—Grade 2 Belting***

Operating Conditions	*Thickness (inches)*
Non-abrasive materials	1/32
Abrasive materials	1/16
Impact loading*	3/32

Note: Cover thicknesses are nominal values subject to manufacturers' tolerances.

* While an increased cover gauge helps protect the carcass, if impact is severe, a correct system design which includes carcass design, top cover thickness, and impact rolls in the conveyor is the preferred method of handling.

Deteriorating Conditions. Table 7-15 establishes the basis for determining cover quality for some deteriorating conditions. The actual cover thickness generally should follow the guidelines for a Grade 2 cover in Table 7-13. For all special materials not listed, or where extreme concentrations of chemical solutions are likely to be encountered, a belt manufacturer should be consulted to determine appropriate cover quality and thickness.

High Temperature Covers. Special cover constructions are available for use in systems where temperatures range from 150 °F to 400 °F. In general, special precautions must be observed when working with materials in this temperature range. Extra care must be taken because the heat will not only affect the cover but will also bake the belt carcass, thereby reducing its overall strength and durability. Heat retention and transmission of heat to the carcass by belt fasteners cause localized loss of strength at mechanical splices. A good rule for hot material service is to limit the stress on the carcass to 75% of its normal service rating.

Molded Covers. For special applications and/or unusual operating conditions, covers with special molded surfaces may be used to advantage. One type has a rough top, or various patterns of molded surface designed primarily for conveying packages up inclines, but also occasionally used for conveying light-weight bulk materials on steep inclines. The second type is a ribbed or cleated cover used in bulk conveying to allow the conveyor incline to be increased without backsliding the load. Also, special designs for handling wet materials or slurries permit drainage or retention of fluids as required.

Frequency factor. The frequency factor indicates the number of minutes for the belt to make one complete turn or revolution. It can be determined by the following formula:

$$F_f = \frac{2L}{V}$$

where L = Center to center length of the belt conveyor, ft.
V = Belt speed, fpm
F_f = Frequency factor, minutes

For a frequency factor of 4.0 or over, minimum top cover thicknesses can be considered based on the loading conditions. For a frequency factor of 0.2, the appropriate top cover thickness should be increased up to twice this minimum amount. For frequency factors between 0.2 and 4.0, vary the top cover thickness accordingly.

Breakers. No breaker is required under covers when the conveyor belt is to handle fine material having no lumps exceeding ¾ inch and has well designed loading conditions. For lump sizes over ¾ inch and up to 2 inches, the use of breakers generally will depend on the ratio of lumps to fines and the character of the loading conditions. For lump sizes of 2 inches and over, breakers generally are required. Breakers are not normally required with reduced-ply belting. For other than RMA Grade 2 belting or for severe operating conditions, consult individual belt manufacturers for recommendations.

Table 7-13 provides a general guide for the proper selection of the thickness and grade of the top cover (carrying side) of a conveyor belt to yield a reasonable performance and belt life. Table 7-14 provides similar guidance for thickness and grade of the pulley-side cover.

Loading Considerations

Loading Conditions Resulting in Normal Cover Wear.

- Material feed is in the same direction as belt travel. See Chapter 11, "Loading the belt."
- Equivalent free fall of material onto the conveyor belt is not over 4 ft. See Tables 7-11 and 7-12.
- Loading area of the belt conveyor is horizontal or has a slope of not more than 8°.
- Properly designed chutes and skirtboards to form, center, and settle the load on the belt. See Chapter 11, "Loading chutes and skirtboards."
- Material temperature is in the range of 30°F to 150°F.
- Material handled contains nothing that will deteriorate the cover or carcass of the belt. See Table 7-15.

Loading Conditions Resulting in Minimum Cover Wear. All the conditions above (Resulting in Normal Cover Wear) plus the following:

- In the process of loading, the material is traveling at approximately the same speed as the belt. See speed-up conveyor, Chapter 11, "Direction of loading."
- Special attention has been paid to the equipment design in the loading area to reduce impact on the conveyor belt to a minimum. By a minimum is meant an equivalent free fall of less than 3 ft. The use of a carefully designed loading chute and the use of closely spaced impact idlers are presumed.
- Provision has been made in loading chute to place the fines on the conveyor belt first, to provide a bed for large lumps. See Chapter 11, "Loading chutes," and Figure 11.3.
- With Loading Conditions Resulting in Minimum Cover Wear as herein defined, top covers of a conveyor belt sometimes can be reduced 1/32 inch to 1/16 inch from the values listed in Table 7-13.

Loading Conditions Resulting in Maximum Cover Wear. Any of the following conditions constitute Loading Conditions Resulting in Maximum Cover Wear:

- Material is loaded 90° transversely (right angle) to the direction of the belt. See Chapter 11, "Direction of loading."
- Material is loaded at more than 90° transversely to the direction of the belt. See Chapter 11, "Direction of loading."
- Loading area has a slope in excess of 8° to the horizontal.
- Equivalent free fall of the loaded material is greater than 4 ft. See Tables 7-11 and 7-12.
- Material loaded has no velocity in the direction of belt travel, or has a negative velocity in the direction of belt travel.

Table 7-15 *Deteriorating Conditions for Conveyor Belt Covers*

Typical materials handled without cover deterioration	
Chemicals	Materials wetted with or containing the following chemicals and not over 150 °F, may be handled satisfactorily on conveyor belts with covers of Grades 1 and 2. Acetone Ammonium hydroxide Black sulfate liquor Butyl alcohol Ethyl alcohol Sulfur, elemental, dry Sulfuric acid
Heat	Hot fine material up to 150 °F Hot lump material up to 150 °F
Fertilizers	Super phosphate Triple super phosphate Phosphate rock or pebbles, acid treated, to produce super or triple super phosphate

- With Loading Conditions Resulting in Maximum Cover Wear, the top cover thickness may have to be increased by 1/16 inch to 3/16 inch above the values listed in Table 7-13 in order to obtain a reasonable life.

Materials Handled Resulting in Deterioration of Covers. Chemicals not listed *may* have a deteriorating effect on the rubber covers of conveyor belting, but, because of considerations of concentration and temperature, do not lend themselves readily to classification. Although certain materials may be classified as not being detrimental of themselves, conditions must always be examined for the presence of processing or incidental chemicals or oils which might have a deteriorating effect. Therefore, when handling chemicals not listed in Table 7-15, oily materials, or ordinary materials having temperatures exceeding 150°F, consult the belt manufacturer for cover quality recommendations.

Economic Considerations

The most economical belt will be that combination of carcass and cover which best meets the job requirements taking the following factors into consideration: tension rating, wear and impact resistance, takeup capability and maintenance, splice-holding ability and splice maintenance, troughability and load support, pulley diameter compatibility, and resistance to special conditions (e.g., heat, cold, water, chemicals, etc.).

Because of the large number of fabrics and constructions that can be employed to satisfy the conveyor belt requirements, the tables in this chapter provide conservative recommendations for each condition under consideration.

Chapter 8

Pulleys and Shafts

Contents

It is accepted engineering practice to consider pulleys and shafts together because they form a composite structure whose operating characteristics are mutually related. Therefore, they are discussed as one topic of belt conveyor design and construction in this chapter.

Conveyor Pulley Assemblies

Conveyor pulley construction has progressed from fabricated wood, through cast iron, to present welded steel fabrication. Increased use of belt conveyors has led industry away from custom-made pulleys to the development of standard steel pulleys with universally accepted size range, construction similarities, and substantially uniform load-carrying capacity for use with belts having a carcass composed of plies or layers of fabric. ''Standard'' drum and wing pulleys are suitable for these applications. The present trend, however, is to use higher tonnage conveyor systems with wider, stronger belts that incorporate a carcass of either steel cables or high-strength tensile members. In these applications, where high tensions are encountered, the use of custom-made ''Engineered'' welded steel pulleys is dictated. See Chapter 7 for a description of the various types of conveyor belts.

Pulley Types

The most commonly used conveyor pulley is the standard steel pulley. See Figure 8.1. They are manufactured in a wide range of sizes and consist of a continuous rim and two end discs fitted with compression type hubs. In most wide-faced conveyor pulleys, intermediate stiffening discs are welded inside the rim. Other pulleys available are self-cleaning wing types, which are used at the tail, takeup or snub locations where material tends to build up on the pulley face, and magnetic types, which are used to remove tramp iron from the material being conveyed. Figures 8.1 through 8.8 illustrate the more common types of conveyor pulleys now in use.

Standard Steel Drum Pulleys: Standard welded steel drum pulleys are defined by American National Standards Institute (ANSI) standard number B105.1. The standard establishes load ratings, allowable variation from nominal dimensions, permissible crown dimensions, and overall dimensions normally necessary to establish clearances for location of adjacent parts. The standard applies to a series of straight and crown-faced welded steel conveyor pulleys which have a continuous rim and two end discs, each with a compression type hub to provide a clamp fit on the shaft.

FIGURE 8.1. *Typical welded steel pulley.*

FIGURE 8.2. *Fabricated curve crown pulley.*

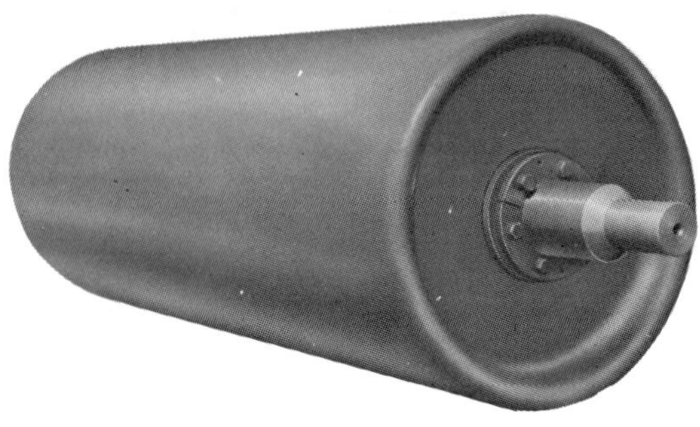

FIGURE 8.3. *Spun-end curve crown pulley.*

FIGURE 8.4. *Lagged welded steel pulley.*

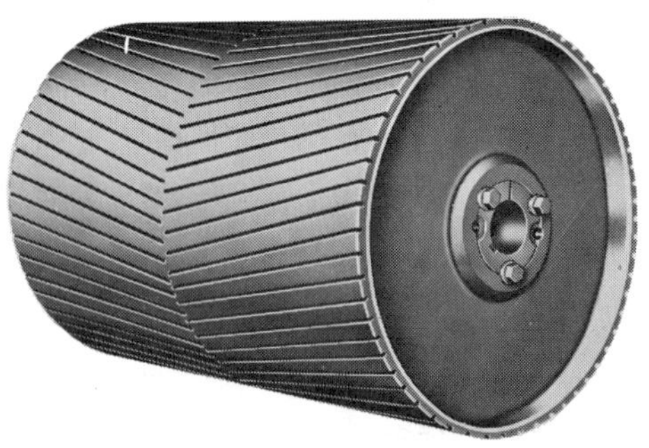

FIGURE 8.5. *Welded steel pulley with grooved lagging.*

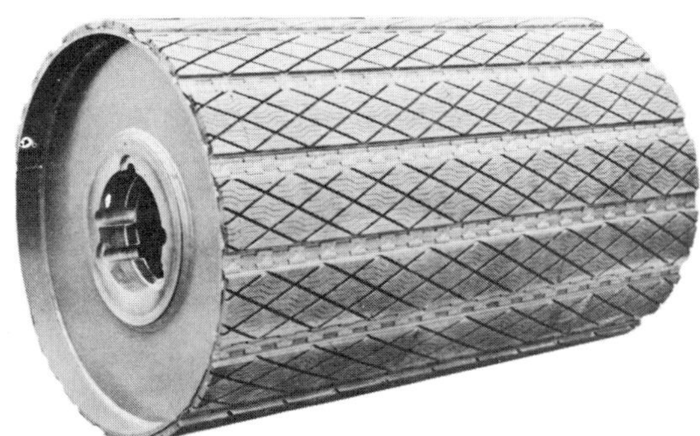

FIGURE 8.6. *Slide-lagged pulley.*

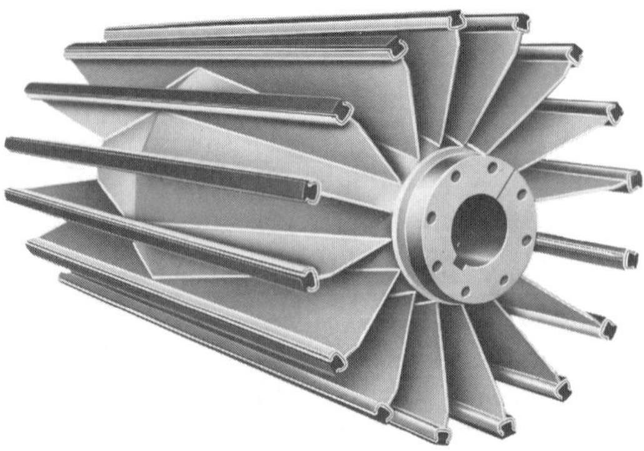

FIGURE 8.7. *Lagged wing pulley.*

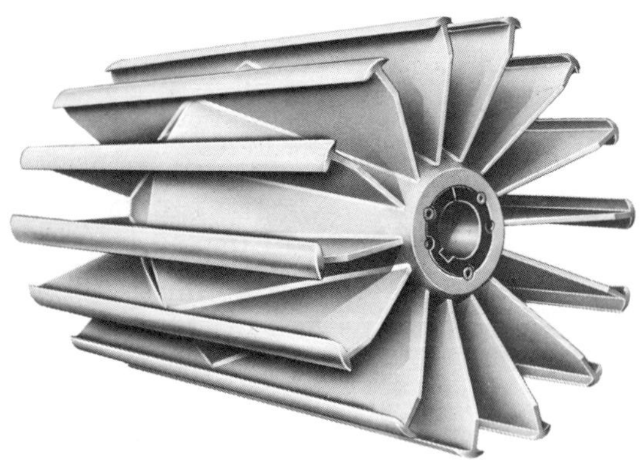

FIGURE 8.8. *Fabricated wing type of pulley.*

The tabulated ratings for drum pulleys and shaft combinations are based on using non-journaled shafting with pulleys centrally located between two bearings. Ratings are based on either a maximum shaft bending stress of 8000 psi or a maximum free shaft deflection slope at the pulley hub of 0.0023 in./in., whichever governs. High strength shafting may be required with drive pulleys to withstand the added shaft stresses from torsional loads, overhung loads or turndowns for bearings. Belt tension limits are also provided and must be checked, especially for pulleys with low arc of contacts such as snubs or bends.

ANSI B105.1 is not applicable to single disc pulleys, wing pulleys, cast pulleys, or pulleys not utilizing compression hubs. The standard is not intended to specify construction details other than those listed above. The standard covers pulleys up to 60″ diameter for shaft diameters up to 10″ and face widths up to 66″ and therefore encompasses the majority of combinations of welded steel pulleys with compression type hubs that are normally used in current belt conveyor and elevator practice.

Welded steel conveyor pulleys covered by ANSI B105.1 should not be used with steel cable or other high modulus conveyor belts because such belts create tension concentrations and demand manufacturing tolerances beyond the capacities of these pulleys. Such conveyor belts require engineered conveyor pulleys.

Standard Steel Wing Pulleys: Standard welded steel wing pulleys are defined by American National Standards Institute standard number 501.1. Like the drum pulley standard, this standard establishes load ratings, allowable variations from nominal dimensions, permissible crown dimensions, and overall dimensions normally necessary to establish clearances for location of adjacent parts. The standard covers pulleys up to 36″ diameter for shaft diameters up to 8″ and face widths up to 66″ and therefore encompasses the majority of combinations of welded steel wing pulleys with compression type hubs that are normally used in current belt conveyor and elevator practice.

The standard applies to a series of straight and crown-faced welded steel wing pulleys which have a number of steel wing plates that extend radially from the longitudinal axis of two compression hub assemblies and are equally spaced about the pulley circumference. The purpose of the compression hubs is to provide a clamp fit on the shaft. The wings are supported or joined by welded steel plates so arranged as to form the shape of two frustums of cones or regular pyramids joined at their bases. A contact bar with generous radii is attached to the outer longitudinal edge of each wing to provide contact area with the belt.

The tabulated ratings for wing pulleys and shaft combinations are based on using non-journaled shafting through the pulley hubs, with pulleys centrally located between two bearings. Ratings are based on either a maximum shaft bending stress of 6000 psi or a maximum free shaft deflection slope at the hub of 0.0023 in./in., whichever governs. Belt tension limits are also provided.

Welded steel wing pulleys covered by ANSI 501.1 should not be used with steel cable or other high modulus conveyor belts because of the eccentricity inherent in the construction of a wing pulley. It is not recommended to operate standard wing pulleys above a belt speed of 450 feet per minute. For higher speeds, manufacturers should be consulted.

Table 8-1. *Welded Steel Drum Pulley Weights, Pounds*

These are representative weights for estimating purposes. Since actual weights may vary, final shaft and pulley designs should be checked with the manufacturer.

Pulley		Belt width (inches)																		
		18			24		30		36	42		48		54		60		72	84	96
		Pulley face width (inches)																		
Dia. (inches)	Max. Bore (inches)	20	22	24	26	30	32	36	38	44	46	51	54	57	60	63	66	78	90	102
6	2½	30	35	35	40	40	45	45	50	60	60	65	65	70	75	80				
8	2½	45	50	50	55	60	65	70	75	80	85	95	100	105	110	115				
10	2½	55	60	60	65	70	75	85	90	100	105	110	120	125	130	135	140			
	3	60	65	70	75	80	85	95	100	110	115	120	125	130	140	145	150			
	3½	75	80	80	85	90	95	105	110	120	125	130	135	140	145	150	155			
12	2½	70	75	75	85	90	100	105	115	120	130	140	150	155	160	165	175			
	3	80	85	90	100	105	110	115	125	130	135	145	155	160	170	175	180			
	3½	95	100	105	110	115	120	125	135	145	150	160	165	170	175	180	190			
14	2½	85	90	95	100	110	120	135	145	165	180	195	205	215	225	235	245			
	3	95	100	105	115	120	130	145	155	180	195	210	220	230	240	250	260			
	3½	105	110	115	130	140	150	160	175	190	205	220	230	240	245	255	265			
	4	130	135	140	150	160	170	185	200	210	220	235	245	255	260	275	280			
16	2½	105	110	115	130	140	150	160	175	200	220	235	240	250	260	270	280			
	3	115	125	130	140	150	165	170	185	220	240	250	260	270	280	290	300			
	3½	130	135	140	155	160	170	180	200	225	240	260	265	275	285	295	310			
	4	160	165	175	185	195	205	215	225	245	260	280	285	290	300	310	325			
18	2½	125	130	140	160	170	185	200	215	240	265	280	300	310	320	330	350			
	3	135	150	160	180	190	205	215	230	250	275	290	305	320	335	345	360			
	3½	145	155	160	180	200	215	225	240	265	275	295	310	330	345	355	375			
	4	175	185	190	215	225	235	250	270	295	310	330	355	375	400	410	430			
	4½	200	205	210	235	250	260	275	290	315	325	345	370	395	410	425	450			
20	2½	145	160	170	190	200	215	230	260	300	310	330	350	370	390	400	420			
	3	160	170	180	200	220	240	250	275	300	320	350	365	380	390	400	420			
	3½	170	180	195	210	225	240	270	285	300	330	360	375	390	400	415	435	500	550	600
	4	190	200	215	240	250	270	285	305	330	360	400	410	425	435	450	470	550	600	660
	4½	210	225	235	265	280	290	305	325	350	375	400	425	440	455	475	500	580	630	700
	5																	600	650	700
	6																	750	825	900
24	3	200	210	230	250	265	285	300	325	400	425	450	475	500	510	525	550			
	3½	215	225	245	265	280	300	315	350	400	435	460	485	500	515	535	560	600	675	750
	4	235	245	265	285	300	320	345	380	410	450	500	520	540	560	580	600	700	750	825
	4½	255	265	280	300	315	335	370	400	460	480	500	550	570	590	610	635	725	800	900
	5	280	290	300	335	365	400	420	440	500	530	560	615	635	655	680	710	900	950	1050
	6																	1100	1200	1300
	7																	1100	1200	1300
	8																	1200	1300	1400

Table 8-1 continued. ***Welded Steel Drum Pulley Weights, Pounds***

Pulley		*Belt width (inches)*																		
		18			24		30		36	42		48		54		60		72	84	96
Dia. (inches)	*Max. Bore (inches)*	*Pulley face width (inches)*																		
		20	22	24	26	30	32	36	38	44	46	51	54	57	60	63	66	78	90	102
	3	315	330	350	380	400	430	450	485	525	550	670	700	750	800	850	900			
	3½	325	340	360	390	410	440	475	500	550	600	660	700	750	800	850	900	925	1000	1100
	4	350	375	400	425	450	475	500	525	575	635	700	750	800	825	860	900	1100	1200	1300
30	4½	370	385	400	435	460	490	520	550	600	650	700	780	815	860	900	950	1100	1200	1400
	5	400	420	440	470	500	525	550	580	630	690	750	840	870	900	950	1000	1200	1300	1400
	6	525	550	575	600	625	650	680	740	775	825	875	900	925	950	1000	1050	1500	1600	1700
	7	550	560	575	600	650	675	725	750	800	850	900	950	975	1000	1050	1100	1500	1700	1800
	8																	1600	1800	1900
	10																	2200	2300	2500
	3½	450	475	500	550	590	625	650	675	800	840	880	925	950	1000	1075	1150	1400	1500	1700
	4	475	500	540	575	600	650	675	700	800	850	900	950	975	1000	1100	1200	1400	1500	1700
	4½	490	500	550	600	625	650	700	750	800	850	900	950	1000	1050	1100	1200	1500	1600	1800
36	5	530	550	575	630	675	700	735	775	850	925	960	1000	1050	1100	1200	1300	1700	1900	2100
	6	635	660	690	735	765	800	835	875	1000	1030	1100	1200	1250	1275	1300	1400	1900	2000	2200
	7	725	750	800	850	900	950	1000	1050	1150	1175	1250	1350	1375	1400	1475	1550	2000	2100	2300
	8	—	—	850	900	950	1000	1050	1100	1200	1250	1300	1400	1450	1500	1550	1600	2100	2300	2500
	10																	2700	2900	3200
	4	625	650	700	750	800	850	900	950	1000	1075	1150	1200	1250	1300	1350	1400	1900	2100	2300
	4½	640	675	700	775	850	875	900	950	1050	1100	1175	1225	1275	1300	1350	1500	2000	2200	2400
42	5	675	700	740	800	850	900	950	1000	1075	1150	1200	1250	1300	1350	1400	1500	2100	2200	2400
	6	800	825	875	925	960	1000	1050	1125	1200	1250	1325	1400	1475	1525	1600	1700	2200	2400	2600
	7	975	1000	1050	1100	1150	1200	1260	1325	1425	1500	1550	1600	1700	1750	1800	1900	2800	3100	3400
	8	1025	1050	1100	1150	1200	1250	1300	1375	1450	1525	1600	1650	1700	1750	1825	1900	3000	3300	3600
	10																	3500	3700	4000
	4	775	800	825	950	1000	1050	1075	1125	1200	1300	1500	1575	1650	1700	1750	1800	2500	2700	2900
	4½	800	825	875	950	1050	1100	1150	1200	1300	1450	1600	1650	1700	1750	1850	2000	2500	2700	2900
48	5	825	850	875	1000	1050	1100	1175	1225	1300	1450	1600	1700	1750	1800	1900	2000	2500	2700	3000
	6	900	950	1000	1100	1175	1225	1275	1350	1450	1600	1750	1825	1900	1950	2000	2100	2700	2900	3200
	7	—	1025	1100	1200	1300	1400	1500	1600	1700	1800	1900	2000	2075	2150	2225	2300	3500	3700	4100
	8	—	1150	1225	1300	1400	1450	1500	1600	1700	1800	1900	2000	2100	2200	2300	2400	3700	4000	4300
	10																	4200	4400	4800
	4½	1050	1100	1200	1300	1375	1450	1525	1600	1700	1800	1925	2000	2100	2200	2300	2400	3000	3200	3500
	5	1100	1150	1200	1300	1400	1500	1600	1700	1800	1900	2000	2100	2150	2200	2300	2400	3100	3300	3700
54	6	1225	1300	1400	1500	1600	1700	1800	1900	2000	2100	2200	2250	2325	2400	2500	2600	3800	4100	4500
	7	1350	1400	1500	1600	1700	1800	1900	2000	2100	2200	2300	2400	2500	2600	2700	2800	4000	4300	4700
	8	1500	1500	1550	1650	1750	1825	1900	2000	2100	2200	2300	2400	2500	2600	2700	2800	4400	4700	5100
	10	—	2000	2050	2130	2230	2330	2400	2500	2625	2750	2850	2950	3025	3100	3200	3300	5000	5300	5700
	4½	1300	1350	1450	1600	1700	1800	1900	2000	2100	2200	2300	2400	2450	2525	2600	2700	4100	4600	5100
	5	1325	1400	1500	1625	1750	1850	2000	2100	2200	2300	2400	2500	2600	2700	2800	2900	4300	4600	5100
60	6	1450	1500	1600	1700	1800	1900	2000	2100	2250	2400	2500	2600	2700	2800	2900	3000	4600	4900	5400
	7	1600	1650	1750	1900	2000	2100	2200	2300	2400	2500	2600	2750	2850	2975	3000	3200	4600	4900	5400
	8	1700	1750	1800	1900	2000	2100	2200	2300	2500	2600	2700	2800	2900	3000	3100	3200	5200	5500	6000
	10	—	2200	2300	2400	2500	2600	2700	2850	3000	3100	3200	3300	3400	3500	3600	3800	5800	6100	6600

TABLE 8-2. *Welded Steel Wing Pulley Weights, Pounds*

These are representative weights for calculating purposes. Since actual weights may vary, final shaft and pulley designs should be checked with the manufacturer.

Pulley		Belt width (inches)																			
		18			24		30		36		42		48		54		60		72	84	96
Dia. (inches)	Max. Bore (inches)	Pulley face width (inches)																			
		20	22	24	26	30	32	36	38	40	44	46	51	54	57	60	63	66	78	90	102
8	1-5/8	40	45	50	55	60	65	70	75	80	85	90	100	105	110	120	125	135	160	190	225
	2-1/2	50	55	55	60	70	75	80	85	90	95	100	110	115	120	125	130	140	160	200	230
10	1-5/8	55	60	65	70	80	85	95	105	110	120	125	135	145	155	160	170	180	210	250	275
	2-1/2	60	65	70	75	85	95	100	110	115	125	130	140	150	160	165	175	180	210	250	275
12	1-5/8	80	85	95	100	115	125	140	145	155	165	175	190	205	215	225	230	240	290	335	380
	2-1/2	80	90	95	100	115	125	140	145	155	165	175	190	205	215	225	235	245	290	335	380
	3	90	100	105	115	130	140	150	160	170	185	195	210	220	235	245	260	270	315	365	410
14	1-5/8	95	105	115	120	140	150	170	175	185	200	215	240	250	265	280	295	310	360	415	470
	2-1/2	100	110	120	125	145	155	170	180	190	205	215	240	250	265	280	295	310	360	415	470
	3	105	115	125	135	155	165	185	195	200	220	235	255	270	285	300	315	330	390	450	500
	3-1/2	115	125	135	145	165	175	195	205	215	235	245	270	285	300	310	325	340	400	450	500
16	2-1/2	115	125	135	145	165	175	200	210	220	240	250	275	290	300	320	335	350	400	460	520
	3	120	130	140	150	175	185	205	215	230	250	260	290	305	320	340	355	375	450	500	570
	3-1/2	130	140	150	160	180	195	215	230	240	260	275	300	315	335	350	365	385	450	520	600
	4	145	160	170	180	200	215	240	250	260	280	300	325	340	360	375	390	410	480	550	575
18	2-1/2	145	160	170	185	210	225	250	265	280	300	320	350	370	390	405	425	450	525	600	680
	3	150	160	175	190	215	230	260	270	285	310	340	370	390	410	430	450	470	550	625	700
	3-1/2	160	170	185	200	225	240	260	280	300	325	340	370	390	410	430	450	470	550	630	710
	4	170	185	200	215	235	250	280	290	300	330	350	375	400	420	440	460	485	570	650	735
	4-1/2	190	200	215	230	255	270	300	315	330	355	370	405	425	450	470	490	510	600	675	760
20	2-1/2	160	175	190	205	230	250	275	290	305	335	350	385	400	430	450	475	500	600	675	760
	3	165	180	200	215	240	255	285	300	315	350	360	400	425	450	470	500	515	600	675	760
	3-1/2	175	190	205	220	250	265	290	310	325	355	370	405	430	450	475	500	520	600	700	780
	4	190	200	220	230	260	280	310	325	340	370	390	420	440	465	490	510	530	625	700	800
	4-1/2	210	225	240	250	280	300	325	350	360	390	400	450	470	490	510	540	555	640	720	810
24	2-1/2	250	275	300	325	375	400	490	530	560	600	640	700	750	780	825	860	900	1000	1200	1400
	3	280	305	335	360	410	435	490	530	560	600	640	700	750	780	825	860	900	1000	1200	1400
	3-1/2	300	315	340	370	420	440	500	530	560	600	640	710	750	780	825	860	900	1000	1200	1400
	4	300	325	345	375	425	450	500	545	570	620	650	710	750	780	825	860	900	1000	1200	1400
	4-1/2	315	340	365	400	440	470	510	550	575	625	650	715	760	800	835	875	900	1000	1200	1400
	5	320	350	370	400	440	470	520	560	585	635	660	725	760	800	835	875	900	1100	1200	1400
30	2-1/2	400	435	470	500	575	610	685	710	785	850	900	950	1050	1100	1150	1200	1200	1500	1700	1900
	3	400	435	470	510	580	610	690	750	790	860	900	975	1050	1100	1150	1200	1250	1500	1700	1900
	3-1/2	400	435	470	510	580	610	690	750	790	860	900	980	1050	1100	1150	1200	1250	1500	1700	1900
	4	410	450	480	520	590	620	700	760	800	875	900	1000	1050	1100	1150	1200	1250	1500	1700	1900
	4-1/2	425	460	500	530	600	635	700	770	800	875	900	1000	1050	1100	1150	1200	1250	1500	1700	1900
	5	430	460	500	530	600	635	700	770	800	875	900	1000	1050	1100	1150	1200	1250	1500	1700	1900
	6	500	530	565	600	660	700	760	800	850	920	955	1040	1100	1140	1200	1250	1300	1500	1700	1900
36	3	540	585	630	680	770	820	910	1000	1050	1150	1175	1300	1375	1450	1500	1600	1700	1900	2200	2500
	3-1/2	540	585	630	680	770	820	910	1000	1050	1150	1175	1300	1375	1450	1500	1600	1700	1900	2200	2500
	4	550	590	635	680	775	820	915	1000	1050	1150	1200	1300	1375	1450	1500	1600	1700	1900	2200	2500
	4-1/2	560	600	650	700	800	830	925	1000	1050	1150	1200	1300	1375	1450	1500	1600	1700	1900	2200	2500
	5	560	600	650	700	800	830	925	1000	1050	1150	1200	1300	1375	1450	1500	1600	1700	1900	2200	2500
	6	625	675	700	750	850	900	1000	1050	1100	1200	1250	1340	1400	1475	1550	1600	1700	1900	2200	2500
	7	675	700	750	800	900	950	1000	1100	1150	1200	1250	1350	1450	1500	1575	1650	1700	1950	2200	2500
	8	700	740	775	825	900	950	1050	1100	1150	1225	1275	1400	1450	1500	1575	1650	1700	1950	2200	2500

Advantages of Use of ANSI Standards: The standard drum and wing pulleys discussed above can be used in most conveyor applications. The designer of belt conveyors will find the ANSI standards listed above invaluable in determining specifications for pulleys and shafting and in finding information that will permit the framework and supporting bearings to be detailed into the design. Suitable pulleys conforming to these standards can be obtained readily from the principal pulley manufacturers. Many of the sizes are considered stock sizes and are in their inventory of finished goods.

Mine Duty Pulleys: Standard size drum and wing pulleys are also available in mine duty construction. A mine duty pulley is one whose material thicknesses have been increased for a very rigid, conservative design. Mine duty pulleys were originally specified and used for underground mining operations where the abusive environment and high cost of installation demanded a more conservative design.

No ANSI standard governs the load ratings or material thickness of mine duty pulleys. Each manufacturer has established its own specifications and should be contacted for complete information regarding this class of pulley. A mine duty pulley will have much lower stress and deflection on the various components then a corresponding standard pulley. The conservative, rigid construction of mine duty pulleys makes them particularly useful for spares or replacements in critical or troublesome locations.

Engineered Pulleys: An engineered pulley is one which has been specifically designed to meet load conditions of a particular conveyor pulley. Specific information is required for proper and economical design, since the pulley designer must allow for sufficient strength in the rim, disc, shaft, and mounting system to carry the belt loads and to assure proper pulley to shaft connection.

Modern conveyor belting is often manufactured with steel cables or other high strength, high modulus tensile members for applications requiring high production or where large volumes of material need to be conveyed. Typical applications include overland conveyors, primary conveyors in open pit mines, large coal fired power plants, and ore processing facilities.

Special considerations in pulley design are required because of the increased loads imposed by these belts. Experience in design and application has helped engineers meet the need for pulleys that will accommodate the higher belt tensions generated under these conditions.

High tension belts, because of their high modulus, low stretch characteristics, require conveyor pulleys which conform to considerably higher standards than those of the pulleys used with normal fabric ply belts. Startup, braking, and other dynamic loads are more directly transmitted to pulleys. Concentricity of the pulley and proper alignment of the pulley and conveyor are of prime importance when high tensions are involved. Special emphasis should be placed on the importance of adequate, properly aligned structural supports for the pulley assembly to prevent load concentrations and overloads on the pulley resulting from misalignment.

Construction details vary widely, ranging from pulleys with flexible end discs and compression type hubs for plain shafting, to very heavy, rigid end discs fitted to specially machined shafts. Pulleys designed for this class of service are subject to more rigid manufacturing specifications, usually requiring accurately machined surfaces and close control of welded joints.

All the components that make up the pulley and shaft assembly must be integrated to provide a fully capable power transmission system. Several factors must be considered for proper application and economical design of conveyor pulleys intended for load conditions exceeding those necessary to conform to ANSI standards. For such engineered pulleys, the following information is required for proper design:

Diameter, face width, straight or crowned
Bearing centers
Location of pulley; head, bend, snub, takeup, etc.
Type of belt takeup; gravity, screw, etc.
Type of conveyor belt
Belt tensions on pulley
Belt wrap angle on pulley
Shaft diameter (at hub, bearing, drive) if predetermined
Lagging specifications
Overhung load with location on drive pulley shaft (if it exists)
Horsepower
Belt speed
Starting mechanism (reduced voltage, fluid coupling, etc.)
Special environmental conditions

Pulley Overloads

Normal running loads for standard pulleys should not exceed ratings in the ANSI B105.1 and 501.1 load tables. Starting and occasional peak loads should not exceed ratings by more than 50%. Overloads may result from such causes as starting, jam-ups, maladjusted take-ups, brakes, misalignment and excess amounts of material on the belt. Overload conditions may result in premature failure of either pulleys, shafting, or bearings.

Normal running loads for engineered class pulleys should not exceed the loads used for design. Starting and occasional peak loads should not exceed design loads by more than 50%. If starting or peak loads are expected to exceed design loads by more than 50%, that information should be provided to the pulley manufacturer to be considered in the pulley design.

Pulley Diameters

Standard steel drum pulley diameters are 6, 8, 10, 12, 14, 16, 18, 20, 24, 30, 36, 42, 48, 54, and 60 inches. Standard wing pulley diameters are 8, 10, 12, 14, 16, 18, 20, 24, 30, and 36 inches. All other diameters are considered special. These nominal diameters apply to straight and crown-face pulleys and are for bare pulleys only; they do not include any increases in diameter due to the application of lagging. Pulley diameters should be selected per belt manufacturer's recommendations, as described in Chapter 7, as well as for other drive or space considerations.

Permissible variations from nominal diameters of standard steel pulleys are based on face width as follows:

Pulley face (inches)	*Variation*			
	Over nominal		*Under nominal*	
	drum	*wing*	*drum*	*wing*
6-26	1/4	1/8	1/8	3/8
over 26-66	5/8	1/8	1/8	3/4

These limitations apply equally to straight-face and crown-face pulleys. The nominal diameter is measured at the midpoint of the pulley face width. The diameter is defined as the bare diameter exclusive of lagging. Listed variations may occur from one pulley to another. The permissible diameter variations listed are not to be construed as runout tolerance. Runout tolerance on diameter is measured at the midpoint of the bare pulley face and is as follows:

Diameter (inches)	*Maximum Total Indicator Reading*
8-24	0.125
30-48	0.188
54-60	0.250

Engineered pulleys to be used with steel cable or high modulus belts are usually machined with a straight face having the following permissible runout tolerance:

Type	*Maximum Total Indicator Reading*
Unlagged pulley	0.030
Lagged pulley (under lagging)	0.030
Lagged pulley (over lagging)	0.030

Pulley Face Widths

Pulley face width is the length of rim, wing or contact bar along the shaft axis. Standard pulley face width is normally equal to the belt width plus 2 inches for belt widths up to 42 inches and belt width plus 3 inches for belt widths over 42 inches. Engineered pulley face widths are generally 6 to 12 inches greater than the belt width to provide greater clearance. The conveyor belt should not wander beyond the edge of the pulley face. Refer to Chapter 7 for more information on pulley face widths.

Standard drum pulley face widths defined by ANSI standards are 20, 22, 26, 32, 38, 44, 51, 57, 63, and 66 inches, with standard wing pulley face widths also including 40, 54, and 60 inches. Face width variations from nominal, according to ANSI standards, are ± 1/8 inch for all size drum pulleys and ± 1/4 inch for all wing pulleys. The listed variations in face width may occur from one pulley to another. The permissible face width variations are not be construed as an edge runout tolerance. Edge runout tolerance is specified by the individual pulley manufacturer.

Pulley Crown

See Chapter 7 for a discussion of the use of crowned conveyor pulleys. Straight-face pulleys are recommended for all installations using reduced ply, high modulus, low stretch belts, such as those with a carcass of steel cables or high strength tensile members. There are currently two types of pulley crowns available, the taper crown and the curve crown.

Taper Crown: On taper crown pulleys, the face forms a ''V'' with the rotating axis larger in diameter in the center of the pulley. This crown is expressed in inches of crown per foot of total face width, by which the diameter at the center of the face exceeds the diameter at the edge. Normal crowns of this type vary from 1/16 to 1/8 inch per foot of total face width.

Curve Crown: Curve crown pulleys have a long, flat surface in the center of the pulley with the ends curved to a smaller diameter. Except on short pulleys, the curved surface extends in approximately 8 inches from the edge.

Trapezoidal Crown: Trapezoidal crown pulleys have a flat surface in the middle portion of the pulley face with the ends tapered. Trapezoidal crown pulleys may be appropriate for wider face width pulleys.

Pulley Weights

Pulley weights must be used to determine pulley and shaft selection. Average weights for standard steel drum pulleys are given in Table 8-1, and weights for standard wing pulleys are listed in Table 8-2. Weights listed are averages. If more accurate values are required, it is necessary to obtain the actual pulley weight from the manufacturer, as there are some variations in manufacturing practices. Engineered pulley weights are dependent on the tensions encountered and can vary widely.

Pulley Lagging

Conveyor pulleys can be covered with some form of rubber, fabric, or other material. Lagging is used on driving pulleys to increase the coefficient of friction between the belt and pulley. Lagging is also used to reduce abrasive wear on the pulley face and to effect a self-cleaning action on the surface of the pulley. Abrasive wear and material buildup can substantially decrease pulley life. Drive pulleys should always be lagged. Non-driving pulleys, especially on the carrying side of the belt, should be lagged whenever abrasive or buildup conditions exist.

Thickness and Attachment: Lagging thickness can vary from a few thousandths of an inch, as with a sprayed-on coating, to thicknesses of one to two inches, as with some solid rubber vulcanized coatings. Common methods of attachment are bolting, cementing, tack welding, and vulcanizing. Vulcanized lagging is generally preferred for heavy duty or severe service applications. Bolted-on and welded-on slide type lagging has the advantage of being replaceable in the field. It can be obtained in various grooved and other specialized surface finish types. Bolted-on lagging usually consists of a rubber cover reinforced with multiple ply fabric construction similar to conveyor belting. The fabric plies are required to provide strength under the bolt heads. Slide-on lagging is constructed of rubber molded and vulcanized to steel backing plates. They are slid into steel retainer strips welded onto the pulley face. Replacement of the pads can be accomplished by sliding the old strip out and the new strip in without removing the pulley from its conveyor location.

Lagging Hardness: Rubber lagging used on drive pulleys normally has a durometer hardness of 60 $\pm$ 5 on the Shore A scale. The lagging used on non-driving pulleys may have a hardness of 45 $\pm$ 5 or 60 $\pm$ 5, depending on the application. For snub and bend pulleys, which contact the carrying side of the belt, softer rubber tends to better resist buildup on the pulley face. With high tension belts, lagging of 70 $\pm$ 5 hardness is sometimes used.

Lagging Grooving: Drive pulleys which operate in wet or damp conditions are commonly grooved. These grooves generally take the shape of a herringbone or chevron shaped pattern cut into the lagging. Generally, the dimensions of the groove are 1/4 inch wide by 1/4 inch deep with a 1/8 inch minimum thickness of material under the bottom of the rubber. Grooves are usually spaced on 1-1/4 to 1-3/4 inch centers. In a chevron pattern, the grooves meet at the center of the pulley face, while in the herringbone pattern the grooves are offset by one-half the groove spacing. Figure 8.5 illustrates the herringbone pattern. In both patterns, the apex points in the direction of belt travel. There are also groove configurations and sizes of rubber lagging which can be furnished to enhance belt tracking or reduce the accumulation of material on the pulley face.

High Tension Applications: With steel cable and other high modulus conveyor belts, lagging is always used on drive pulleys and is preferred on pulleys which contact the carrying side of the belt. For maximum pulley life and belt safety lagging can be used on all non-drive pulleys. Lagging for high tension applications should be solid vulcanized rubber machined concentric with the shaft.

Shafting

Suitable shafting to be used with a steel pulley cannot be selected independently of the pulley load rating. In fact, the load capacity of a given pulley is a function of the shaft that is installed in that pulley. The shaft and pulley must be treated as a composite structural assembly. This is because the structural rigidity of the assembly depends upon both the shaft and the pulley and their interaction.

The shaft diameter required for a pulley assembly is a function of two criteria, shaft diameter required for strength and shaft diameter required for deflection. Depending on the exact pulley assembly, either strength or deflection can be the determining factor for shaft diameter selection.

Shaft Materials: Pulley design is based upon the use of any commercial or standard shafting material, such as AISI C1018 or C1045 steel. Load ratings of standard or mine duty pulleys are not increased when higher strength shafting is used. High strength shafting is of value in cases where it permits the shaft ends to be turned down so that smaller diameter, high capacity antifriction bearings can be used. It is also sometimes of value on drive shafts to withstand the added torsional stresses. While the use of high strength steel increases the strength of the shaft, its use does not decrease deflection.

Resultant Radial Load: The resultant pulley radial load is the vector summation of the belt tensions, pulley weight, and weight of shaft. The force from weights always acts downward, and the forces from the belt act in the path of the belt and away from the pulley. A graphical representation, as illustrated in Figure 8.9, or trigonometric methods are simple means of obtaining the resultant load. When a drive pulley shaft has an overhung load outboard of the pillow block bearings, such as with a shaft mount reducer or chain drive, refer to the ANSI drum pulley standard B105.1 for inclusion of the effects of the overhung load.

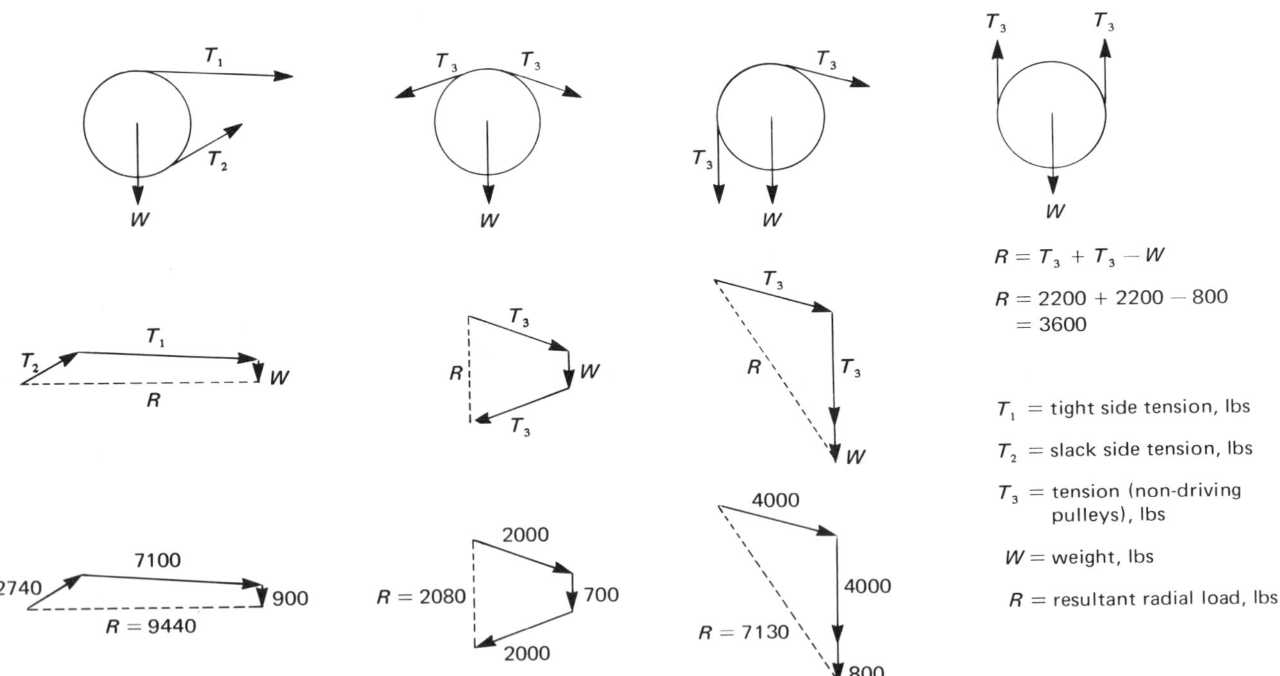

FIGURE 8.9. *Graphical means of obtaining resultant radial load.*

Shaft Sizing by Stress Limit: The equation given in standard B105.1 for the diameter of a pulley shaft loaded in bending and torsion (drive pulley with no overhung load) is:

$$D = \sqrt[3]{\frac{32F.S.}{\pi}\sqrt{\left(\frac{M}{S_f}\right)^2 + \frac{3}{4}\left(\frac{T}{S_y}\right)^2}}$$

The recommended values are:

F.S., Factor of Safety = 1.5

S_f, Corrected shaft fatigue limit =

$k_a \; k_b \; k_c \; k_d \; k_e \; k_f \; k_g \; S_f^*$

Where:

k_a, surface factor = 0.8 for machined shaft

k_b, size factor = $(D)^{-0.19}$

k_c, reliability factor = 0.897

k_d, temperature factor = 1.0 for −70 to +400 degrees F

k_e, duty cycle factor = 1.0

k_f, fatigue stress concentration factor =

0.63 for steel (BHN < 200) with profiled keyway

0.77 for steel (BHN < 200) with sled runner keyway

0.50 for steel (BHN > 200) with profiled keyway

0.63 for steel (BHN > 200) with sled runner keyway

(The steels listed can be considered to be under 200 BHN)

k_g, miscellaneous factor = 1.0 for normal conveyor service

S_f^*, = 29,000 for C1018

= 41,000 for C1045

= 47,500 for C4140 (annealed)

(S_f^* = 0.5 tabulated ultimate tensile strength)

S_y, yield strength = 32,000 for C1018

45,000 for C1045

60,500 for C4140 (annealed)

M = bending moment, lb-inches

T = torsional moment, lb-inches

For drive pulley shafts with overhung loads outboard of the pillow block bearings, such as a shaft mount reducer or chain drive, refer to the ANSI drum pulley standard B105.1 or vector methods for inclusion of the effects of the overhung load on M_B.

The method shown above can also be used to develop allowable bearing turndown sizes by adjusting M_B for the correct moment arm and multiplying M_B and M_T by the stress concentration factors for shaft turndowns, available in many engineering texts.

For non-drive pulleys set T = 0. In the ANSI drum pulley standard B105.1, a value of 8000 psi is used for S_f for non-drive pulleys. In the ANSI wing pulley standard 501.1, a value of 6000 psi is used for S_f.

Shaft Sizing for Deflection: The pulley assembly is a structural unit, and the strengths of different components are interdependent. The shaft diameter, the bearing centers, the thickness of the end disc, the resultant loads on the pulley, the hubs, and the tapered bushings are all interconnected in this respect. A steel pulley with very thin end discs allows the shaft to deflect almost as it naturally would by virtue of the two loads applied to it through the hubs. A pulley made with very thick end discs has minimal shaft deflection. For this reason, only the pulley manufacturer can determine the actual deflection of the shaft at the end disc.

When selecting pulley assemblies, a shaft deflection limitation is required and varies for different types of pulley to shaft attachment methods. For preliminary calculations, the free shaft deflection, where the shaft is treated as a simple beam without any stiffening effects from the pulley, can be used for calculation. ANSI B105.1 and 501.1 specify free shaft deflection limits of .0023 in./in. for standard drum and wing pulleys. This represents a maximum deflection angle of 8 minutes at the pulley end disc.

For pulleys that must withstand higher belt tensions and higher resultant loads than those given in standards B105.1 or 501.1, pulley manufacturers design an Engineered pulley. For Engineered pulleys used in critical applications, such as steel cable belts or high tension belts for mining or power plant usage, the free shaft deflection should be limited to .0015 in./in. This represents a maximum deflection angle of approximately 5 minutes at the pulley end disc. The system designer often specifies a particular deflection limit.

The equation given in standard B105.1 for the free shaft deflection for any pulley assembly is:

$$\text{Tan } \alpha = \frac{RA(B\text{-}2A)}{4E_yI}$$

Where: A = Moment arm for pulley (inches)
B = Bearing centers (inches)
R = Resultant pulley load (pounds)
E_y = Youngs modulus in psi (29×10^6 for steel)
I = Area moment of inertia of shaft in inches^4 ($.049087\ D^4$)
D = Diameter of shaft
tan α = Tangent of the angle made by the deflected shaft and its neutral axis before bending, at the pulley end disc.

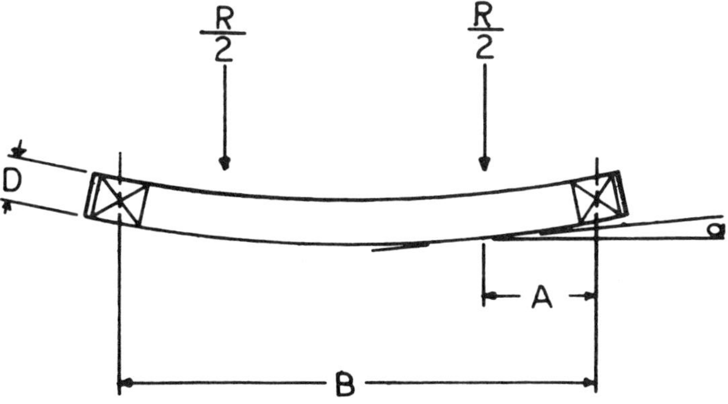

FIGURE 8.10. *Shaft deflection.*

For more information on shaft selection and allowable shaft deflection, consult a CEMA pulley manufacturer. It is suggested that complete pulley loading information be supplied with all pulley orders or requests for quotations.

Chapter 9

Vertical Curves

Contents

Vertical curves in belt conveyors are used to connect two tangent portions which are on different slopes. They are of two basically different types: *concave* vertical curves, where the belt is not restrained from lifting off the idlers; and *convex* vertical curves, where the belt is restrained by the idlers.

Disregarding what may be the theoretically perfect shape for either curve, it is commercially satisfactory to consider them as arcs of a circle.

These curves occur on both the carrying and the return runs of the belt and in a variety of tension conditions. The reader is referred to Chapter 6, "Tension relationships and belt sag between idlers," especially Figures 6.8 through 6.16, for various common belt conveyor profiles and tensions.

For simplification, the text and diagrams in this chapter are principally concerned with vertical curves on the carrying run of the belt conveyor.

Concave Vertical Curves

A conveyor belt is said to pass through a concave vertical curve when the center of curvature lies above the belt. (See Figure 9.1.) In such cases, the gravity forces of the belt and the load (if present) tend to hold the belt down on the idlers while the tension in the belt tends to lift it off the idlers. It is necessary to proportion the vertical curve so that the vector sum of these forces acts in a direction which allows the belt to stay down on the idlers and insures that the load will not be spilled. It is preferable that the belt does not lift off the idlers under any condition, including the starting of the empty belt.

FIGURE 9.1. *Concave vertical curve.*

If this is not practical, it is permissible to let the empty belt lift off the idlers if the following conditions are met. (1) Nothing above the belt will damage it (i.e. headroom of the structure, tunnel, skirtboards, guard rails, belt cover or machinery, etc.). Sometimes the empty belt can be protected from such sources of damage by locating flat idlers above the carrying strand. (2) Wind will not affect the proper training of the belt. (3) Lack of troughed support will not result in spillage as the loaded portion of the conveyor belt approaches the vertical concave curve.

Design of Concave Vertical Curves

Because of the considerations mentioned above, it is good practice to design vertical concave curves with sufficient radius to allow the belt to assume a natural path on the troughing idlers under all conditions.

The illustration in Figure 9.2 makes it clear that the location of the beginning of the concave vertical curve, point c, tangent point of the curve, is indeterminate until the minimum radius is known. However, a close approximation can be made by assuming that the beginning of the concave vertical curve is at point c_1 in Figure 9.2. After determining the minimum radius from the following formulas, a second and exact calculation should be made.

The following formulas all involve point c, the start of the concave curve. But, for the first approximation, point c_1 will be used.

To prevent the belt from lifting off the idlers with the belt conveyor running, the formula is:

$$\text{Minimum radius, } r_1 = \frac{1.11 T_c}{W_b} \tag{1}$$

where: r_1 = minimum radius, ft, to prevent belt from lifting off the idlers.

T_c = belt tension, lbs, at point c (or c_1).

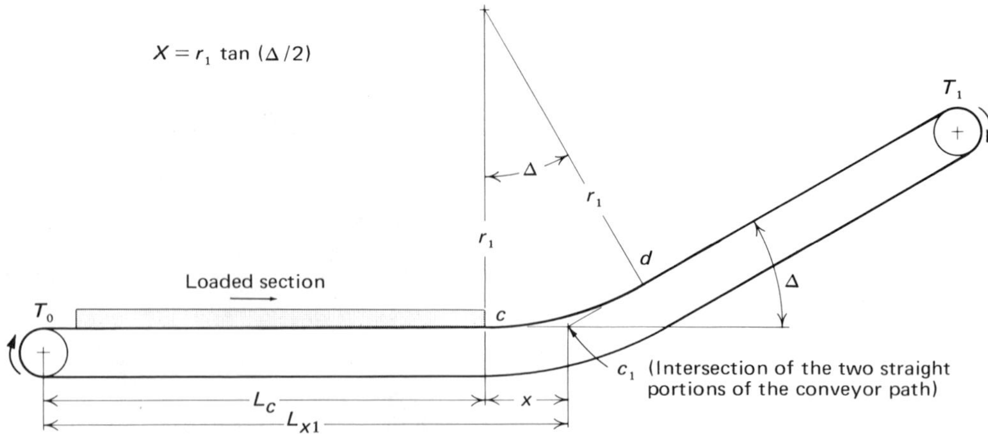

FIGURE 9.2. *Profile of conveyor with concave vertical curve.*

W_b = weight of belt per foot, lbs
1.11 = constant, based on maximum conveyor incline of 25° to horizontal.
See Figure 9.3.

However, two hazards may exist. These require checking. The first involves the tendency of the belt edges to buckle when the tension in the belt is too low. The second is the possibility that the tension in the center of the belt may exceed the allowable tension in the belt.

To insure that the belt tension is sufficiently high to avoid zero tension in the belt edges at a concave curve, a check of the curve radius should be made by the use of the following formula for fabric constructions:

$$\text{Minimum radius, } r_1 = \frac{(\text{Factor } A)(b^2)(B_m)(p)}{T_c - 30b} \quad \textbf{(2)}$$

For steel-cable belts, however, this radius can be reduced to permit a controlled buckling which experience has shown neither harms this type of belt or its splices nor causes excessive spillage. For steel-cable belts, use the following formula:

$$\text{Minimum radius, } r_1 = \frac{(\text{Factor } A)(b^2)(B_m)(p)}{(T_c - 30b)(2.5)} \quad \textbf{(3)}$$

To prevent stressing the center of the belt beyond the rated tension of the belt, check the radius of the concave curve by using the following formula for both fabric and steel-cable construction:

$$\text{Minimum radius, } r_1 = \frac{(\text{Factor } B)(b^2)(B_m)(p)}{T_r - T_c} \quad \textbf{(4)}$$

In these formulae:
r_1 = minimum radius of concave curve, ft
b = belt width, inches

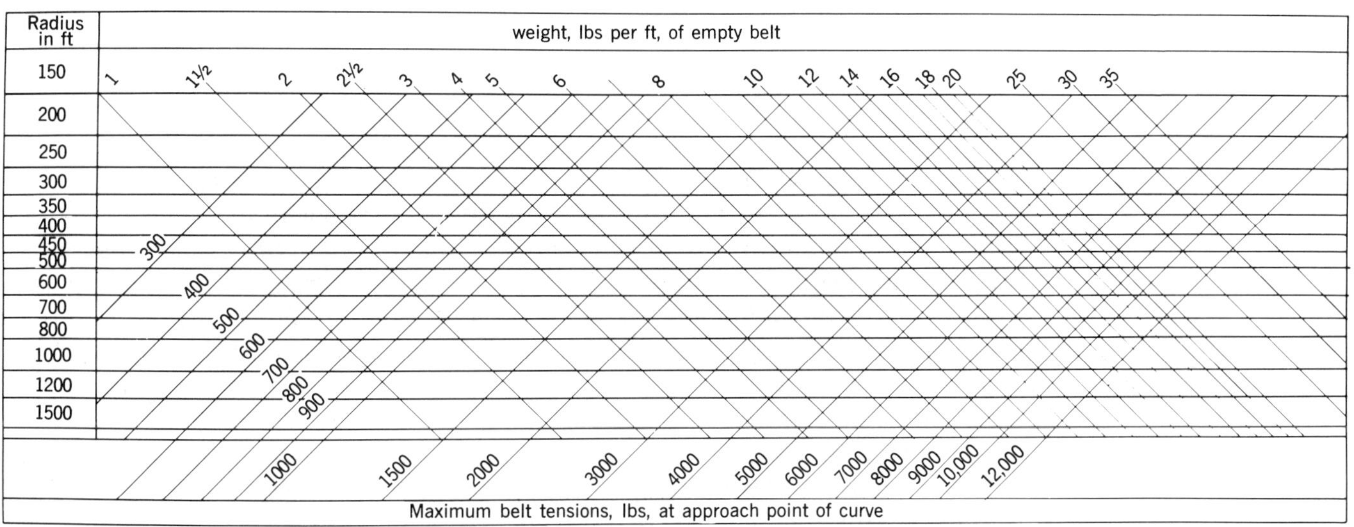

Chart is based on gradual acceleration of belt when starting the belt conveyor.

FIGURE 9.3. *Recommended minimum radii for concave vertical belt conveyor curves.*

p = number of plies in the belt
T_c = tension in belt at point c (or c_1), lbs
T_r = rated belt tension, lbs
B_m = modulus of elasticity of the conveyor belt, lbs per inch width per ply

Belt moduli vary widely among belt manufacturers because of the different fabric types used and the various ways of processing the fabrics and building the belt carcass. The modulus values calculated from the table below may vary considerably from specific values given by manufacturers but, in most cases, they will be conservatively high and can be used for preliminary or estimating work.

Longitudinal or Warp Reinforcement	*B_m Approximate Belt Modulus*
Cotton	50 times rated tension
Nylon	70 " " "
Polyester	100 " " "
Rayon	100 " " "
Steel cable	400 " " "
(Rated tension is rating in pounds per inch width per ply)	

For the final design, accurate values should be obtained.

Factor A and Factor B depend upon the trough angle of the belt conveyor carrying idlers, as indicated below:

Trough angle	*Factor A*	*Factor B*
20°	0.0063	0.0032
35°	0.0106	0.0053
45°	0.0131	0.0065

Formulas (2) and (3) are used for avoiding zero tension in the belt edges, and should be applied to the operating empty belt.

Formula (4) is used to prevent stressing the center of the belt beyond its rated tension. It should be applied to the condition where the belt is loaded from the tail pulley to the start of the curve and power is employed to start the belt from rest. Under starting conditions the allowable rated tension of the belt may be increased. See "Starting and stopping maximum tensions" in Chapter 6, page 102.

Use the largest of the three radii calculated in formulas (1), (2) or (3), and (4), above. (See "Example Problem of Concave Vertical Curve," page 227.) If formula (2) or (3) governs, investigate the possibility of increasing T_c by providing additional takeup weight.

Calculation of T_c Tension. The T_c belt tension can be determined by making the necessary additions to the tail tension, T_t, or subtractions from the head pulley tension, T_1. Refer to Chapter 6, "Belt tension at any point X on conveyor length," page 106, and Problem 4 in Chapter 6.

The decision to work forward from T_t or to work backward from T_1 depends upon the complexity of the path of the conveyor belt from these points to point c (c_1 for the first approximation).

For the purposes of illustration, the following works forward from T_t. The value of the tail tension, T_t, can be computed for various conveyor configurations having concave vertical curves from the formulas given in Chapter 6, associated with Figures 6.8B, 6.9C, 6.10C, 6.11B, 6.12C, 6.13B, 6.14B, 6.15C, and 6.16B.

Having obtained T_t, tension T_c then is determined as follows:

$$T_c = T_t + L_c[K_t(K_x + K_{y1}W_b) + K_{y1}W_m] \pm H_c(W_b + W_m)$$

where:

T_c = belt tension, lbs, at point c (or c_1)
T_t = belt tension, lbs, at tail pulley
L_c = length of belt, ft, from tail pulley to point c (or c_1)
K_t = temperature correction factor (see Chapter 6, "Factor K_t," Figure 6.1)
K_x = idler friction factor (see Chapter 6, "K_x—idler friction factor," pages 80-81)
K_{y1} = K_y factor for the particular belt path from the tail pulley to point c (or c_1)(refer to Chapter 6, "K_y—Factor for calculating the force of belt and load flexure over the idlers," Tables 6-1 and 6-2)
W_b = weight of belt, lbs per ft
W_m = weight of material, lbs per ft
H_c = vertical distance, ft, if any, from the tail pulley to the point c (or c_1)

The formula above covers the condition where the belt is most likely to lift while running. When H_c is positive, this occurs when the belt is loaded from the tail pulley to point c and is empty forward of point c (i.e., there is no load forward of point c to hold the belt down on the troughing idlers). When H_c is negative, this could occur when the belt is empty.

Calculation of T_{ac} Tension at point c during acceleration. The effect of acceleration of the belt conveyor when starting from rest must be considered, as the tension in the belt at point c will be increased over the running tension T_c.

To prevent the belt from lifting from the idlers during acceleration at the start-up, it is necessary to calculate the acceleration forces and determine the total belt tension at the beginning of the curve. Refer to Chapter 6 for the effect of acceleration.

Where motors with higher than required horsepower are used, care should be taken in the calculation of acceleration forces to prevent underestimating the tension force in the belt at point c. If this is not done, the conveyor belt may lift off the idlers.

$$T_{ac} = T_c + T_a$$

where: T_{ac} = total tension, lbs, at point c during acceleration
T_c = tension, lbs, at point c during normal running
T_a = tension, lbs, induced in the belt by accelerating forces at any given point (in this instance, at point c)

The accelerating force at any point on the conveyor is in direct proportion to the mass being accelerated. Since the mass is the weight divided by the gravity acceleration, the accelerating force is also in direct proportion to the weights accelerated. Therefore,

$$T_a = F_a \left(\frac{W_c}{W_t}\right)$$

where: T_a = tension, lbs, induced in the belt by accelerating forces at any given point
F_a = total accelerating force, lbs., for concave vertical curve calculations, conveyor loaded from tail to point c only
W_c = total weight, lbs, to be accelerated by the belt at point c

$$= LW_b + W_{ri}N_{ri} + L_c\left(\frac{W_{ti}}{S_i}\right) + L_c(W_b + W_m)$$

+ equivalent pulley weights, lbs

W_{ri} = equivalent weight, lbs, of moving parts of a single return idler
N_{ri} = number of return idlers
W_{ti} = equivalent weight, lbs, of the moving parts of a single troughing idler
S_i = troughing idler spacing, ft
L = total centers length, ft, of conveyor
L_c = length of conveyor, ft, from tail pulley to point c
L_2 = $L - L_c$ = remaining length, ft, of the conveyor from point c forward
W_b = weight of belt, lbs per ft
W_m = weight of material, lbs per ft
W_t = total equivalent wt., lbs., of all moving conveyor parts, excluding drive and drive pulley, plus loaded portion from tail to point c $= W_c + L_2W_b + L_2\left(\frac{W_{ti}}{S_i}\right)$

Like the formula for T_c, the above formulas apply to that condition where the belt is loaded from the tail pulley to point c, and where there is no load from point c to the terminal pulley. When the takeup is not near the discharge, the effect of the length of return run belt and the effect of the number of return run idlers should be reduced accordingly.

When the minimum radius has been calculated, based upon point c (or c_1 for the first approximation), the location of point c can be determined from the chart in Figure 9.4.

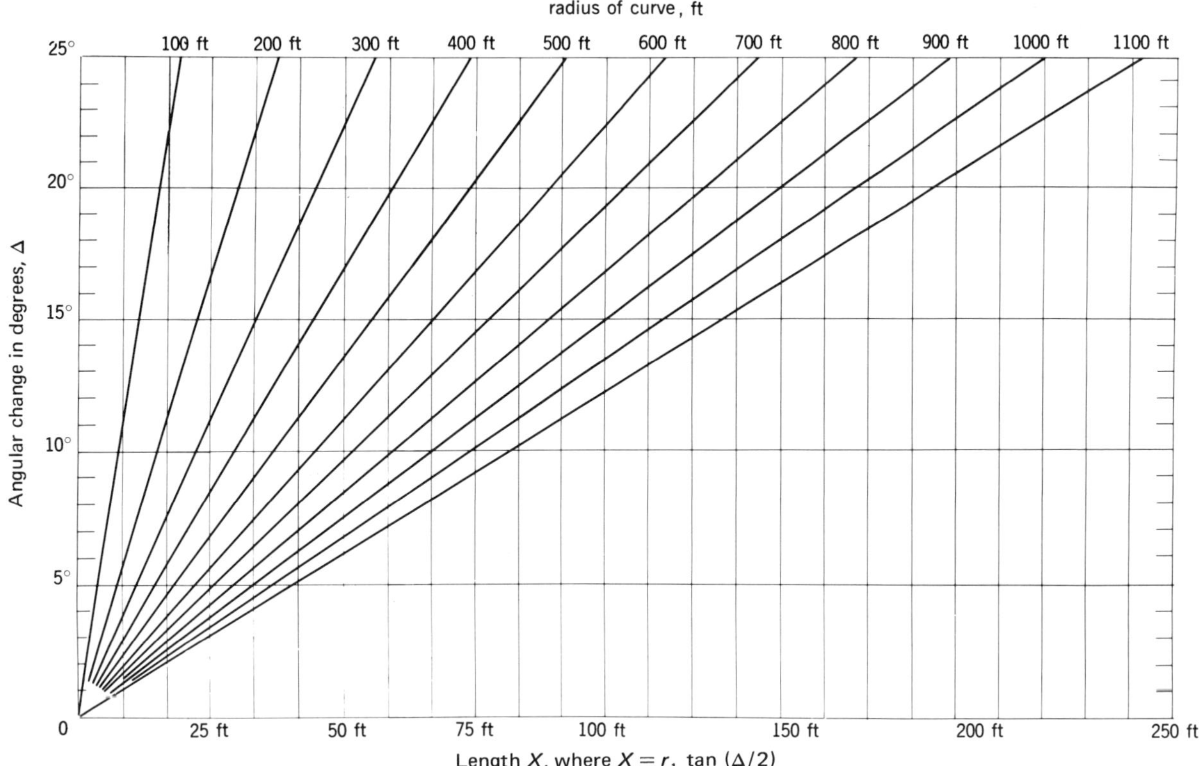

FIGURE 9.4. *Length X for concave vertical curves.*

Problem ***Determining Minimum Radius of a Concave Vertical Curve***

To illustrate the method of determining minimum radius of the curve, the following problem is offered. (This is the same as Problem 4 in Chapter 6.) The profile is as indicated in Figure 9.5.

Conveyor Specifications:
Belt width = 36 inches, 7 ply, *MP* 70 nylon
Belt modulus = B_m = 4,900 lbs per inch width per ply
Length = L = 4,000 ft
Belt weight = W_b = 10 lbs per ft
Capacity = Q = 800 tph of material weighing 85 lbs per cu ft
Weight of material = W_m = 66.6 lbs per ft
Speed = V = 400 fpm
Idlers = Class C6, 6-inch dia., 20° trough

FIGURE 9.5. *Profile of concave vertical curve.*

Weight of the Moving Parts of the Idlers (refer to Tables 5.13 and 5.14):
Troughing = W_{ti} = 43.6 lbs
Return = W_{ri} = 37.6 lbs
Idler spacing, troughing = S_i = 4 ft, return = 10 ft

$K_x = 0.427$
$K_y = 0.0255$ for 3,000-ft horizontal section
$K_y = 0.016$ for 800-ft inclined section
$K_y = 0.016$ for 200-ft horizontal section

From Problem 4, Chapter 6, T_t = 1,287 lbs and T_{fcx} = 7,141 lbs
T_c (tension at curve) = 1,287 + 7,141 = 8,428 lbs
Therefore,

$$r_1 = \frac{1.11T_c}{W_b} = \frac{(1.11)(8{,}428)}{10} = 936 \text{ ft}$$

Now, check against belt lifting during acceleration at the start,

$$W_c = LW_b + W_{ri}N_{ri} + L_c\left(\frac{W_{ti}}{S_i}\right) + L_c(W_b + W_m) + \text{pulley equivalent weights}$$

Inspection of the profile shows that six non-driving pulleys must be accelerated. Assume that these pulleys weigh 3,600 lbs, as in Problem 4, Chapter 6. Then,

$$\begin{aligned} W_c &= LW_b + W_{ri}N_{ri} + L_c\left(\frac{W_{ti}}{S_i}\right) + L_c(W_b + W_m) + \text{pulley equivalent weights} \\ &= (4{,}000)(10) + (37.6)\left(\frac{4000}{10}\right) + 3{,}000\left(\frac{43.6}{4}\right) + 3{,}000(10 + 66.6) + 3{,}600 \\ &= 40{,}000 + 15{,}040 + 32{,}700 + 229{,}800 + 3{,}600 \\ &= 321{,}140 \text{ lbs} \end{aligned}$$

$$\begin{aligned} W_t &= W_c + L_2 W_b + L_2\left(\frac{W_{ti}}{S_i}\right) = 321{,}140 + (1{,}000)(10) + 1{,}000\left(\frac{43.6}{4}\right) \\ &= 342{,}040 \text{ lbs} \end{aligned}$$

The accelerating force, F_a, may be determined by assuming that the motor and controls will deliver an average accelerating torque of 180% of full load torque of the two motors, as stated in Problem 4, and the drive efficiency will be .94. See Table 6-11.

Then, the accelerating force at the belt line becomes

$$\frac{\text{Effective horsepower} \times 33{,}000}{V} = \frac{(1.8)(200)(.94)(33{,}000)}{400} = 27{,}918 \text{ lbs}$$

Again referring to Problem 4, Chapter 6, the effective tension when the conveyor is running fully loaded is T_e = 14,055 lbs. To determine T_e when only the horizontal portion is loaded, deduct resistance of load moving over inclined and upper horizontal sections and resistance to lift load

$$\begin{aligned} &= (L - L_c)K_y W_m \pm HW_m \\ &= (1{,}000)(.016)(66.6) + 70(66.6) \\ &= 1{,}066 + 4{,}662 \\ &= 5{,}728 \text{ lbs} \end{aligned}$$

Therefore, T_e, for the conveyor loaded only on the horizontal portion, is 14,055 — 5,728 = 8,327 lbs.

The total equivalent force acting at the belt line and available for accelerating is 27,918 — 8,327 = 19,591 lbs.

However, a portion of this will be necessary to overcome the inertia of the drive. This effect can be compensated for by converting the WK^2 of the drive to the equivalent weight at the belt line, and adding it to W_t. Referring to Problem 4, in Chapter 6, this equivalent weight of the drive is 55,615 lbs (see page 151).

Because the accelerating force is directly proportional to the total weight being accelerated, the actual accelerating force available to accelerate the conveyor then is calculated as:

$$F_a = 19{,}591\left(\frac{342{,}040}{342{,}040 + 55{,}615}\right) = 16{,}851 \text{ lbs}$$

and:
$$T_a = F_a\left(\frac{W_c}{W_t}\right) = 16{,}851\left(\frac{321{,}140}{342{,}040}\right) = 15{,}821 \text{ lbs}$$

Therefore:
$$T_{ac} = T_c + T_a = 8{,}428 + 15{,}821 = 24{,}249 \text{ lbs}$$

The minimum radius to prevent the belt from lifting during the calculated acceleration of starting the conveyor (loaded only from the tail to point c_1) may be found by substituting T_{ac} for T_c in the formula for the minimum radius:

$$r_1 = \frac{1.11T_{ac}}{W_b} = \frac{(1.11)(24{,}249)}{10} = 2{,}692 \text{ ft}$$

If a more accurate determination is required, recalculate the radius, based on the new T_{ac}, for the exact location of point c.

Checking for belt buckling during running conditions, and for overstressing the center of the belt during starting conditions, with the conveyor loaded only from the tail end to point c (or c_1), the following two conditions can occur.

For buckling of the belt, T_c for empty belt = 8,428 − (.0255) (3,000) (66.6) = 3,333 lbs.
Therefore,

$$r_1 = \frac{(\text{Factor } A)b^2(B_m)(p)}{T_c - 30b} = \frac{(0.0063)(36)^2(4{,}900)(7)}{3{,}333 - (30)(36)} = 124 \text{ ft.}$$

For overstressing the center of the belt,

$$r_1 = \frac{(\text{Factor } B)b^2(B_m)(p)}{T_r - T_c} = \frac{(0.0032)(36)^2(4{,}900)(7)}{31{,}752 - 24{,}249} = 19 \text{ ft.}$$

It is assumed that, for the operating conditions in this example, the approximate value of $T_r = 17{,}640 \times 1.8 = 31{,}752$ lbs.

Therefore, the minimum concave radius requirement is 2,692 ft.

Graphical Construction of Concave Vertical Curve

After the minimum radius has been calculated and point c located, the concave curve can be graphically constructed, as indicated in Figure 9.6 and Tables 9-1 and 9-2.

Example Radius of curve decided upon is 300 ft and the angle Δ is 20°. After locating the working point, which is the intersection of the horizontal and inclined runs of the conveyor if extended to meet, the tangent points of the curve will be

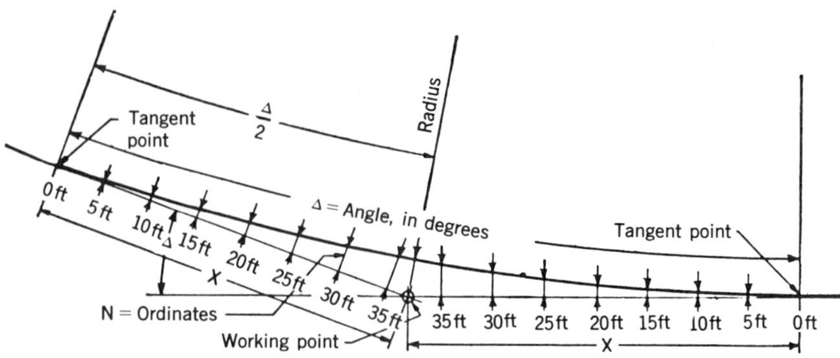

FIGURE 9.6. *Method of plotting vertical curves. After the proper radius of the vertical curve is determined, and the exact angle of inclination of the belt conveyor decided upon, the conveyor can be laid out by using Tables 9-1 and 9-2 below.*

found to be 52 ft, 10¾ inches (Dim, "X", Table 9-1) from the working point. By laying off points, starting from each tangent point, 5 ft, 0 inches apart towards the working point and then drawing ordinates through these points at 90° from the lines representing the continuation of the horizontal and inclined conveyor runs and then measuring off the distance given in Table 9-2 on each respective ordinate, the curve may then be drawn through these points.

Precautions for the Design of Vertical Concave Curves

With the trend toward stronger fabrics and new types of belt construction, the belt conveyor designer should consider the possibility of a lighter weight belt being used as a replacement at some future date. Because such a lighter belt would require a larger minimum radius, it is wise to design for the largest radius possible, considering economics and physical space requirements.

In general, the minimum radius of the vertical concave belt conveyor curve should not be less than 150 ft.

Convex Vertical Curves

A conveyor belt is said to pass through a convex vertical curve when the center of curvature lies below the belt. (See Figure 9.7.) In such cases, the gravity forces of belt and of load (if present), and the belt tension itself, press the belt onto the idlers.

FIGURE 9.7. *Convex vertical curve.*

TABLE 9-1. *Location of Tangent Points on Concave, Vertical Curves*

Δ *Angle of Inclination (degrees)*	*Dimension "X"—Distance from Tangent Point to Working Point, in ft and inches*							
	Radius (ft)							
	150	200	250	300	350	400	450	500
5	6- 6 5/8	8- 8 3/4	10-11	13- 1 1/4	15- 3 3/8	17- 5 1/2	19- 7 3/4	21-10
6	7-10 3/8	10- 5 3/4	13- 1 1/4	15- 8 5/8	18- 4 1/8	20-11 1/2	23- 7	26- 2 1/2
7	9- 2 1/8	12- 2 3/4	15- 3 1/2	18- 4 1/4	21- 4 7/8	24- 5 1/2	27- 6 1/4	30- 7
8	10- 5 7/8	13-11 7/8	17- 5 3/4	20-11 3/4	24- 5 3/4	27-11 5/8	31- 5 5/8	34-11 1/2
9	11- 9 5/8	15- 8 7/8	19- 8 1/8	23- 7 1/4	27- 6 1/2	31- 5 3/4	35- 5	39- 4 1/4
10	13- 1 1/2	17- 6	21-10 1/2	26- 3	30- 7 1/2	35- 0	39- 4 1/2	43- 9
11	14- 5 3/8	19- 3 1/8	24- 0 7/8	28-10 5/8	33- 8 3/8	38- 6 1/4	43- 4	48- 1 3/4
12	15- 9 1/4	21- 0 1/4	26- 3 1/4	31- 6 3/8	36- 9 1/2	42- 0 1/2	47- 3 1/2	52- 6 5/8
13	17- 1 1/8	22- 9 1/2	28- 5 3/4	34- 2 1/4	39-10 1/2	45- 6 7/8	51- 3 1/4	56-11 5/8
14	18- 5	24- 6 3/4	30- 8 3/8	36-10	42-11 3/4	49- 1 3/8	55- 3	61- 4 3/4
15	19- 9	26- 4	32-11	39- 6	46- 1	52- 8	59- 2 7/8	65-10
16	21- 1	28- 1 1/4	35- 1 5/8	42- 2	49- 2 1/4	56- 2 5/8	63- 3	70- 3 1/4
17	22- 5	29-10 3/4	37- 4 3/8	44-10	52- 3 1/4	59- 9 3/8	67- 3	74- 8 3/4
18	23- 9 1/8	31- 8 1/8	39- 7 1/8	47- 6 1/4	55- 5 1/4	63- 4 1/4	71- 3 1/4	79- 2 1/4
19	25- 1 1/8	33- 5 5/8	41-10	50- 2 1/2	58- 6 3/4	66-11 1/4	75- 3 5/8	83- 8
20	26- 5 3/8	35- 3 1/4	44- 1	52-10 3/4	61- 8 1/2	70- 6 3/8	79- 4 1/8	88- 2
21	27- 9 5/8	37- 0 3/4	46- 4	55- 7 1/4	64-10 1/2	74- 1 5/8	83- 4 7/8	92- 8

TABLE 9-2. *Ordinate Distances of Points on Concave Vertical Curves*

Distance from Tangent Point (ft)	*Dimensions "N"—Length of Ordinates in ft and inches, at intervals of 5 ft from Tangent Point*							
	Radius (ft)							
	150	200	250	300	350	400	450	500
5	0-1	0-0 3/4	0-0 5/8	0-0 1/2				
10	0-4	0-3	0-2 3/8	0-2	0-1 3/4	0-1 1/2	0-1 3/8	0-1 1/4
15	0-9	0-6 3/4	0-5 3/8	0-4 1/2	0-3 7/8	0-3 3/8	0-3	0-2 3/4
20	1-4	1-0	0-9 5/8	0-8	0-6 7/8	0-6	0-5 3/8	0-4 7/8
25	2-1 1/4	1-6 7/8	1-3	1-0 1/2	0-10 3/4	0-9 3/8	0-8 3/8	0-7 5/8
30		2-3 1/8	1-9 3/4	1-6	1-3 1/2	1-1 5/8	1-0	0-10 3/4
35		3-1	2-5 1/2	2-0 1/2	1-9	1-6 1/2	1-4 1/4	1-2 3/4
40			3-2 5/8	2-8 1/8	2-3 1/2	2-0	1-9 3/8	1-7 1/4
45			4-1	3-4 3/4	2-10 3/4	2-6 1/2	2-3	2-0 3/8
50				4-2 3/8	3-7 1/8	3-1 5/8	2-9 3/8	2-6 1/8
55				5-1	4-4 1/4	3-9 1/2	3-4 1/2	3-0 1/2
60					5-2 1/4	4-6 3/8	4-0 1/4	3-7 3/8
65						5-3 3/4	4-8 3/4	4-3
70						6-2	5-5 3/4	4-11 1/8
75							6-3 5/8	5-7 7/8
80							7-2	6-5 3/8
85								7-3 3/8
90								8-2 1/8

When a troughed conveyor belt passes around the convex curve, the tension stress present is distributed across the belt so that the belt edges, being on a larger radius, are more highly stressed than is the belt center, where the radius of curvature is less. Similarly, the troughing idlers on a convex curve are more heavily loaded by radial pressures from the belt than those idlers not on the curve. A curve of sufficiently large radius holds these extreme stresses and loads within acceptable limits.

If a convex vertical curve is located where the belt tension is low, the distribution of stress across the belt may result in less than zero tensile stress at the center of the belt. This can produce buckling in the belt and possible spillage of the load.

Design of Convex Vertical Curves

The following equations are used to determine the minimum radius to use to prevent undesirable conditions such as belt buckling and load spillage:

$$\text{Minimum radius, } r_2 = \frac{(\text{Factor } C)\, b^2(B_m)(p)}{T_r - T_c} \qquad \textbf{(5)}$$

(to prevent overstress of belt edges)

$$\text{Minimum radius, } r_2 = \frac{(\text{Factor } D) b^2(B_m)(p)}{T_c - 30b} \qquad \textbf{(6)}$$

(to prevent buckling of the belt)

$$\text{Minimum radius, } r_2 = 12\left(\frac{b}{12}\right) \qquad \textbf{(7)}$$

where:
r_2 = minimum radius of convex curve, ft
b = belt width, inches
p = number of plies in the belt
T_c = tension in the belt at point c (or c_1), lbs
T_r = rated belt tension, lbs
B_m = modulus of elasticity of the belt, lbs per inch width per ply. For values of B_m, see discussion of concave vertical curve design, page 222.

Factor C and Factor D depend upon the trough angle of the carrying idlers, as indicated below:

Trough angle	*Factor C*	*Factor D*
20°	0.0063	0.0032
35°	0.0106	0.0053
45°	0.0131	0.0065

Equation (5) should be applied to the condition where the belt is being started from rest, with the belt loaded from the tail pulley to the convex curve. Under starting conditions, the allowable rated tension of the belt may be increased. See Chapter 6, "Starting and stopping maximum tensions," (page 102).

Equation (6) should be applied to the condition where the belt is operating empty.

Always use the largest of the three values of the minimum convex curve radii determined by formulae (5), (6), and (7), above. See the problem, "Determining minimum radius of a convex vertical curve," page 234. If formula (6) governs, investigate the possibility of increasing T_c by providing additional takeup weight.

Idler Spacing On Convex Curves

Both the carrying and return idlers should be spaced so that the sum of the belt load, plus the material load, plus the radial resultant of the belt tension does not exceed the load capacity of the idlers.

The radial resultant of the belt tension can be calculated approximately as follows:

$$F_r = 2T_c \sin\left(\frac{\Delta}{2n}\right)$$

where:

- F_r = resultant force, lbs, on idlers at convex vertical curve, produced by the belt tension at the curve
- T_c = tension in belt, lbs, at point c or c_1
- Δ = change in the angle of the belt, degrees, between entering and leaving the curve
- n = number of spaces between the idlers on the curve (must be an integral number)

$$\text{Arc length of curve} = 2\pi r_2 \left(\frac{\Delta}{360}\right), \text{ ft}$$

Troughing Idler Spacing on Convex Curves. The troughing idler spacing on a convex curve can be determined in the following manner:

$$\text{Maximum troughing idler spacing, } S_{ic} = \frac{(I_{lr} - F_r)}{(W_b + W_m)}$$

where:

- S_{ic} = maximum troughing idler spacing, ft, on the curve
- I_{lr} = allowable load per troughing idler (i.e., troughing idler load rating, lbs); see Chapter 5
- F_r = resultant force, lbs, on idlers at convex vertical curve, produced by belt tension at curve
- W_b = weight of belt, lbs per ft
- W_m = weight of material, lbs per ft

The above formula for maximum troughing idler spacing on the curve is subject to the following three conditions: (1) If the formula results in a troughing idler spacing on the curve greater than the normal idler spacing adjacent to the curve, S_{ic} is limited to values no greater than the normal troughing idler spacing. (For normal idler spacing, see Chapter 5, "Idler spacing," page 67). (2) If the formula results in a troughing idler spacing greater than one-half of the normal idler spacing adjacent to the curve, but less than such normal idler spacing, S_{ic} is limited to values no greater than the value given by the formula. (3) If the formula results in a troughing idler spacing less than one-half of the normal idler spacing adjacent to the curve, S_{ic} is limited to no less than one-half normal idler spacing adjacent to the curve. Solve for a new F_r. If possible, increase the radius of the curve to that based on this new F_r value.

There is also a practical limitation in determining the S_{ic} value. The idler spacing on the curve should be in integral and equal increments to simplify structural frame details. This further limits the actual value of S_{ic}.

If the length of arc of the curve (arc) is given in ft,

$$n = \frac{(\text{arc})}{S_{ic}}$$

where: n = number of spaces between idlers on the curve. Use the next largest integer.

Problem

Determining Minimum Radius of a Convex Vertical Curve

To illustrate the method of determining the minimum radius of a convex curve and the troughing idler spacing, the following problem is offered. (This is the same as Problem 4 in Chapter 6.) A profile of the conveyor is shown in Figure 9.5.

Conveyor Specifications:

Belt width = 36 inches, 7-ply, *MP* 70 nylon
Belt modulus, B_m = 4,900 lbs per inch width, per ply
Belt weight, W_b = 10 lbs per ft
$T_r = (b)(p)(70) = (36)(7)(70) = 17{,}640$ lbs
Capacity, Q = 800 tph
Speed, V = 400 fpm
Material weight, W_m = 66.6 lbs per ft
Idlers = Class C6, 6-inch diameter, 20° trough,
Idler spacing, S_i = 4 ft
Maximum allowable idler load, I_{lr} = 900 lbs
Tension at curve, T_c = 15,112 lbs (see Problem 4, Chapter 6)
Assume 30,892 lbs during acceleration.

Using Equation (5) during acceleration of the belt:

$$r_2 = \frac{(\text{Factor } C)\ b^2\ (B_m)(p)}{T_r - T_c}$$

$$= \frac{(0.0063)(36)^2(4{,}900)(7)}{(17{,}640 \times 1.8) - 30{,}892}$$

$$= 326 \text{ ft}$$

Using Equation (6) when belt is running empty and T_c = 4,388 lbs:

$$r_2 = \frac{(\text{Factor } D)\ b^2\ (B_m)(p)}{T_c - 30b}$$

$$= \frac{(0.0032)(36)^2(4{,}900)(7)}{4{,}388 - (30)(36)}$$

$$= 43 \text{ ft}$$

Using Equation (7):

$$r_2 = 12\left(\frac{b}{12}\right) = 36 \text{ ft}$$

Because Equation (5) yields the largest minimum radius, use 326 ft for the minimum radius of the convex curve.

Length of arc of curve, ft:

$$\text{arc} = 2\pi r\left(\frac{\Delta}{360}\right) = 2\pi(326)\left(\frac{5}{360}\right) = 28.4 \text{ ft}$$

Number of spaces between idlers:

$$n = \frac{\text{arc}}{S_i} = \frac{28.4}{4} = 7.10$$

however, the next greater integer = 8.

Resultant idler load:

$$F_r = 2T_c \sin\left(\frac{\Delta}{2n}\right)^{\circ} = (2)(15{,}112)\sin\left(\frac{5}{16}\right)^{\circ}$$

$$= (2)(15{,}112)\sin .3125^{\circ} = (2)(15{,}112)(.00545)$$

$$= 165 \text{ lbs (in round numbers)}$$

Troughing idler spacing:

$$\text{Maximum } S_i = \frac{(I_{lr} - F_r)}{(W_b + W_m)} = \frac{(900 - 165)}{(10 + 66.6)} = 9.5 \text{ ft, maximum}$$

The limitation for troughing idler spacing on convex curves, page 233, applies. The normal idler spacing adjacent to the curve is 4 ft. Therefore, the 4-ft idler spacing on the curve will be maintained.

Return Idler Spacing on Convex Curves. The spacing of return idlers can be determined similarly to the method used for troughing idlers. Use the resultant return idler load plus belt weight and then compare this value with the allowable load rating table in Chapter 5.

Use of Bend Pulleys for Convex Curves

A convex curve employing troughing idlers is recommended for all installations where space will permit for two reasons. First, the belt edge stress in a troughed belt is reduced by a properly designed convex curve. Second, there is less disturbance of the material on the belt as it passes through the change in belt profile, thereby reducing wear on the belt and idlers and preventing spillage over the edges of the troughed belt.

Bend pulleys on the carrying runs of troughed belts, in place of convex curves, are not generally recommended. A bend pulley should be used only in special cases, when space will not permit a properly designed convex curve and the belt conveyor is not sufficiently loaded to cause spillage of material over the edges of the flattened belt as it passes over the bend pulley.

Under these conditions, the diameter of the bend pulley should be large enough to insure retention of the material on the belt as the belt changes direction. The diameter required varies with the cosine Δ (angle of change in direction) and V^2 (square of the belt speed). This becomes quite large for belt speeds greater than 500 fpm. Naturally, this is another reason why troughing idlers are preferable.

The minimum diameter of the bend pulley, for a given belt velocity or speed, should be as listed below:

Minimum diameter of bend pulley (inches)	*Belt velocity or belt speed (fpm)*
16	200
20	300
36	400
54	500

In no case should the diameter be less than the minimum value shown in Tables 7-5, 7-6, and 7-7 in Chapter 7.

Chapter 10

Belt Takeups, Cleaners, and Accessories

Contents

Belt Takeups

All properly designed belt conveyors require the use of some form of takeup device for the following reasons: (1) To insure the proper amount of slack-side tension, T_2, at the drive pulley to prevent belt slippage. (2) To insure proper belt tension at loading and other points along the conveyor (necessary to prevent loss of troughing contour of the belt between idlers, thus avoiding spillage of the material from the belt). (3) To compensate for changes in belt length. (4) To allow belt storage for making replacement splices (without which storage, small sections of new belt would have to be added, requiring two splices for each splice repair).

Belt Stretch or Elongation

Any conveyor belt can be expected to have several types of stretch or elongation.

Elastic stretch. This is that part of the stretch which occurs in a conveyor belt during starting acceleration or braking deceleration. This stretch is almost entirely recovered when the applied pull or stress is removed.

Constructional stretch. This is due more to the type of fabric weave than to the textile material used. In a conventionally woven fabric, the warp strands which are crimped tend to straighten out as the load is applied. This results in belt growth, a portion of which is nonrecoverable.

Permanent length change. This includes changes in length caused by elongation in the basic fiber structure. It also includes that portion of the elastic stretch and constructional stretch which is nonrecoverable.

Takeup Movement

The required length of takeup movement depends upon several factors:

1. Type of starting or braking. Across-the-line starting or braking will require considerably more allowable takeup movement than controlled acceleration or deceleration.
2. Frequency of starts and stops of a fully loaded belt.

3. Whether a run-in period with metal fasteners can be utilized. If so, this would conveniently allow for removal of any nonrecoverable length change before making a final vulcanized splice.
4. Belt stretch and elongation characteristics of the belting being used.

The takeup should provide sufficient movement to accommodate acceleration or deceleration surges without having the takeup strike against its stops. It should also allow for some "live" storage of belting so that, in case of an accident, it may not be necessary to splice in a short length of belt, which would require two splices. In addition, takeup movement should provide for changes in belt length due to stretch or shrinkage.

Manual Takeups

Manual takeups have the advantages of compactness and low cost; however, because with this type the takeup of the belt is at best periodic, the resulting belt tensions are almost always too high or too low. Therefore, manual takeups are recommended only where an automatic takeup is not practical because of space limitations or in the case of relatively short, light duty belt conveyors, where takeup considerations are not as critical.

In high-tonnage inclined conveyors, such as stackers, the use of manual takeups for any length of belt is not normally recommended. Such cases should be referred to a CEMA member.

The most commonly used manual takeup is the screw takeup, which is illustrated in Figure 10.1. Other common types include ratchet and jack-operated takeups. Screw takeups are generally available with 12, 18, 24, 30, and 36 inches of takeup travel.

The main problem with the use of a manual takeup is that it requires a vigilant and careful operator to observe when takeup is required, and then adjust the takeup just to the point where the proper tension is provided. The problem is made difficult by the absence of an accurate index for judging how much tension is needed, or how much has been provided by a particular adjustment.

With new belts, it is necessary to provide a large amount of slack-side tension to prestretch the belt. This prevents slack in the belt from ac-

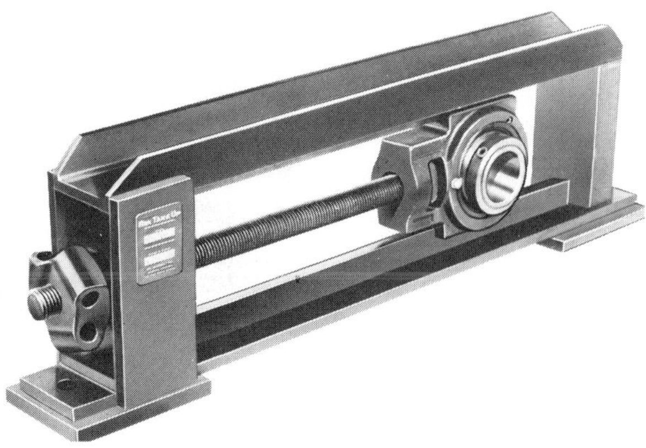

FIGURE 10.1. *Manually adjusted screw takeup assembly.*

cumulating behind the drive pulley and causing belt slippage when the conveyor operates. With manual takeups, the operator must rely on personal judgment in making the suitable adjustment.

Manual Takeup Location. Manual takeups normally are located at the end of the conveyor, opposite the drive end. This is the most convenient and least costly arrangement as it involves no extra pulleys. However, it is possible to locate manual takeups elsewhere on the return run of the belt conveyor, if required.

Automatic Takeups

Automatic takeups are the more desirable type for use on any belt conveyor. They can be installed horizontally, vertically, or on an incline. They can be either gravity-operated or power-operated by hydraulic, electric, or pneumatic means. The most common type is the gravity takeup.

Other types of automatic takeups can be given consideration when special conditions, such as space limitations or portability, are involved. Shown in Figures 10.2 and 10.3 are a horizontal automatic gravity takeup at a tail pulley and a vertical automatic gravity takeup for use near a head pulley. Figure 10.3 illustrates a vertical automatic gravity takeup on an inclined conveyor.

Recommended Takeup Movement. The foregoing points have been considered in making recommendations for the amount of required takeup movement and for the position of the takeup in making the final splice.

FIGURE 10.2. *Horizontal automatic gravity belt takeup. Counterweight and cables not shown.*

FIGURE 10.3. *Vertical automatic gravity takeup on an inclined conveyor*

The values for takeup movement listed in Table 10–1 are generally suitable for most conveyor applications. A reduction or increase in these values will depend upon several factors which include belt selection and environmental, as well as operating, conditions. Therefore, it is advisable to consult the manufacturer of the belt under consideration before takeup requirements are determined.

Automatic Takeup Location. Automatic takeups may be located at any place in the return run of the belt conveyor. The prime consideration is where

TABLE **10-1.** *Recommended Takeup Movement**

Conveyor centers (ft)	*Takeup movement (ft)*			
	*Multiple-ply belt***	*Reduced-ply belt*	*Steel-cable belt*	*Solid-woven belt****
50 or less	1.5	1.5	—	
100	3	3	—	
200	5-6	6	—	
300	7-8	8	—	
500	10-14	14	—	
700	13-18	18	—	
1,000	18-25	25	7	
1,500	25-34	34	8	
2,000	30-40	40	10	
2,500	35-47	47	12	
3,000	39-54	54	15	
3,500	42-59	59	17	
4,000	45-64	64	20	
4,500	48-70	70	22	
5,000	50-75	75	25	

* Allows for ¾ of the total movement for belt stretch.
** The larger allowance is recommended for belts having 35-50 RMA ratings.
*** Consult belting manufacturer.

the automatic takeup will work best in relation to the drive, to keep belt tensions at a minimum. Other considerations, such as available space, maintenance conditions, and the economics of the location, should also be taken into account.

Generally, the most inexpensive location for an automatic takeup is at the tail of an inclined conveyor. At this point, no additional pulleys will be involved. On steeply inclined conveyors, the weight of the takeup pulley assembly and belt may provide sufficient slack-side tension to prevent drive pulley slippage, without the need for additional counterweight.

On long, horizontal, or slightly inclined conveyors, the automatic takeup should be located near the drive, where it will act quickly enough to prevent slippage of the belt on the drive pulley during acceleration at start-up. If the takeup is located elsewhere, its movement must be calculated to be sure that it exceeds the rate at which the belt will be deposited in the takeup. Refer to Chapter 6, "Belt tensions for various profiles," for usual takeup locations on various conveyors.

Automatic Takeup Force Requirements. An automatic gravity takeup must provide a force equal to twice the required belt tension, at the place where the takeup is installed. This force usually is supplied by a counterweight composed of steel, cast iron, concrete, or some other heavy material equal to the force required. Or the force may be somewhat less in magnitude and multiplied appropriately by the mechanical advantage of a system of ropes and sheaves. Some adjustment to the weight force should be provided, as actual operating conditions may change the force requirements originally calculated in the design of the belt conveyor.

To calculate the required force of the automatic takeup or the weight force of a gravity takeup, the following formula can be used:

$$W_g = \frac{2T + W_f - W_p}{R_1}$$

where:

W_g = required force, lbs, provided by the takeup; in gravity takeups, it is the weight force, lbs

T = belt tension, lbs, at the point where the takeup is located

W_f = force, lbs, to overcome friction of the takeup carriage, ropes, sheaves, or other frictional resistance

W_p = component of weight force of takeup carriage, wheels, pulley, shaft, shaft bearings, etc., acting in direction of resultant pulley load; where elements move horizontally, W_p becomes zero

R_1 = mechanical advantage ratio, if any mechanical advantage is provided.

In an hydraulically or pneumatically operated automatic takeup, the force is calculated as above. As with a gravity automatic takeup, the force should be adjustable to meet unforeseen operating conditions. This can be done by varying the hydraulic or pneumatic pressure to suit the actual operating conditions.

Cleaning Devices

Many materials conveyed on belts are sticky. Portions cling to the conveying surface of the belt and are not discharged with the rest of the material at the discharge points. Material is carried back on the return run, where it may cause excessive wear, build-up on return idlers, misalignment of the belt, and possible damage by forcing the belt against some part of the supporting structure. The material that is carried back on the return run eventually drops off the belt, causing maintenance and housekeeping problems. It is, therefore, desirable to clean the belt before it contacts any snub pulleys or return idlers.

Materials which will stick to the belt usually will stick to any snub pulleys which contact the dirty side of the belt. Therefore, pulley-cleaning devices may be as necessary as the belt cleaners.

There are several types of belt cleaners. Selecting one which should be used for a particular material is difficult, since such factors as temperature, moisture content, material size, etc. vary with each application. Because these factors determine the effectiveness of the cleaning device, each job requires individual consideration. Even after a belt cleaner is installed, adjustments will be required on the job to meet the characteristics of the material. Sometimes, more than one belt cleaner is required.

Belt cleaning is simplified by the use of vulcanized splices, especially with those cleaning devices which employ blades in contact with the belt surface. Improperly installed mechanical fasteners can catch on cleaning devices and cause them to jump and vibrate. Recessed mechanical fasteners will help minimize this problem.

Proper maintenance and adjustment of the belt cleaner will help prevent belt damage, reduce wear on the belt and cleaner blades, and assure efficient cleaning action. To minimize the attention that must be given to belt-cleaning devices to compensate for wear, they usually are mounted on a pivoted, counterweighted arm, or on a spring-tension arm.

Types of Belt Cleaners

Single- or Multiple-Blade Belt Scrapers. These are designed for scraping material from the belt surface. A typical single-blade scraper is shown in Figure 10.4.

One or more blades are held in contact with the belt surface by counterweight or spring tension. A single-blade scraper consists of one blade across the width of the belt. A multiple-blade scraper consists of two or more parallel blades across the width of the belt.

A scraper blade can be made from any one of several materials. A combination of materials sometimes is used on multiple-blade scrapers to improve cleaning results. For example, a rubber blade and a special steel blade may be used in combination.

Strips of belting should not be used for scraper blades because fine particles of the material being handled on the conveyor may become embedded in the fabric of the blade, causing excessive wear on the cover of the conveyor belt. Any rubber scraper blade should be made of solid rubber with no fabric or fiber reinforcement.

Scraper blades are made in three designs:

1. Straight blade across the belt. Refer to Figure 10.4. When used on contact with the belt on a crowned pulley, the blade wears to conform to the shape of the belt and pulley. The blade usually is made so that it can be adjusted for the most effective cleaning action.

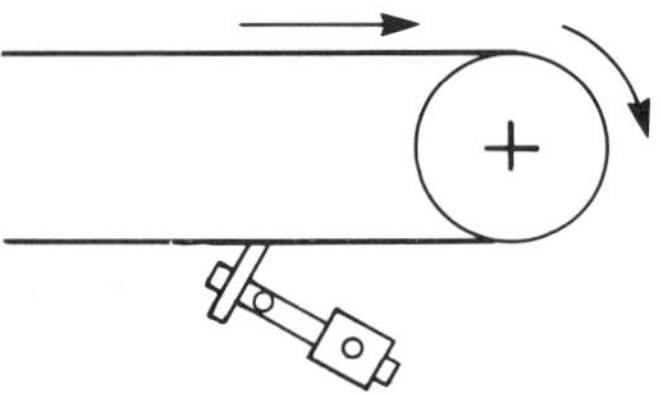

FIGURE 10.4. *Example of single-blade counterweighted belt scraper.*

2. Sectionalized blades. Figure 10.5 illustrates one form of sectionalized blades. Here, the blades are sectionalized to conform to the crown of the pulley. This design does not require a break-in period to wear the blade to the shape of the belt on a crowned pulley. The blade effectively cleans the belt from the time of initial installation.

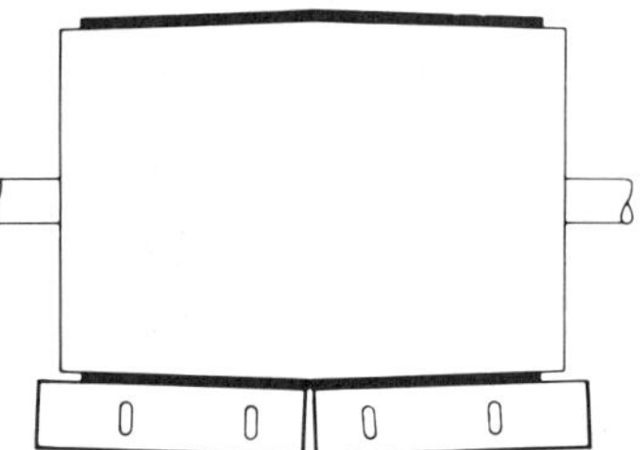

FIGURE 10.5. *Typical pattern of sectionalized scraper blades.*

3. Articulated blades. This design consists of a series of short blades, pivoted on arms which are spring-loaded to maintain contact with the belt surface. The blades are arranged in an overlapping pattern, so that the entire width of the belt is cleaned. This type of belt cleaner must be inspected frequently and adjusted so that the blades are free to pivot and to contact the belt properly. See Figure 10.6.

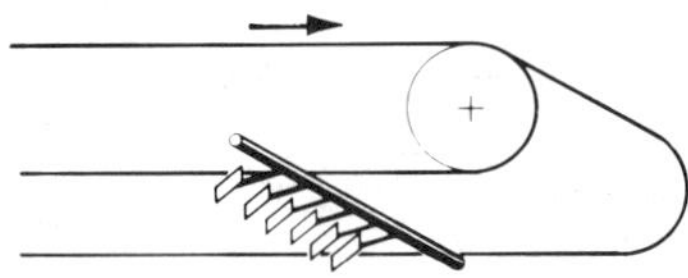

FIGURE 10.6. *Example of articulated blade scraper*

Rotary Belt Cleaners. These consist of power driven shafts or tubes to which brush bristles or blades are attached. These brushes or blades are made wider than the nominal belt width. The brush bristles are made of nylon or bassine. The blades are made of rubber. See Figures 10.7 and 10.8.

The rotary brushes, which have a flicking action, usually have bristles in a helical or parallel pattern. They are of two types:

1. Low-speed rotary brush. This type of rotary brush is designed to operate at peripheral speeds of 400 to 600 fpm, and is most effective on dry granular materials. It gives a longer life because the lower speed causes less wear on the bristles.

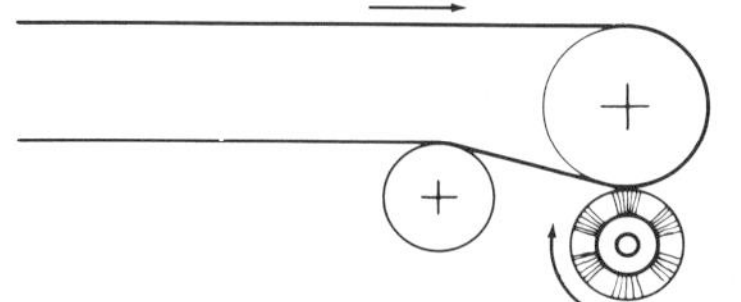

FIGURE 10.7. *Typical rotary belt-cleaning brush*

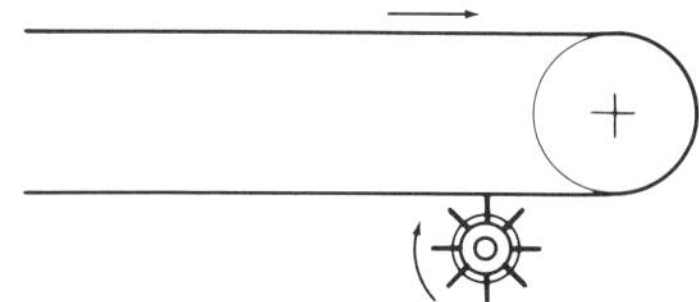

FIGURE 10.8. *Typical example of rotary rubber blade cleaner*

2. High-speed rotary brush. The high-speed rotary brush operates at a peripheral speed of 1,000 to 1,500 fpm. It is most effective on damp granular materials. The high speed produces centrifugal dislodgement of material from the bristles.

Rotary-blade belt cleaners use rubber blades arranged parallel to, or helically on, the shaft. These rubber blades have a squeegee or scraping action. They are also made in two types:

1. Low-speed rotary blade cleaner. The peripheral speed of this type of rotary blade cleaner is 1,000 fpm. It is used on either dry or damp materials. The relatively low speed gives longer life to the rubber blades.
2. High-speed rotary blade cleaner. The peripheral speed is 1,400 fpm. This high-speed rotary blade cleaner is adapted to cleaning the belt of wet, sticky materials which would adhere to and build up on the bristle brushes.

The direction of motion of rotary brush and rotary blade cleaners is such that the periphery of the brushes or blades moves opposite to the direction of the conveyor belt.

Rotary brushes and parallel or helical rotary blade belt cleaners can be driven by chain from the adjacent head pulley shaft, or by a separate drive.

Water spray and wipers. The use of a high-pressure water spray has been effective as a belt cleaner on certain difficult applications. The high-pressure spray is directed against the surface of the conveyor belt by means of nozzles with control valves. A rubber-bladed scraper is installed behind the water spray so that after the water spray washes and cleans the belt, the rubber wiper acts as a squeegee to remove the excess water.

While this type of cleaning is quite effective on certain materials, it has two disadvantages. Provision has to be made to dispose of the wash water, and freezing weather makes the whole scheme inoperable.

Location of Belt Cleaner

The belt cleaner should be located so that the material which is removed from the belt can fall into the discharge chute or can be collected for practical disposal. Single- or multiple-blade scrapers, with spring or counterweighted construction, should be located at points around the contact surface of belt and pulley or immediately after the belt leaves the pulley.

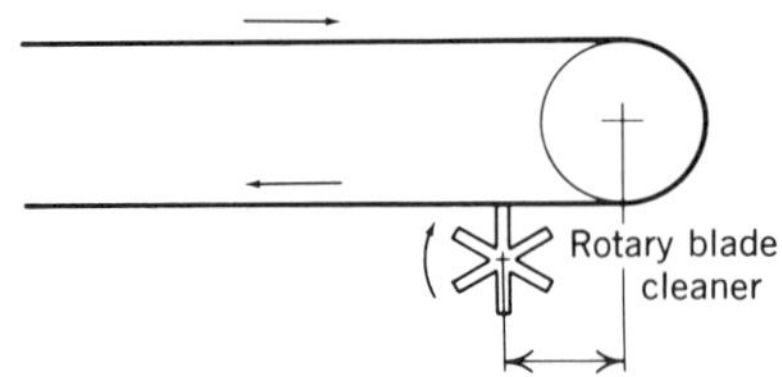

FIGURE 10.9. *Position of rotary blade cleaner on belt.*

Design of the discharge chute often determines the specific location. The articulated-blade cleaner is located on the return run of the belt, just after the belt leaves the pulley. This is shown in Figure 10.6.

The rotary blade cleaner is usually located behind the point where the belt breaks contact with the pulley, as illustrated in Figure 10.9. The rotary brush cleaner should be located as described for the rotary blade cleaner. However, if necessary for reasons of chute design, snub pulley location, etc., the rotary brush can also be positioned to brush the belt clean while the belt is still in contact with the pulley.

Belt Turnover Scheme

To eliminate the problems caused by a dirty belt in contact with return idlers, the belt can be twisted 180° after it passes the discharge point. This brings the clean surface of the belt in contact with the return idler rolls. The method of accomplishing this is illustrated in Figure 10.10. The belt must be turned back again 180° before entering the tail section, to bring the conveying side of the belt up at the loading point.

Because abnormal tensions are induced in the carcass of the belt, a CEMA member should be consulted for the proper location of the snub pulleys which turn the belt. These are not shown in Figure 10.10. The distance required to accomplish the 180° turn of the belt is approximately 12 times the belt width.

FIGURE 10.10. *Belt turnover scheme.*

Pulley Wipers

Snub or bend pulleys which contact the dirty side of the belt must also be cleaned. A common method is to mount a bar scraper on the ascending side of the snub pulley, about 45° below the horizontal center line, so that the material scraped from the pulley will fall free. This is shown in Figure 10.11.

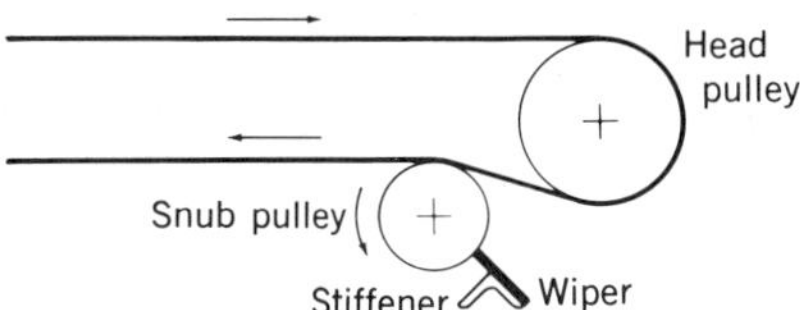

FIGURE 10.11. *Snub pulley wiper.*

The scraper bar should have a hardened bevel scraping edge. Also, it should be adjustably fixed in position on a stiffener so that the beveled edge of the scraper blade just clears the pulley face and works like a chisel to shave off any material that clings to the pulley.

Return–Run Belt Cleaning

Even though a belt conveyor is carefully designed, spillage from the belt may occur at the loading point and elsewhere along the belt. If such spillage occurs, it might land on the return run of the belt and become trapped between the belt and tail pulley, causing possible damage to, or misalignment of, the belt.

Two devices used to prevent this difficulty are deck plates and return–belt scrapers.

Deck Plates. Made up of flat, bent, or curved plates or metal sheets, deck plates or decking can be placed between the carrying and the return run of the belt. This construction, which is conventional on all well-designed belt conveyors, provides excellent protection for the return run of the belt, both from any spilled material and from the weather.

The three styles of decking shown in Figure 10.12 are used more or less according to the designer's preference and the nature of the supporting structure.

Sometimes decking is applied only at the loading point and forward of it for about the next 50 feet. If this design is used, then a return–belt scraper, as described below, should be provided.

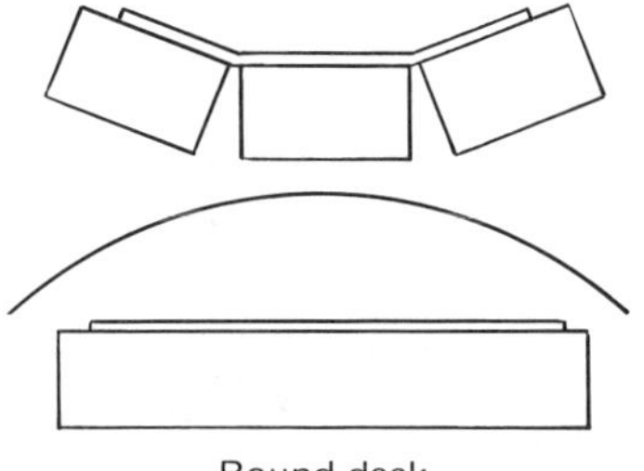

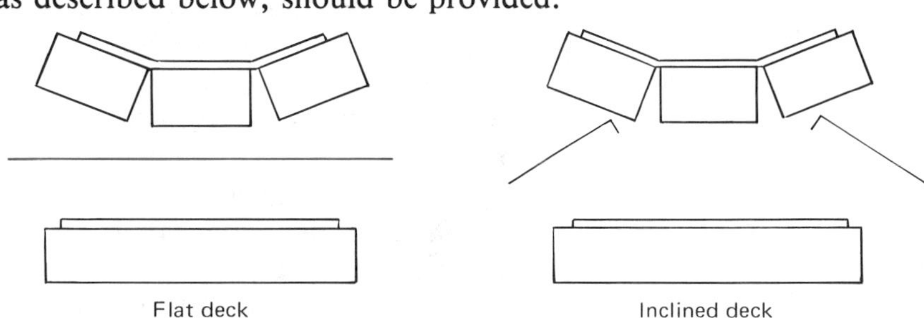

FIGURE 10.12. *Three typical styles of belt conveyor decking.*

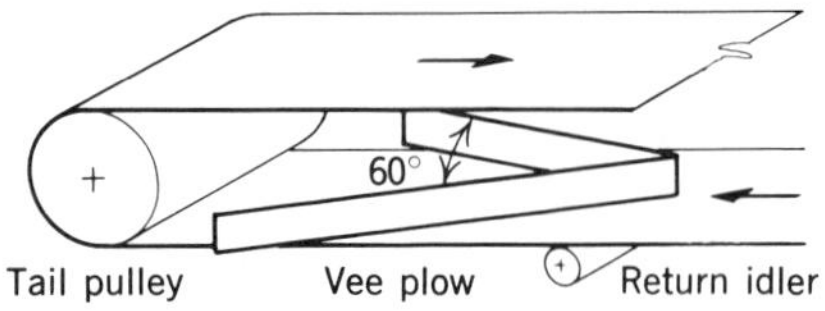

FIGURE 10.13. *Typical V-type return-belt scraper.*

Return-Belt Scraper. Located on the upper side of the return belt, just forward of the tail pulley, the return-belt scraper will scrape from the belt any material that may have spilled on the return run and will thus prevent such spilled material from becoming trapped between the belt and the tail pulley. This scraper should be a V-type with a 60° included angle. The vertex should point toward the head end of the conveyor. The scraper should be located between the tail pulley and the first return idler from the tail pulley, and it should also ride on and be supported by the conveyor belt. See Figure 10.13.

Belt Conveyor Accessory Equipment

After selecting and designing the major components of a belt conveyor system, it is still necessary to consider a number of secondary items which are discussed here as belt conveyor accessory equipment. This includes wing-type tail pulleys, tramp iron detectors, in-process weighing and sampling devices, and weather and spillage protection devices such as housings, wind breaks, wind hoops, and decking.

Weather Protection

The need to protect belt conveyors from the weather varies with the climate, the material handled, and the type of operation. The material being handled may be affected by rain, or, if the material contains moisture, by freezing temperatures. Rain on the pulley side of the belt or on the drive pulleys may cause slippage between the belt and the pulley. Ice and snow on these same surfaces can completely stop a belt conveyor, if the drive cannot move the belt. Also, the life of rubber covers may be shortened by the effect of intense sunlight.

Strong winds can lift conveyor belts off their idlers and cause severe belt training problems. On conveyors with narrow belts, and on those handling light materials, it is possible for the wind to lift the belt, spill the load, and cause serious damage to the belt. This is particularly true of tripper belts, where the belt leaves the troughing idlers and rises on a rather long incline to the tripper head pulley. Wind may also create a nuisance by blowing fine materials from the conveyor belt.

When the belt conveyor is not contained within an enclosed gallery (which obviously offers the most complete protection for the conveyor parts and maintenance personnel), one or more of the following belt protections can be used.

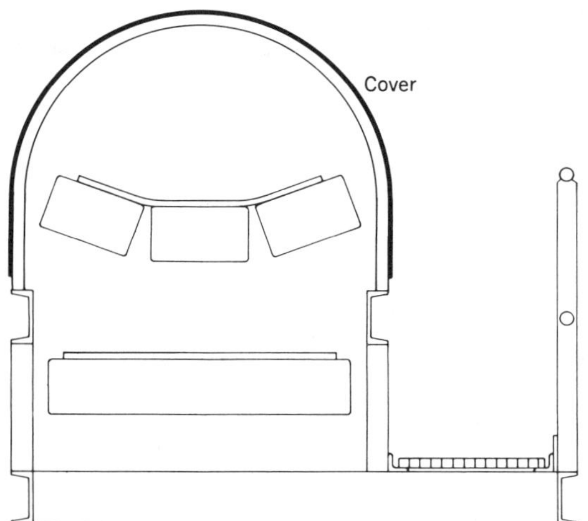

FIGURE 10.14. *Half covers over conveyor belt.*

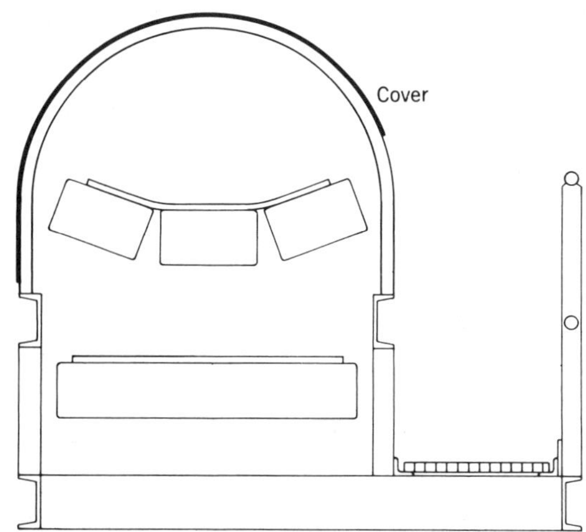

FIGURE 10.15. *Three-quarter covers over conveyor belt.*

Half-Covers. Half-covers usually are semicircular in shape and are made of flat or corrugated sheets, commonly either of galvanized iron or aluminum. They are fastened to the stringers and are situated over the belt, as shown in Figure 10.14. The arrangement and fastening of the covers should permit convenient access to permit servicing the belt and idlers. The belt conveyor covers frequently are hinged at one side.

Three-quarter covers. As the name implies, these covers are less than a complete semicircle. On the walkway or access side of the belt conveyor, to facilitate inspection, the lower edge of the cover may terminate slightly above the high edge of the troughed belt. Figure 10.15 shows three-quarter covers over a conveyor belt.

Wind Breaks. Where protection is needed from wind alone, it can be provided by installing a suitable reinforced metal sheet on the windward side of the conveyor stringers. This sheet should extend above and below the stringers to protect both carrying and return runs of the belt. See Figure 10.16.

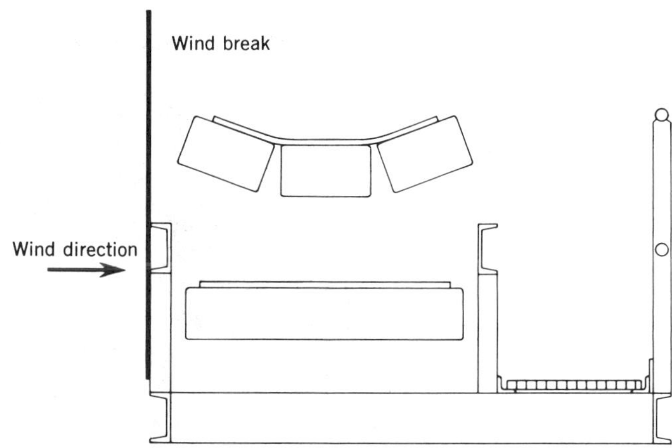

FIGURE 10.16. *Wind break on belt conveyor.*

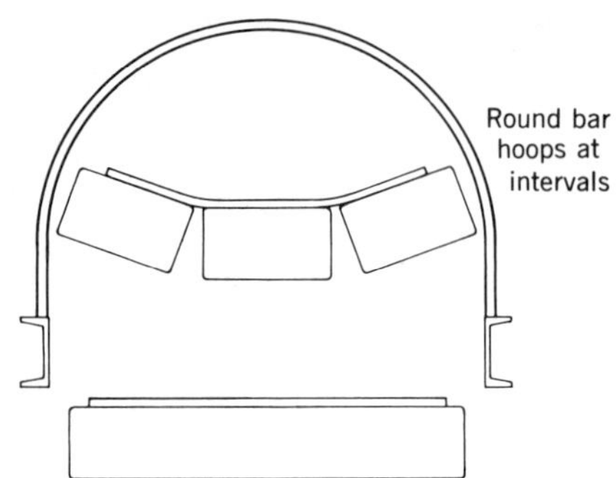

Figure 10.17. *Example of wind hoops on a belt conveyor.*

Wind hoops. In some instances, it is necessary only to prevent the empty or lightly loaded belt from being blown off the idlers. This can be accomplished by providing suitable hoops, spaced at regular intervals along the belt conveyor, as shown in Figure 10.17.

In some operations, the conveyor may be shut down for long intervals of time. If the belt is not otherwise protected it can be lashed to the stringers at frequent intervals by light rope ties.

Spillage Protection

Wing-Type Pulleys. A wing-type pulley is a cast or fabricated pulley with a discontinuous surface (see Figures 8.7 and 8.8). Between the cross bars are inclined valley-shaped recesses that prevent fine or granular material from being caught between the tail pulley and the return belt. The material gathers in the valley-shaped recesses and falls out of the open ends as the pulley revolves.

If a conveyor handling such fine or granular material is likely to spill some of it on the return belt, the wing pulley is an effective device for removing spillage without belt damage. Wing-type pulleys also have been used as vertical gravity takeup pulleys, with the same benefits.

Tramp Iron Detectors

In Chapter 8, the use of magnetic pulleys for removing tramp iron is indicated. There are other methods of removing or detecting tramp iron. One is to suspend a stationary magnet over the belt, alone or in combination with a magnetic pulley.

Another method is to arrange a short conveyor above and at right angles to the belt conveyor. Here, a stationary magnet, between the runs of the cross belt, extends beyond one side of the conveyor below. Material that is picked off the belt conveyor is held against the underside of the cross belt by the magnet. As soon as the belt is removed from the magnetic field, the material drops into a container. This device, although employing a stationary magnet, is self-cleaning.

A third approach, sometimes preferred, is to pass the belt through a magnetic field of adjustable intensity. The device is electrically arranged so that a disturbance of this magnetic field, by passage of a preselected size of iron particle, will shut down the conveyor belt.

Conveyor Belt Scales

For in-plant process functions, the continuous weighing of a wide variety of bulk materials on a moving conveyor belt is accomplished through the use of belt scales. Three types of scales are manufactured specifically for this purpose: the all-mechanical system, the electronic scale, and the nuclear scale.

The first two types operate by weighing the actual material as it passes over a weigh bridge, measuring the belt speed, and integrating these measurements into a rate output. Figure 10.18 illustrates a typical electronic load cell type of belt scale.

The nuclear scale measures the relative density of the material on the moving belt and compares this with the known density of the material. A conversion of density to weight is computed and integrated with belt speed to achieve a rate output.

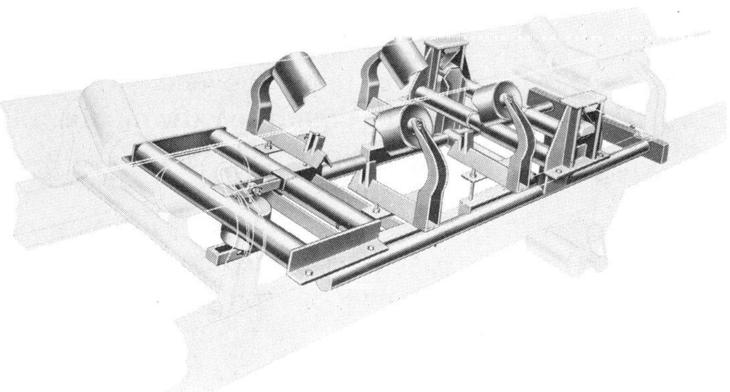

FIGURE 10.18. *Typical electronic load cell type of belt scale.*

The various manufacturers of belt scales have designed their equipment for compatability with standard belt conveyor design practices, so that scales are easily adaptable to both existing and newly designed conveyor systems. Scale accuracy is generally considered to be within ½ of 1% if properly installed in accordance with the manufacturer's recommendations.

Sampling Devices

Sampling is a process of obtaining a small portion of a material which is representative of the whole. Samples are taken both to determine acceptable quality and to control operations and inventory. The most accurate method of sampling is to stop the loaded belt, insert a template to fit the curve of the

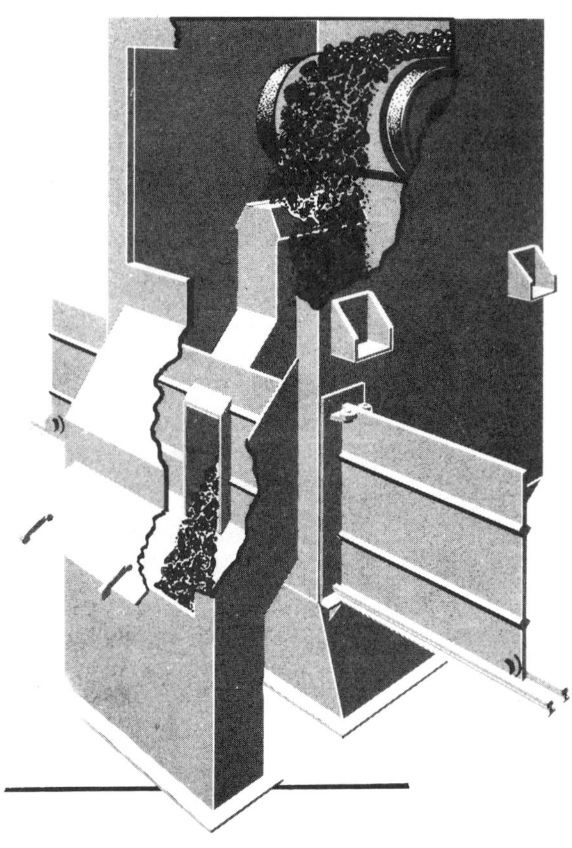

Figure 10.19. *Typical sampling arrangement.*

belt, and shovel or sweep off a precalculated quantity of material. This must be done a specific number of times, depending upon the size of the lot being sampled. This large sample is then reduced and sent to the laboratory for analysis.

However, for most operations, this method would prove impractical if performed continually. An acceptable alternative is to cut the stream of material completely at a right angle to the flow of the stream at a suitable discharge point in the conveying system. Figure 10.19 illustrates a typical sampling arrangement. A cutter device traverses the stream at a right angle and comes to park completely out of the stream. The cutter is driven across the stream by means of a hydraulic cylinder. Electric motor and pneumatic drives are also sometimes used.

Chapter 11

Conveyor Loading and Discharge

Contents

The successful operation of a belt conveyor requires: first, that the conveyor belt be loaded properly; second, that the material carried by the belt be discharged properly. These two requirements are very important and must be given most careful consideration if the belt conveyor is to function as intended.

Loading the Belt

While the loading of material onto a belt conveyor involves many considerations, of prime importance is the placing of the material centrally on the belt in such a manner than the material velocity in the direction of belt travel is, as nearly as possible, equal to the velocity of the belt itself.

If the material isn't delivered onto the belt at the belt speed, there will be a turbulence in the mass of material at the loading point. A build-up in volume may form at this point. This material turbulence is a function of the velocity difference.

Provided that the material is delivered centrally onto the belt, the ideal loading is attained when the forward velocity in the direction of the belt is exactly the same as the velocity of the belt. When this condition exists, there is minimum wear on the belt cover, the minimum power is required to operate the belt, the material takes the proper load shape quietly without spillage, and the minimum degradation or dusting of the material is assured. While this ideal condition is difficult to reach in actual practice, it is well worth the effort to attain it as closely as possible.

Other factors of loading, listed below, also must be considered.

Direction of Loading

There are only two possible directions—in the direction of belt travel, or transverse to the direction of belt travel.

Loading in the Direction of Belt Travel. This type of loading really is best, as it presents the simplest and easiest design problem. The flow of material down a loading chute onto the moving belt surface can be arranged so that the forward velocity of the material is nearly the same as the belt velocity.

The material flow can be directed centrally onto the belt so that the shape of the load is symmetrical. The skirtboards at the loading point can be of minimum length. Proper loading of the belt is not appreciably affected by fluctuations in the feed rate. The loss of height at the transfer from one conveyor to another is at a minimum with this direction of loading.

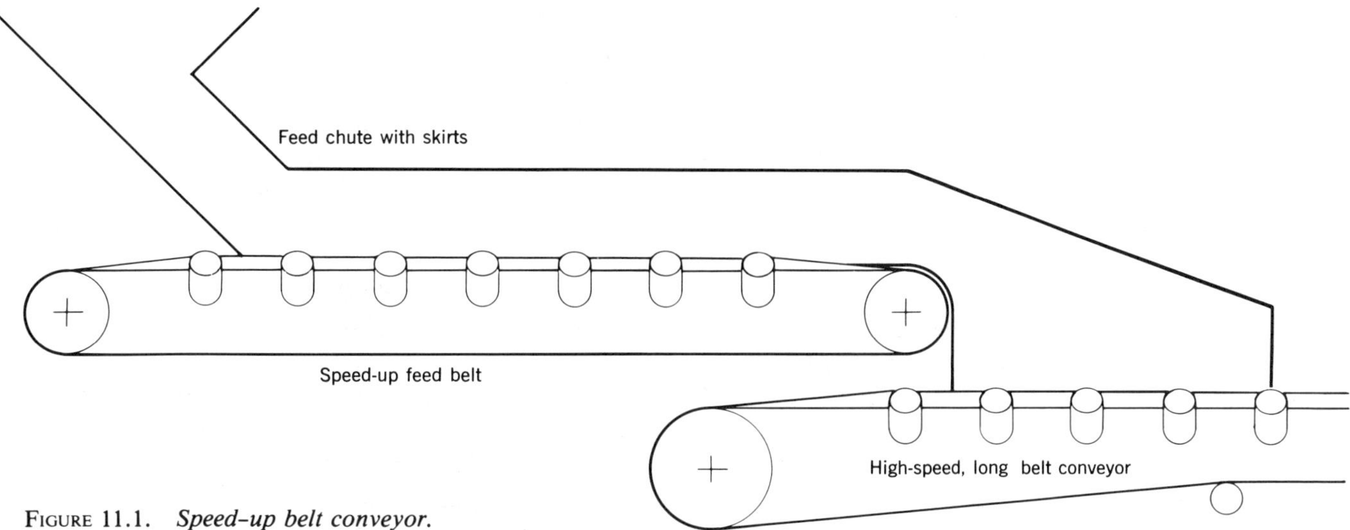

FIGURE 11.1. *Speed-up belt conveyor.*

Unfortunately, loading of belt conveyors in the direction of belt travel is not one of the usual layout configurations.

A short, speed-up belt conveyor can be used to advantage between the loading chute and the loading point of a long, high-speed belt conveyor, as shown in Figure 11.1. Use of the short, high-speed conveyor avoids the wear that otherwise would occur to the cover of the long and costly high-speed belt because of the acceleration of the loaded material.

The length of this speed-up conveyor should be sufficient to bring the speed of the material very close to the speed of the long belt conveyor. The cover of the speed-up belt should be thick enough to accept the wear caused by the acceleration of the material leaving the loading chute.

If the impact of the material leaving the loading chute is high, the speed-up conveyor should be flat. It should run on flat, impact-absorbing idlers. Continuous skirtboards should be provided. This flat belt can be made of the required number of plies and thickness of cover to meet the impact, without consideration of belt-troughing capabilities.

If the impact condition is not severe, a lighter-bodied belt can be used, with a cover thick enough to take the wear. This lighter belt can be carried on troughing idlers, as illustrated in Figure 11.1.

Flat or troughed, speed-up conveyor belts must be considered expendable and therefore must be periodically replaced.

Loading Transverse to the Direction of Belt Travel. This type of loading is found very frequently in belt conveyor layouts. The horizontal angle that a belt conveyor makes with the succeeding belt may be less or more than 90°.

Any angularity at the transfer presents the problem of turning the flow of material, so that finally, as the material passes to the succeeding belt conveyor, its velocity in the direction of the succeeding belt travel is as close to the speed of this belt as possible. Obviously, as the angularity of transfer increases from a small angle to 90° or more, the design of the loading chute becomes a more difficult problem, from the standpoint of both desirable material velocity and central loading of the receiving belt.

With angular transfers up to 90°, the height required at the transfer point increases. Also, it is increasingly difficult to load the belt centrally at all rates of material flow. With large angularity, in spite of careful loading chute design, field adjustments sometimes are required to center variable loads on the receiving belt. Skirtboards at the loading point may have to be higher and longer, in order to retain the material on the belt without spillage until the material is accelerated to the belt speed and assumes the proper load shape. There is more wear on the belt cover, and there is likely to be increased wear on the loading chute itself.

With careful consideration of loading chute design, a compromise of all the factors can be effected which will nearly balance the disadvantages of this direction of loading.

Horizontal angularities of transfer greater than 90° should be avoided. While it is possible to effect a transfer of material from one belt conveyor to another where the angularity exceeds 90°, the design of the loading chute becomes very difficult. Considerable height must be lost at the transfer point. And, it seldom is possible to obtain an acceptable material velocity in the direction of belt travel. The hazards of belt cover wear and loading chute wear are thus accentuated.

Transverse Belt Displacement

If the material being loaded on a belt conveyor has a general direction of flow at the point of engagement with the belt surface, other than directly in the line of belt travel, the belt is likely to be displaced transversely on its supporting idlers.

Likewise, if the material is not uniformly spread on the belt and piles up against one of the skirtboards, the belt can be displaced transversely on the idlers. Such belt displacement causes belt-training difficulties and may result in spillage of the material over the edge of the belt beyond the skirtboards.

Loading an Inclined Belt Conveyor

If the belt conveyor is inclined (slanting upward) at the loading point, it is difficult to attain the minimum difference between the velocity of the material and the velocity of the belt. The greater the incline and belt speed, the greater the difficulty. Thus, the material loaded on an inclined portion of a belt conveyor is subject to much turbulence before it can be accelerated to the speed of the belt. Consequently, the loading skirts may have to be higher and longer to retain the material and avoid spillage. The wear on the belt cover is great if the material is at all abrasive or has sharp-edged particles or lumps.

Conveyors declined (slanting downward) also present loading problems. Careful consideration of the material and belt velocities is essential to satisfactory loading.

Where possible, it is desirable to make the loading portion of the belt conveyor horizontal. This horizontal loading portion is then connected to the succeeding inclined or declined portion by means of a suitable vertical curve.

Impact at Loading Point

Because—in a vertical plane—the direction of flow of the material being loaded on a belt conveyor is almost never exactly in the direction of belt travel, some impact is encountered as the material engages the surface of the belt. Heavy impact tends to damage the belt cover and weaken the belt carcass.

Very fine materials, even if heavy, do not cause much impact. But the belt can be deflected between the idlers unless the idlers are closely spaced under the loading point. Such belt deflection can cause leakage under the skirtboards with resultant spillage of material over the belt edges, at or beyond the loading point.

Lumpy materials, especially if the lumps are heavy, do cause appreciable impact on the belt. If the lumps are sharp, the belt cover may be nicked or cut. Heavy lumps also tend to crush the belt carcass, thus weakening it.

Impact-absorbing idlers should be used. They should be located and spaced under the loading point of the belt so that the principal portion of the lumpy material flow engages the belt between the supporting idlers rather than over any one of them.

Where the material consists of a mixture of fines and lumps, the loading chute can be arranged, first, to deposit the fines on the belt and, second, to deposit the lumps. By so doing, the impact of the lumps is cushioned by the layer of fines. For an illustration of such a loading chute, refer to Figure 11.3, as well as Figure 11.2.

Impact problems, especially with lumpy materials, are accentuated when the loading portion of the belt conveyor is inclined.

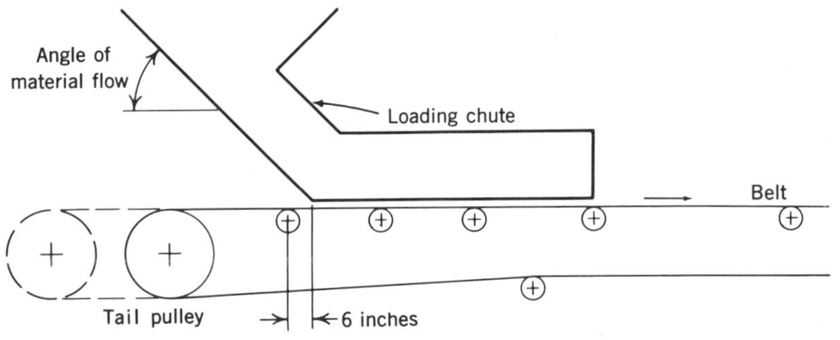

FIGURE 11.2. *Typical loading chute.*

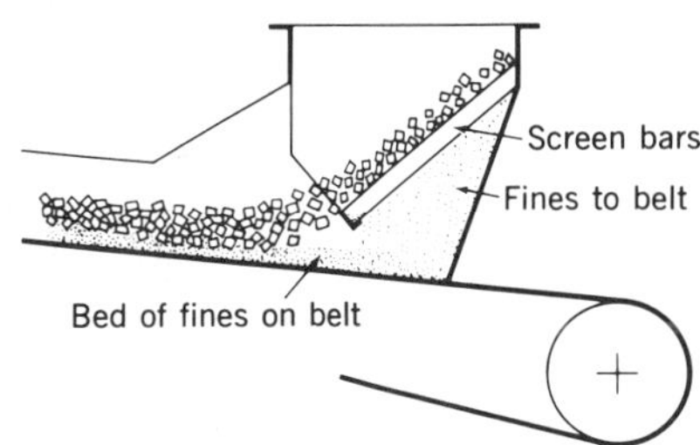

FIGURE 11.3. *Grizzly or screened loading chute (after Hetzel).*

Loading Chutes and Skirtboards

Loading chutes and skirtboards can seriously damage conveyor belts if not properly designed and installed. Therefore, their design should be given careful consideration. Chutes and skirtboards should be fastened securely, to position them accurately with respect to the belt. Make the skirtboards no longer and the bottom edge no closer to the belt than necessary. Where belt conveyors with trippers are arranged with inclined loading sections, the skirtboards should extend to the bend pulley or to the first of the group of idlers at the convex vertical curve. This is done to maintain the shape of the material load on the belt, right up to the beginning of the curve.

Skirtboards should extend to a point where the material has stabilized at the belt speed. Greater distances are required to stabilize the material when loading on an incline. Therefore, longer skirtboards are required for the loading portions of inclined belt conveyors.

Loading Chutes

The design of loading chutes is worthy of careful study. Where the loading of the belt is to be accomplished in the direction of the belt travel, loading chutes can be quite simple. Where the loading of the belt must be accomplished in an angular direction, the design becomes more complex. The design of successful transverse loading chutes and skirtboards for high-speed belt conveyors requires much care and ingenuity.

Obviously, the loading chute must be inclined in order to give the material flow a desirable forward velocity. If the material is fine and contains some moisture, the chute must be made steep enough so that the material will slide rapidly. However, if the material is lumpy, the steepness of the chute is limited to that angle at which the material will slide satisfactorily, but not bounce and tumble. High lump velocities may be controlled by the use of baffle bars or chains hung in the path of the lumps.

Multiple angle chutes, curved chutes, and sometimes covered chutes can be employed to impart a uniform sliding action to the material. If sufficient velocity cannot be imparted to the material—and in the proper direction—it may be necessary to reduce the speed of the receiving belt conveyor. This is done to obtain the minimum difference between the forward velocity of the material flow and the belt velocity. However, such a compromise may result in a wider, more costly belt.

Loading chutes can be made of metal or other materials. Metal chutes are the most common. For abrasive materials, the chute can be lined with abrasion-resisting, removable plates or other material, such as ceramic liners. For corrosive materials, corrosion-resistant metal coatings, rubbers, synthetics, or fused-on glass linings can be used.

Width of Loading Chutes. The width of a loading chute should be no greater than two-thirds the width of the receiving belt. On the other hand, the inside width of the loading chute should be at least two and a half to three times the largest dimension of uniformly sized lumps, when they represent a considerable percentage of the material flow. Where lumps and fines are mixed, the inside width of the chute may be made two times the maximum lump size.

These proportions are essential to the proper loading of the belt and to the prevention of interlocking and jamming of lumps in the chute. Thus, the width of the loading chute could, in some cases, determine the width of the belt on the receiving conveyor.

Grizzly or Screened Loading Chutes. Where a mixture of lumps and fines is to be handled, heavy loading impact on the belt can be minimized in one of two ways: (1) by arranging the chute to place a layer of fines on the belts ahead of the loading of the lumps, or (2) by using curved or perforated chute bottoms, or grizzly chutes. The latter sometimes are called screened chutes. Figure 11.3 shows a grizzly or screened chute for loading a slightly inclined belt conveyor.

When loading a declined belt conveyor, it is best to load the fines ahead of the lumps for another reason. If the fines make a layer on the belt in which the lumps subsequently may be imbedded, the lumps will be prevented from bouncing and cascading down the declined, moving conveyor belt.

Details of a Loading Chute. The back or bottom plates of chutes used for loading should be fitted rather closely to the belt and should be provided with adjustable rubber edging to prevent leakage of fines. Such rubber edging also prevents lumps from getting under or behind the back plate and jamming between the back plate and belt.

Stone Box Loading. If the material to be handled is severely abrasive and the speed of the receiving belt is slow, it is possible to arrange the chute bottom so that it acts as a box in which some of the material is retained. The flow of the abrasive material across this box portion is on the retained material. Thus, wear on the chute bottom is avoided. This arrangement commonly is called a ''stone box'' and is used in gravel, rock, and ore handling. Figure 11.4, below, shows a typical stone box loading chute.

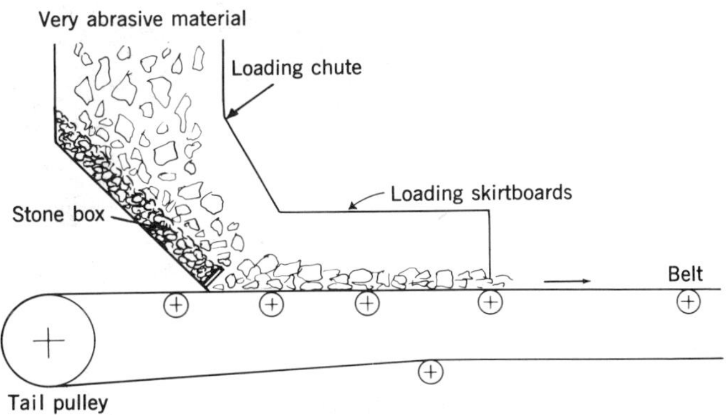

FIGURE 11.4. *Stone box loading chute.*

Skirtboards

To retain the material on the belt—after it leaves the loading chute and until it reaches belt speed—skirtboards are necessary. These skirtboards usually are an extension of the sides of the loading chute and extend parallel to one another for some distance along the conveyor belt. The skirtboards normally are made of metal, although wood sometimes is used. The lower edges of the skirtboards are positioned some distance above the belt. The gap between the skirtboard bottom edge and the belt surface is sealed by a rectangular rubber strip, attached or clamped to the exterior of the skirtboard. See Figure 11.5. Figures 11.6 and 11.7 show typical skirtboard arrangements for flat and troughed belts.

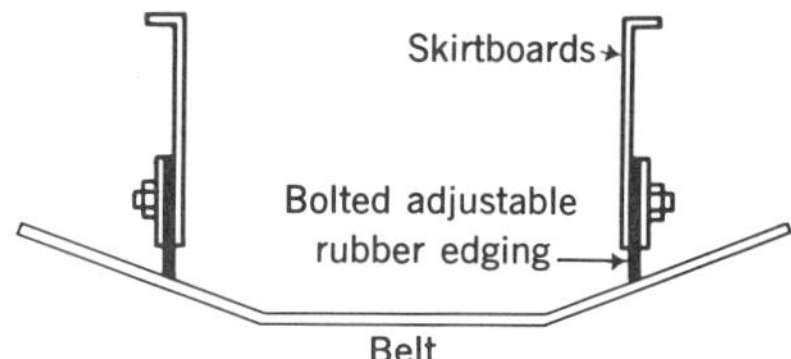

FIGURE 11.5. *Typical application of skirtboard on troughed belt.*

If the material conveyed contains hard lumps, particularly lumps with sharp edges, the gap between the bottom of the skirtboards and the belt should be made to increase uniformly in the direction of belt travel. Any lump forced under the skirtboard edge will quickly free itself as the belt moves forward, and thus not abrade the belt.

When handling a mixture of large lumps and fines or only sized lumps, the skirtboards sometimes are not made parallel to each other but are

FIGURE 11.6. *Continuous skirtboards on a flat belt. Here, frequent skirtboard adjusting is required so that the edging just touches the belt surface. Improper pressure of the rubber skirtboard edging may overload the belt conveyor driving motor.*

FIGURE 11.7. *Skirtboards on a troughed belt. Usually, when the loading is in the direction of the troughed belt travel (above), the skirtboard length is a function of the difference between the loading material velocity and the belt velocity.*

splayed out in the direction of the belt travel. Such an arrangement prevents lumps from jamming between the skirtboards. Frequently, the back plate of such a set of skirtboards is made in a curve instead of straight across the belt. Splayed skirtboards should be kept as short as possible because the rubber edging is difficult to fit to the contour of the troughed belt, both initially and during subsequent maintenance.

Commonly used proportions and details of skirtboards and rubber strip edgings are as follows:

Spacing of Skirtboards. The maximum distance between skirtboards customarily is two-thirds the width of a troughed belt, (0.666*b*). However, it is desirable, when possible, to reduce this spacing to one-half the width of the troughed belt (0.500*b*), especially for free-flowing materials, such as grain.

On flat belts, depending on how well the belt is trained centrally, how well it is supported by idlers or a loading plate beneath the belt, and how effectively the rubber edging seal is maintained, the space between the skirtboards may be only a few inches less than the belt width. Such spacing commonly is used when handling damp or prepared molding sand, or similar materials which do not tend to slump much upon leaving the end of the loading area.

Length of Skirtboards. Usually, when the loading is in the direction of the troughed belt travel the skirtboard length is a function of the difference between the loading material velocity—at the moment the material reaches the belt—and the belt velocity. For the installation where this difference is small, the length of the skirtboards can safely be 2 ft for each 100 fpm of belt speed, but not less than 3 ft. Skirtboards preferably should terminate above an idler rather than between idlers.

Table 11-1 *Minimum Uncovered Skirtboard Height*

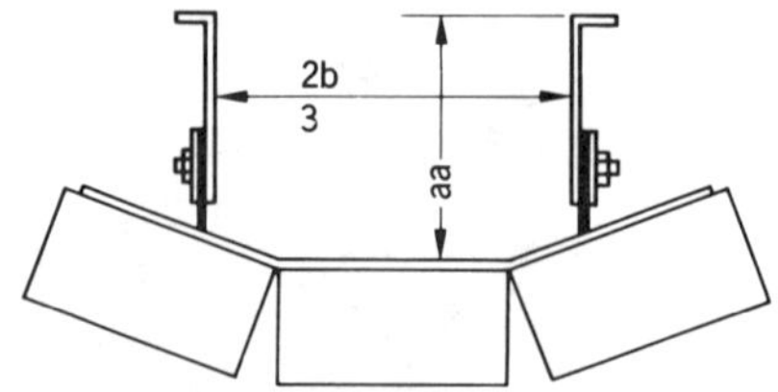

20° Three-Equal-Roll Troughing Idlers

Belt width b	aa, Skirtboard height, inches — Lump size, inches: 2*	4	6	8	10	12	14	16	18
18	5.0	5.0	5.0						
24	5.5	5.5	6.0	6.6					
30	5.8	6.3	7.0	7.6	8.3				
36	6.7	7.3	8.0	8.6	9.3	10.0			
42	7.7	8.3	9.0	9.6	10.3	11.0	11.6		
48	8.7	9.3	10.0	10.6	11.3	12.0	12.6	13.3	
54	9.7	10.3	11.0	11.6	12.3	13.0	13.6	14.3	15.2
60	10.7	11.3	12.0	12.6	13.3	14.0	14.6	15.3	16.2
72	12.7	13.3	14.0	14.6	15.3	16.0	16.6	17.3	18.2
84	14.7	15.3	16.0	16.6	17.3	18.0	18.6	19.3	20.2
96	16.7	17.3	18.0	18.6	19.3	20.0	20.6	21.3	22.2

35° & 45° Three-Equal-Roll Troughing Idlers

Belt width b	aa, Skirtboard height, inches — Lump size, inches: 2*	4	6	8	10	12	14	16	18
18	7.0	7.0	7.0						
24	7.5	7.5	8.0	8.6					
30	8.8	8.8	9.5	10.1	10.8				
36	9.7	10.3	11.0	11.6	12.3	13.0			
42	11.2	11.8	12.5	13.1	13.8	14.5	15.1		
48	12.7	13.3	14.0	14.6	15.3	16.0	16.6	17.3	
54	14.7	14.8	15.5	16.1	16.8	17.5	18.1	18.8	19.7
60	15.7	16.3	17.0	17.6	18.3	19.0	19.6	20.3	21.2
72	18.7	19.3	20.0	20.6	21.3	22.0	22.6	23.8	25.0
84	21.7	22.3	23.0	23.6	24.3	25.0	25.6	27.2	28.7
96	24.7	25.3	26.0	26.6	27.3	28.0	28.6	30.6	32.5

*For material that is all fines, use skirtboard heights in 2-inch lump column

Height of Skirtboards. The height of skirtboards must be sufficient to contain the material volume as it is loaded on the belt. Table 11-1 lists accepted, reasonable skirtboard height for 20°, 35°, and 45° three-equal-roll troughing idlers.

Skirtboard Clearance over Belt. The metal (or wood) portion of the skirtboards should not come closer than 1 inch to the belt surface. This clearance preferably should increase uniformly in the direction of belt travel. As previously explained, this steadily increasing gap permits lumps or foreign objects to move forward without becoming jammed between the lower edge of the skirtboards and the belt.

Greater clearance can be used, but this necessitates wider and thicker rubber-edging strips, particularly for long skirtboards.

Skirtboard Rubber Edging. To prevent leakage of fines through the clearance between the lower edge of the skirtboards and the moving belt, it is common practice to edge the skirtboard exterior with long, flat strips of ¼-to ½-inch thick solid rubber. These strips can be bolted or clamped to the skirtboards in such a manner as to permit the rubber strip to be adjusted to rest lightly on the belt surface, both initially and after wear has taken place. See Figure 11.5.

Such edging should be of solid rubber of at least 60 to 100 durometer hardness. It should contain no fabric to pick up and retain abrasive particles, thus avoiding the abrasion of the belt cover. Strips of old rubber belting should *never* be used for edging skirtboards. The width of the rubber-edging strips will depend upon the manner in which they are attached to the skirtboard and upon the wear allowance.

Rubber edging can be installed vertically or at an angle. Edging installed at an angle provides a better seal between idlers as the belt flexes under load. However, care must be exercised in design to combine good sealing with minimum belt cover wear.

Where the characteristics of the material permit—such as uniform lump size greater than 1 inch and no fines—the rubber skirtboard edging can be safely omitted, but only if the skirtboards are not too close to the edge of the belt. Omission of rubber skirtboard edging does eliminate some wear and grooving of the belt cover.

When splayed skirtboards are used, the rubber edging bears against a wider portion of the belt cover, thus reducing the tendency to form grooves in the belt cover.

Rubber skirtboard edging should be adjusted frequently so that the edging just touches the belt surface. Forcing the edging hard against the belt cover will not only groove the belt cover but also require additional power to move the belt. On conveyors with continuous skirtboards, improper pressure of rubber skirtboard edging may overload the belt conveyor driving motor.

Skirtboard Covers. Suitably high skirtboards can be covered to minimize dusting. The top edges of the skirtboards can be flanged, externally, and the cover fastened to these flanges.

If skirtboard covers are used, especially on skirtboards for inclined belt conveyors, the portion of the cover adjacent to the feed chute should be generously slanted to meet the chute. This is necessary to make room for material which is not yet moving at belt speed and thus avoid a jam in the material flow at the end of the feed chute.

Skirtboards for Intermediate Loading Points. Where a belt is loaded at more than one point in the belt length, care must be used in the arrangement of the skirtboards at the intermediate loading points. Because the load of material tends to flatten and spread on the belt, skirtboards at these intermediate loading points must be designed to let the previously loaded material pass freely. Usually, intermediate-loading-point skirtboards are spaced closer together, with a generous clearance above the previous load surface. Rubber edging is omitted from intermediate loading-point skirtboards as it serves no purpose.

Spillage may occur at intermediate loading points, even with the most careful design of the skirtboards, because of fluctuating initial loading. Dusting at intermediate loading points is almost inevitable.

Often, when intermediate loading points are relatively close together, it is better to continue the skirtboards between the loading points than to hazard the use of relatively short lengths at the intermediate loading points. Continuous skirtboards are good insurance against spillage.

Sometimes the use of a wider-than-normal belt or a more deeply troughed belt will facilitate loading without spillage at intermediate loading points.

Friction Against Skirtboards. The additional belt pull required to overcome the friction of the material against the skirtboards and to overcome the friction of the rubber skirtboard edging is discussed in Chapter 6.

Feeders

The cross-sectional areas and capacities given in Chapter 4, Tables 4-2 through 4-5 are based on a continuous, even flow of material. Some variations in the flow may be permitted, provided these variations do not appreciably affect the average flow.

However, intermittent or irregular feeding of the material to the belt will result in alternate empty and overloaded portions of the belt. Such a condition usually causes a loss of capacity and, very likely, spillage of material over the edges of the belt along the overloaded portions.

Some means of feed regulation must be employed, particularly when a belt conveyor is to be loaded from hoppers, bins, or piles. Material discharged to a belt conveyor from other conveyors requires only a suitable transfer chute. However, the rate of feed must be established somewhere in the conveying system.

Feeders can be of various types, such as screw, belt, drag–scraper, apron, reciprocating plate, vibrating, rotary vane or drum, rotary disc, or table feeders.

Apron, drag–scraper, and vane feeders provide a slightly pulsating feed, unless the spacing of the apron pans, drag–scraper bars, or vanes is small in comparison with the volume of material being fed. However, chutes from these feeders to the belt conveyor usually smooth out the pulsations. Retarding devices, suspended in such chutes, help to reduce the pulsation peaks.

The choice of feeders depends upon the characteristics of the material handled, the manner in which the material is stored, and the tonnage rate of feed. Ten basic feeders which are commonly used are briefly discussed below and illustrated in Figures 11.8 through 11.19.

Screw Feeders

A conveyor screw can be located at the bottom of a storage bin to control and regulate the flow of most materials uniformly and continuously, except

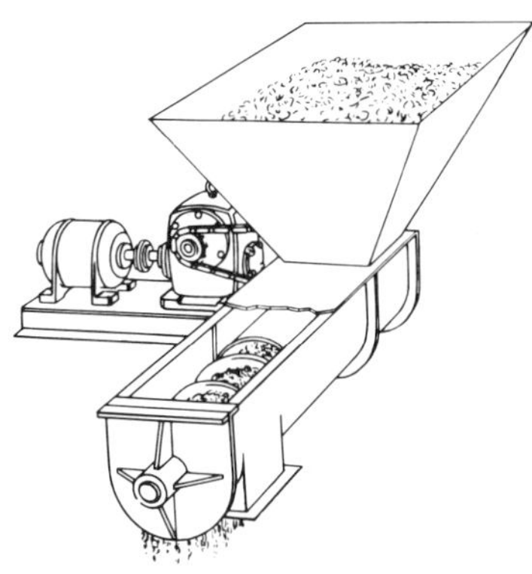

FIGURE 11.8. *Screw feeder.*

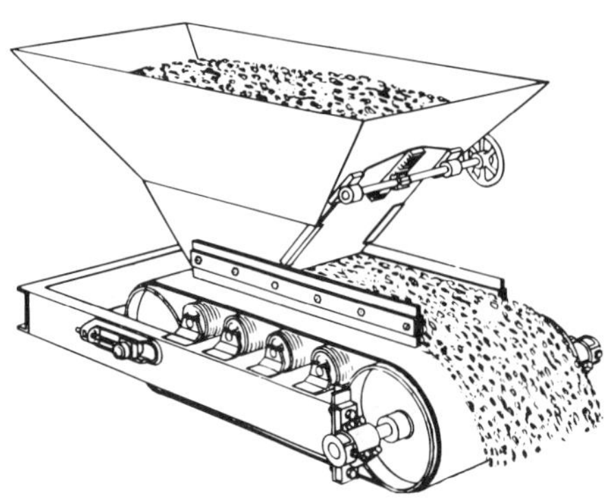

FIGURE 11.9. *Typical belt feeder.*

those materials containing large lumps or highly aerated fines, or those which tend to pack. See Figure 11.8.

Belt Feeders

A belt feeder is a very short belt conveyor, installed under a storage facility. Generally, the belt is flat and is supported on closely spaced idlers or on a smooth slide plate. Belt feeders are used extensively for handling fine, free-flowing, abrasive, and friable materials. See Figure 11.9.

Drag–Scraper Feeders (Bar Drag Feeders)

A drag–scraper feeder consists of a succession of plates or bars mounted between two strands of conveyor chain. The plates or bars drag along the bottom of a trough. A drag–scraper feeder is a simple and compact arrangement for controlling the feed of fine or small-lump material. See Figure 11.10.

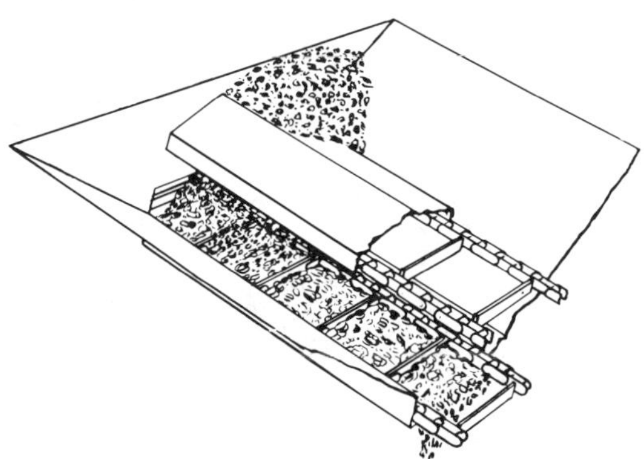

FIGURE 11.10. *Typical drag–scraper feeder.*

Apron Feeders

Apron feeders are used for handling abrasive, heavy, and lumpy materials. The material is carried on overlapping metal plates or "pans," mounted on or between strands of conveyor chain. The chains usually are fitted with rollers which ride on metal tracks. See Figure 11.11.

Reciprocating Plate Feeders

The reciprocating plate feeder is one of the oldest devices for feeding materials from bins or hoppers. This feeder consists of a reciprocally driven plate or pan, operating horizontally or slightly declined, and positioned under a head of material in a bin. The reciprocating element can be double-ended, arranged so that each end is under a separate opening in the same bin. Plate feeders will handle fines, a mixture of lumps and fines, or small-sized lumps. See Figure 11.12.

Vibrating Feeders

A vibrating feeder consists of a pan or trough to which is imparted a vibrating motion so that the material is impelled in a definite, controlled flow. Normally, it is positioned under the opening in the bottom of a bin or a hole under a storage pile. See Figure 11.13.

Vibrating feeders successfully handle a wide range of materials. However, their use should be avoided where the material is of such nature that it will stick to and build up on the surface of the pan or trough.

Rotary–Vane Feeders (Pocket Feeders)

A rotary–vane feeder consists essentially of a shaft-mounted vane or pocket within a snugly fitted, enclosed case. Rotary–vane feeders, also called pocket

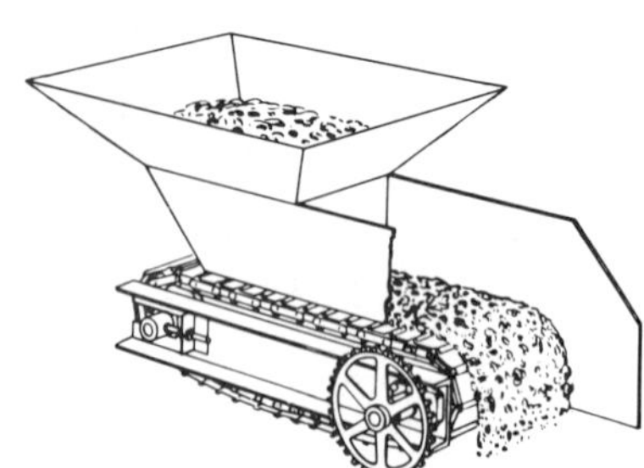

FIGURE 11.11. *Apron feeder.*

FIGURE 11.12. *Single–plate feeder.*

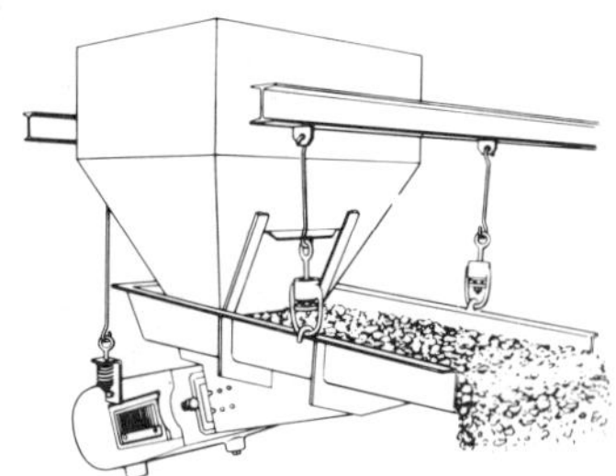

FIGURE 11.13. *Typical electrical vibrating feeder.*

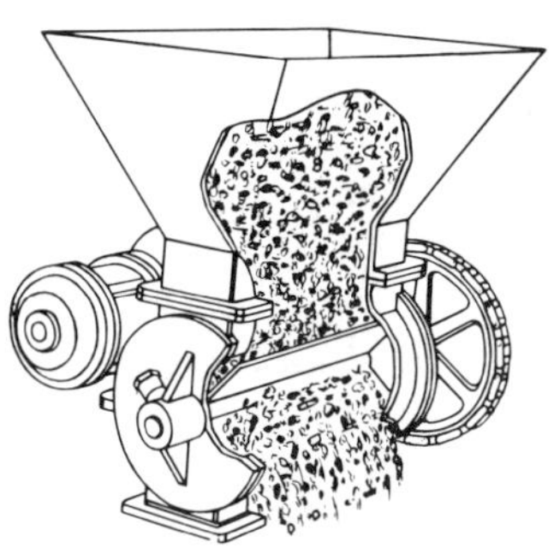

FIGURE 11.14. *Typical vane or pocket feeder.*

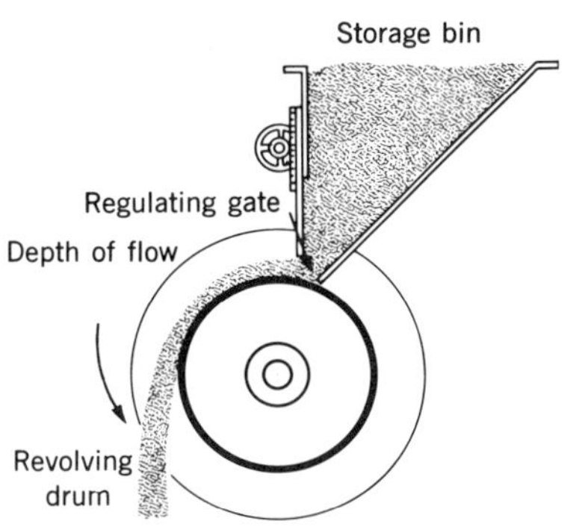

FIGURE 11.15. *Rotary drum feeder.*

FIGURE 11.16. *Rotary-table feeder.*

feeders, provide an intermittent feed of free-flowing fine or small-lump material. See Figure 11.14.

Rotary-Drum Feeders

Rotary-drum feeders provide a fairly accurate control of the rate of feed (in limited applications) of fine or small-lump material. See Figure 11.15. Where the material handled is sticky or otherwise not free-flowing, the use of drum feeders should be avoided.

Rotary-Table Feeders (Disc Feeders)

A rotary-table, or disc, feeder consists of a horizontal circular plate, with or without a central cone, rotating under a circular opening in the bottom of a conical bin or hopper. The material on the plate is plowed off by an adjustable blade and falls over the edge of the table or disc. Rotary-table feeders are used for materials that have a tendency to arch. See Figure 11.16.

Traveling Rotary-Plow Feeders

Traveling rotary-plow feeders are suitable for use in tunnels, under storage piles, or under long storage bins. The plowing mechanism consists of a number of curved arms, operating on a vertical axis, and arranged to sweep material off a narrow shelf running the length of the storage pile or bin. See Figure 11.17.

Feed-Control Gates

If the material is fine and very free-flowing, the judicious use of a simple, adjustable cut-off gate usually will satisfactorily control the flow of such material. These gates are used in reclaiming tunnels beneath storage piles or

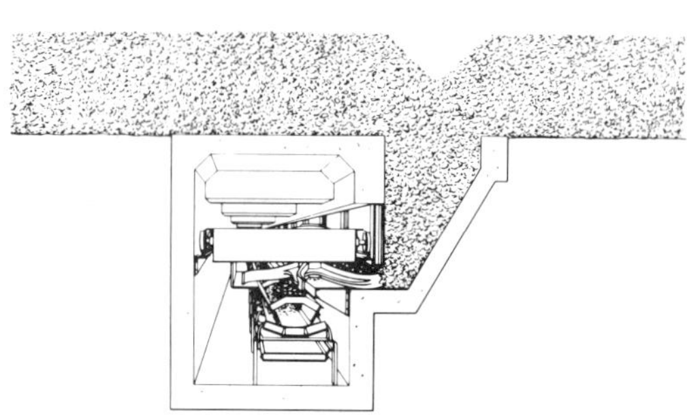

FIGURE 11.17. *Traveling rotary plow.*

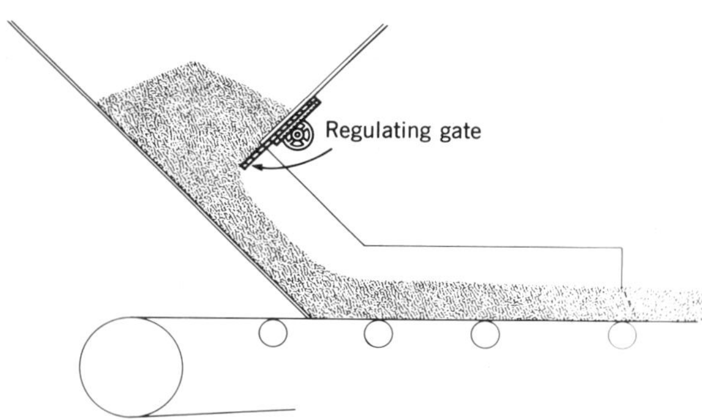

FIGURE 11.18. *Typical regulating gate feeder.*

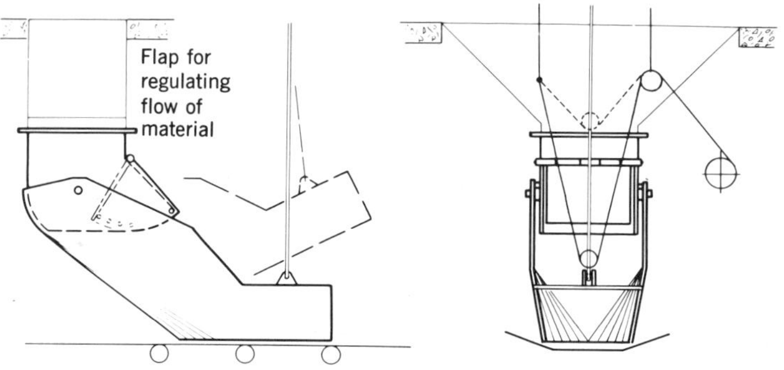

FIGURE 11.19. *Tunnel gate.*

under overhead bins. They are made in several designs, plain or with attached loading chute and skirtboards, and with or without separate cut-off or regulating gates. Several tunnel gates and chutes can be arranged in succession to blend different grades of material on the belt conveyor. See Figures 11.18 and 11.19. Feed-control gates should not be used if the character of the bulk material is variable, as it will be practically impossible to maintain the proper gate adjustment.

Methods of Discharging from the Belt

Materials carried by a belt conveyor can be discharged from the belt in different ways to effect certain desired results. The discharge can be accomplished at a definite point, or points, or it can extend alongside the belt conveyor, on one or both sides, for considerable distances. The flexibility of discharge arrangements of belt conveyors facilitates their use in the maximum fill of long bins and in the creation of large and variously shaped storage piles.

The simplest method of discharge from a conveyor belt is to let the material pass over an end pulley and fall onto a pile. By adding a suitable chute, the discharge can be directed, as desired, to a pile, a bin, or another conveyor. A fork in the discharge chute, with a gate, will permit the material to flow simultaneously in two directions or alternately in either direction.

If several specific points of discharge are required, the conveyor belt can be passed over fixed trippers, which will effect the discharge at these points. A movable tripper can be used to discharge material at many points on either side of the belt, or to one side, continuously or intermittently.

When a flat belt is used, it is possible to position plows at the desired discharge points and direct the material from the belt to either side or simultaneously to both sides of the belt. Such plows can be fixed in position, or can be movable to effect a wider range of discharge points. A troughed belt can be temporarily flattened to permit discharge by a plow.

Whenever the material is discharged over an end pulley, the speed of the belt and the diameter of the end pulley are factors which determine the path of the discharged material. This path is called a trajectory. The shape of discharged material trajectories is important when designing discharge chutes, and when the material falls freely without a chute onto a stockpile.

Care in the design of discharge chutes will pay off handsomely in the sucessful operation of a belt conveyor. Ingenious combinations of discharge chutes and gates will facilitate diversion of the material in desired directions, collection of the material adhering to the belt, avoidance of spilled material, and control of dusting in the case of dry or powdery and fine materials.

Various common discharge arrangements are illustrated in Chapter 2, Figures 2.11 through 2.18.

Discharge Over-the-End Pulleys

This simple discharge is commonly used. The trajectory of the discharged material should be investigated to ascertain where the material will fall. Provision must be made so that the pile will not rise to the end pulley and cause the vulnerable cover of the belt to scrub against the top of the pile.

Dusting of dry discharged material may be a problem. Figure 11.20 shows a simple over-the-end pulley discharge.

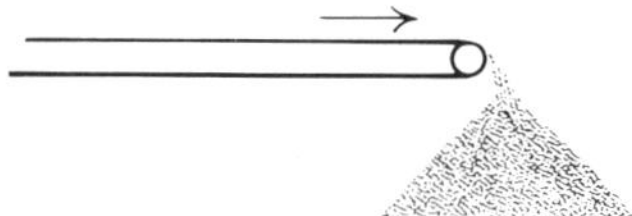

FIGURE 11.20. *Discharge over-the-end pulley.*

Discharge Chutes

The character of discharge chutes is limited only by the ingenuity of the belt conveyor designer. Discharge chutes may be extremely simple or very complex. The prime requirement is that the chute collect all the material discharged and cleaned from the belt. The carrying surface of the belt should be free from any adhering material as it passes back over the return idlers.

Some materials will part freely from the surface of the belt as the belt flexes in passing over the end pulley. Other materials will adhere to the belt. For a discussion of belt-cleaning devices, see Chapter 10.

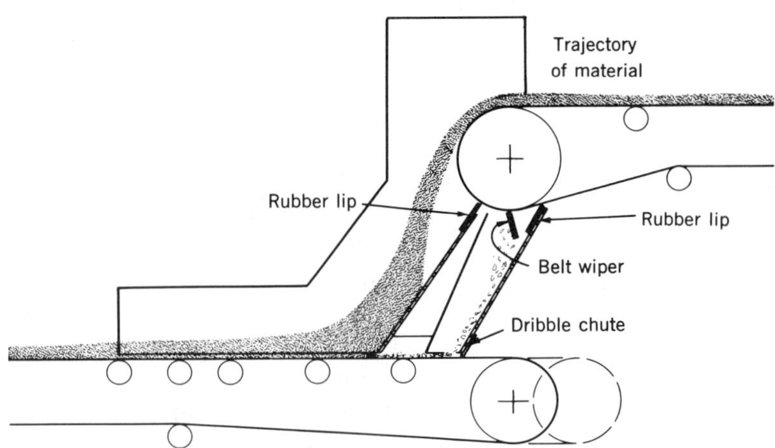

FIGURE 11.21. *Typical simple discharge chute.*

The upper end of the discharge chute must enclose the cleaning device and catch the material cleaned from the belt. Otherwise, a separate dribble chute can be employed.

Gates frequently are provided in the discharge chute to divert the flow of the discharged material.

If the material handled is abrasive, provision must be made in the discharge chute for renewing the wearing surfaces. This is particularly true if the material is heavy and has sharp-edged lumps. Corrosion-resistant materials may be required for certain chemicals, to avoid both damage to the chute and contamination of the chemical.

Some of the most common forms of discharge chutes are illustrated and described below.

A typical simple discharge chute is shown in Figure 11.21. The top of the chute can be provided with a cover, or even an air suction vent, if dusting of the material is a problem. The upper edge of the bottom plate, or the "lip," should be far enough under and back of the pulley to catch all the material, even when the belt is barely moving. To avoid abrading or gouging the belt cover, the arrangement must be such as to eliminate the danger of a hard lump becoming wedged between the lip and the belt on the pulley.

If hard, jagged lumps are contained in the material, it is good practice to provide a strip of rubber on the lip, as shown in Figure 11.21. Should the belt run to one side of the pulley, the sides of the chute should be flared, where the belt enters, to minimize damage to the edges of the belt.

The chute design must be integrated with the trajectory of the material leaving the belt, so that the material will fall on the sloping bottom of the chute and not on a succeeding conveyor belt.

Serious injury to a conveyor belt may result from failure of the material to fall freely at the discharge point. Such failure can be caused by overfilling a bin; by stoppage of the succeeding conveyor or apparatus receiving the discharged material; or by plugging of the discharge chute, as when sticky material refuses to slide on the sloping surfaces. Whenever such a condition is a possibility, it is advisable to provide an automatic protective device, such as a paddle switch. This should be located so that any undesirable accumulation of material will deflect the paddle to shut down the system, or signal a warning to the operator.

A correctly designed chute allows the free, smooth flow of the discharged material. It avoids abrupt changes of direction which invite material build-up and subsequent plugging.

Lowering Chutes

Where dusting and degradation are objectionable, lowering chutes are used for directing materials to storage. Four common forms of lowering chutes are illustrated and described as follows:

Spiral Lowering Chutes. Figure 11.22 shows the usual form of a spiral lowering chute used for gently lowering fragile and/or dusty bulk materials.

Bin Lowering Chutes. Straight, declined, inverted channel chutes run from the belt conveyor discharge point down into the bin, and are secured to the sloping side of the bin near the bin bottom. The slope of the chute to the horizontal must be 10° to 15° greater than the angle of repose of the material. Material will slide down these chutes quietly, without dusting, until it meets with the bin side or the surface of the material in the bin. There, the material will leave the chute and spread out conically. Refer to Figure 11.23.

When properly arranged, these bin lowering chutes will eliminate dusting and fill the bin to its practical capacity.

Rock Ladders. Many specifications, particularly federal specifications, do not permit breaking, crumbling, or degrading rock as it is delivered to a

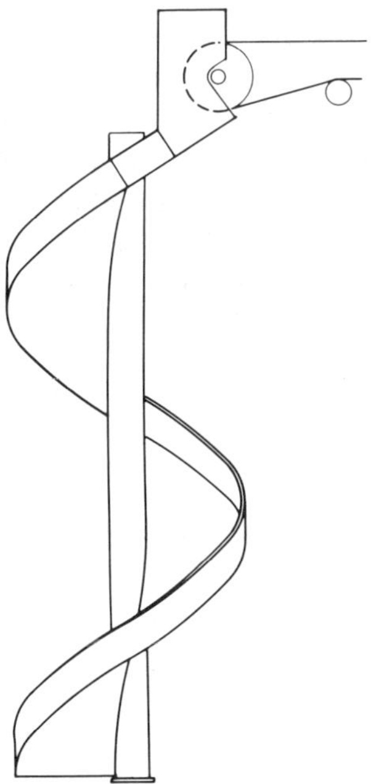

FIGURE 11.22. *Spiral lowering chute.*

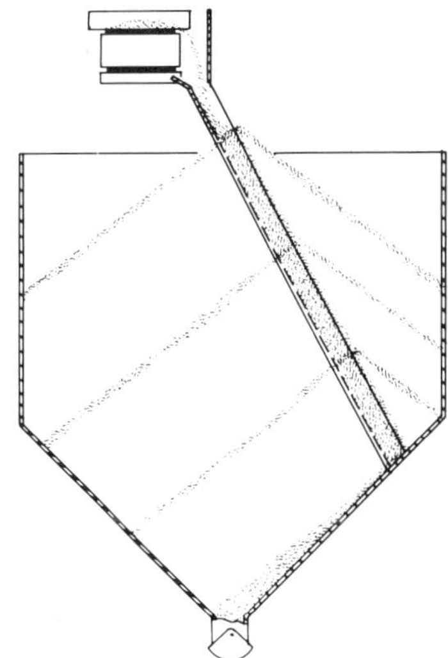

FIGURE 11.23. *Bin lowering chute.*

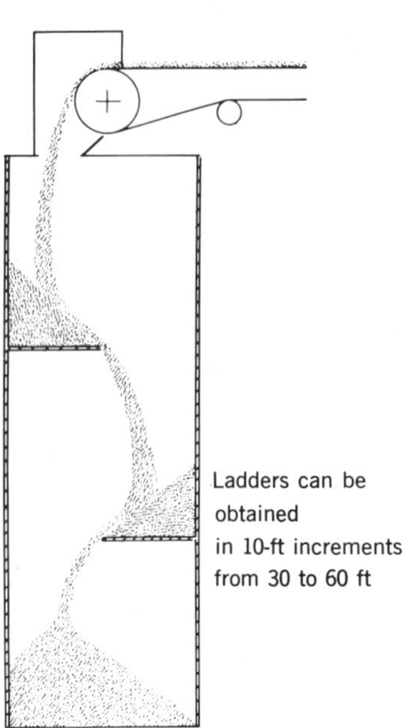

FIGURE 11.24. *"Rock ladder" lowering chute.*

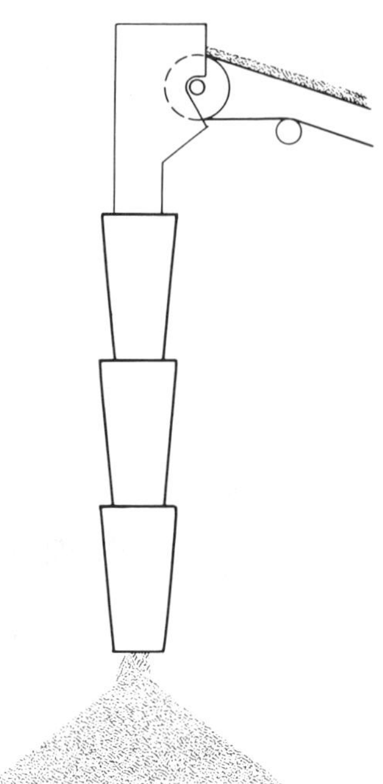

FIGURE 11.25. *Telescopic chute.*

stockpile by the belt conveyor. Good practice calls for the installation of a "rock ladder" under the discharge point of the belt conveyor. Figure 11.24 illustrates one form of the rock–ladder lowering chute.

The rock ladder is a structural steel (or wood) tower having a series of baffles arranged in such a manner that rock discharged from a belt conveyor never has a free drop of more than 5 ft.

If the material is heavy, abrasive, and lumpy, the baffles can be arranged in the form of rock boxes, as illustrated in Figure 11.4.

Telescopic Chutes. The telescopic discharge chute is used to minimize dusting when discharging to a pile. Telescoping sections usually are cable-connected in such a manner that a winch will successively lift the sections to keep the lower end of the chute just clear of the top of the storage pile. Coal is often delivered to storage piles through telescopic chutes. Figure 11.25 illustrates a typical arrangement.

Trippers

Trippers are devices used to discharge bulk materials from a belt conveyor at points upstream from the head pulley. Essentially, a tripper consists of a frame supporting two idling pulleys, one above and forward of the other. The conveyor belt passes over and around the upper pulley and around and under the lower pulley. The belt usually inclines to the upper pulley and may run horizontally or it may then incline again from the lower pulley.

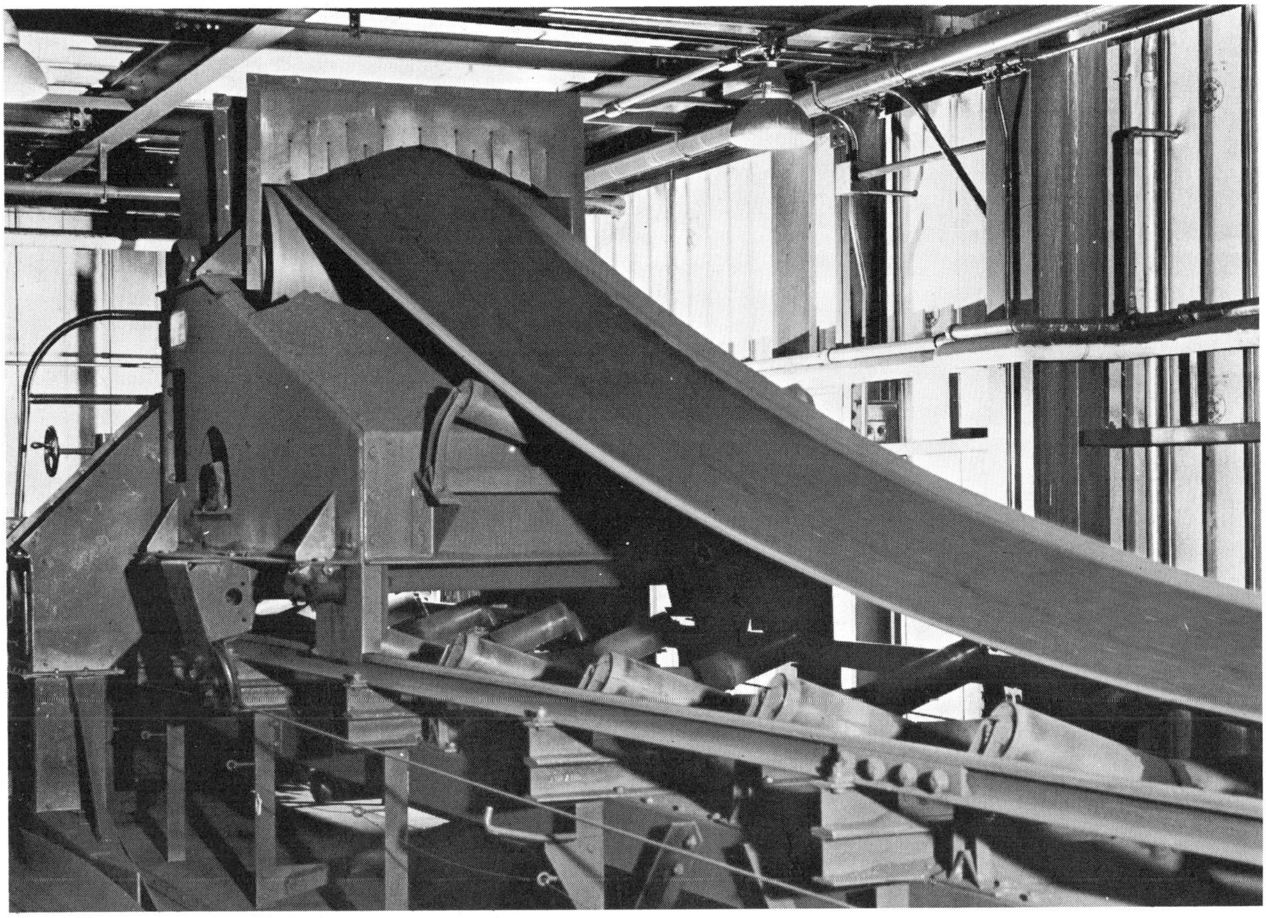

FIGURE 11.26. *Typical motor-driven belt conveyor tripper.*

With this construction, the material handled by the belt is discharged to a chute as the belt wraps around the upper pulley. The chute can be arranged to catch and divert the discharged material in any desired direction. The chute arrangement, augmented by movable gates, can be such as to discharge the material to either or both sides of the belt conveyor, or even back again onto the conveyor belt beyond the tripper. See Figure 11.29.

Trippers can be stationary (fixed) or movable (Figures 11.27 and 11.28). Stationary trippers are used where the discharge of material is to occur at a specific location. More than one stationary tripper can be used on a belt conveyor, either to discharge material from the belt at definite locations or to direct the material back again to the conveyor belt for discharge over the succeeding fixed tripper or the head pulley.

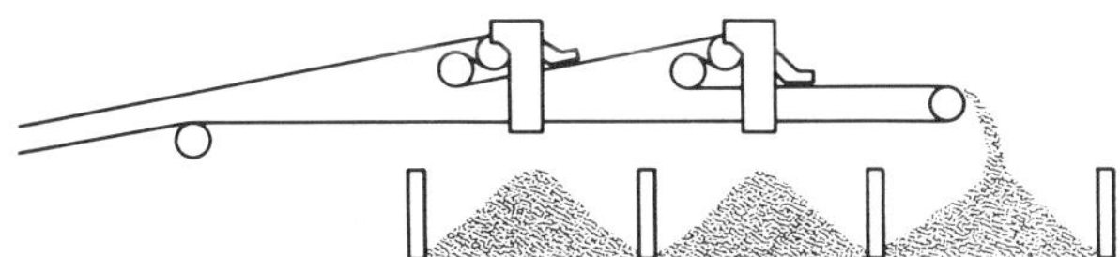

FIGURE 11.27. *Stationary trippers.*

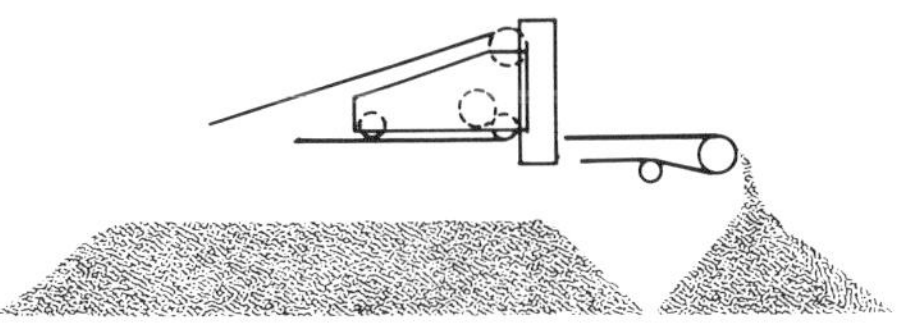

FIGURE 11.28. *Movable tripper.*

Movable trippers have the tripper frame mounted on flanged wheels which engage parallel rails supported on either side of the belt, usually from the conveyor frame. Such movable trippers can be moved by a cable and winch, by the belt itself, or by an electric motor mounted on the tripper. These movable trippers are reversible. The belt and motor-driven trippers frequently are arranged to move continuously back and forth along chosen portions of the belt conveyor, reversing automatically at the ends of their travel. All movable trippers can remain in specific locations for a short time or can be locked in position for longer periods.

Often, movable trippers have a platform to transport an operator, or a crosswalk permitting the operator to cross from one side of the conveyor belt to the other. Controls for the movement of the tripper are located for convenient access by the operator.

When handling dusty material, such as dry, fine, bituminous coal, conveyor belt and motor-driven trippers are provided with seals near the lower ends of the discharge chutes. These seals prevent the escape of dust from the covered bins or hoppers into which the chutes discharge.

There are several forms of seals, including a moving "carpet" of rubber-covered fabric belt, a pair of overlapping rubber–covered fabric belt strips, and a stationary carpet of rubber-covered fabric belt. With the stationary carpet type of seal, the tripper carries pulleys which pick up the carpet and then pass it around the discharge chute.

To increase the utility of the tripper in forming long piles or windrows of material on one or both sides of the belt conveyor, the material discharged over the upper pulley of the movable tripper is directed either to a transverse, horizontal, reversible belt conveyor carried by the tripper, or to one of two transverse, inclined belt conveyors also carried by the tripper.

All trippers absorb a certain amount of power from the belt conveyor drive, because the conveyor belt flexes over the tripper pulleys. Movable trippers that are actuated by the conveyor belt itself absorb greater amounts of power from the conveyor drive. For horsepower requirements of trippers, refer to Chapter 6. Figures 11.27 through 11.32 portray the various tripper arrangements and their uses.

Stationary (Fixed) Tripper

Stationary or fixed trippers pile material on either or both sides of the conveyor belt at specific locations, or return the material to the belt. See Figure 11.27.

Typical Movable Tripper

Driven by the conveyor belt itself, by an electric motor, or by a cable and winch, this tripper moves in a forward and reverse direction to fill a bin from end to end or to make a long pile on each or one side of the belt conveyor.

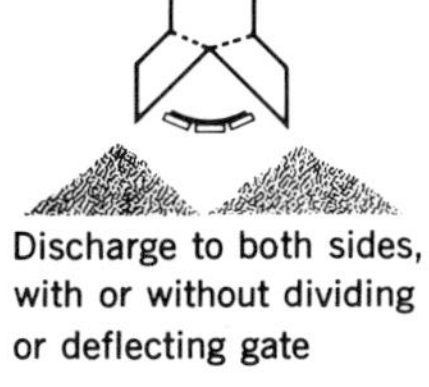

FIGURE 11.29. *Two typical movable trippers.*

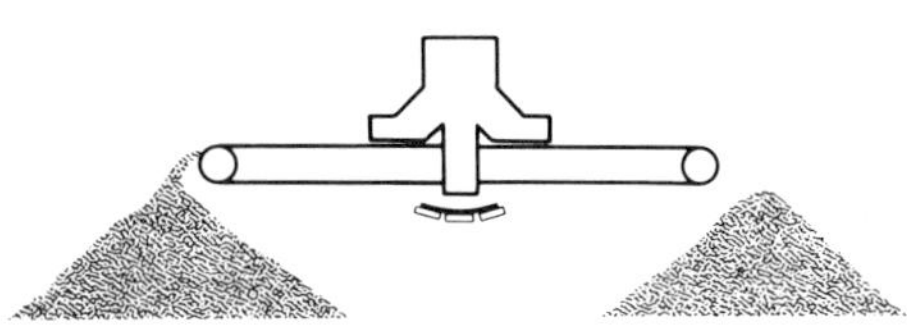

FIGURE 11.30. *Movable tripper with reversible cross belt.*

FIGURE 11.31. *Typical tripper with two transverse stacker belts.*

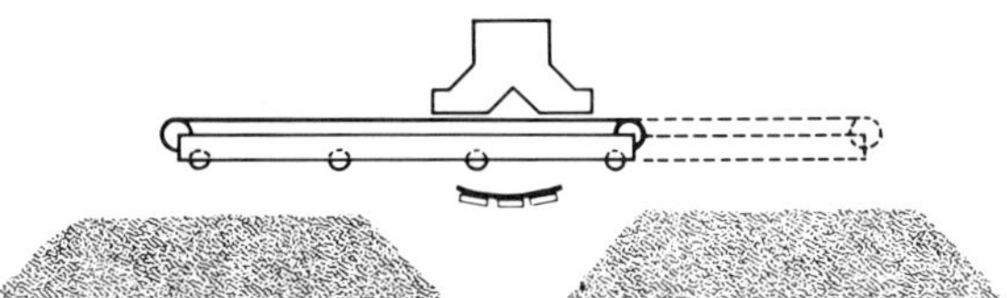

FIGURE 11.32. *Typical movable tripper with reversible shuttle belt.*

The movable tripper can be locked in any number of specific locations for spot discharge. Typical movable trippers are shown in Figures 11.28 and 11.29.

Tripper Discharge Through Auxiliary Arrangements

Three auxiliary tripper arrangements are described.

Tripper with Reversible Cross Belt. With this arrangement, a movable tripper carries a horizontal reversible cross belt. By means of a gate in the bifurcated tripper chute, and a corresponding travel direction of the cross belt, material discharged from a belt conveyor by the tripper can be placed in long piles or windrows on either side of the tripper. Or, if desired, individual piles of various materials or grades of the same material can be stockpiled on either side of the tripper. See Figure 11.30.

Tripper with Two Transverse Stacker Belts. To form higher piles than is possible with a horizontal cross belt, the tripper can be outfitted with two inclined stacker belt conveyors. With this arrangement, continuous high piles can be built on either side of the tripper or spot stockpiles can be built as desired or required. See Figure 11.31.

Tripper with Reversible Shuttle Belt. This arrangement is similar to the tripper with the plain cross belt. However, here the cross belt is a shuttle or movable belt conveyor and is reversible in direction, as shown in Figure 11.32. With this arrangement, flat-topped piles with a large top area can be built as continuous windrows on either side of the tripper, or as spotted stockpiles.

Plows

Discharge of fine materials from flat, horizontal belt conveyors can be accomplished by plows, either stationary or movable, with cable and winch. It is also possible to plow material from flat, inclined belt conveyors, if the angle of incline is not too great and the material is not too free-flowing.

Essentially, discharge plows are metal plates set at an angle not exceeding 35° to the belt center line, fitted at the bottom edge with an adjustable, all-rubber edging. There are a number of common arrangements of plow blades to accomplish discharge to one side, to both sides simultaneously, or to both sides simultaneously but with a portion of the material left on the central part of the belt.

The portion of the belt under the plow customarily is supported on a flat metal slide plate. This keeps the belt flat so that the plow can effect a reasonably clean discharge. Plows seldom remove 100% of the material from a belt, so a dribble of material is likely to pass over the end pulley of the conveyor. A means to collect this dribble must be provided.

Figures 11.33 through 11.38 illustrate various plow discharge arrangements.

Plows Discharging to One Side

Plows used to discharge to one side of a flat belt conveyor can swing horizontally across the belt on a vertical pivot or lift vertically on a horizontal pivot.

Single Horizontal Swing Plow. By a horizontal movement of the plow, all or part of the material can be removed from the belt. Plows removing all of the material from the belt can be manually operated or powered by a pneumatic or hydraulic cylinder. See Figure 11.33.

Single Lift Plow. This type of plow is lowered to discharge material from the belt or raised to avoid discharge. Again, all or part of the material can be removed from the conveyor belt, and the plow can be manually or power operated. Refer to Figure 11.34.

Plows Discharging to Both Sides

Plows which discharge simultaneously to both sides of a flat belt conveyor are called "V" plows. Two blades moving horizontally on vertical pivots can form the V, or a fixed V can be raised or lowered on a horizontal pivot.

Horizontal V-Plow. Figure 11.35 shows the horizontal V-plow. The blades swing on vertical pivots and join at the center. This arrangement makes possible several combinations of discharge: part of the material to one side, part to the other side, or all of the material from the belt. The blades can be manually or power operated.

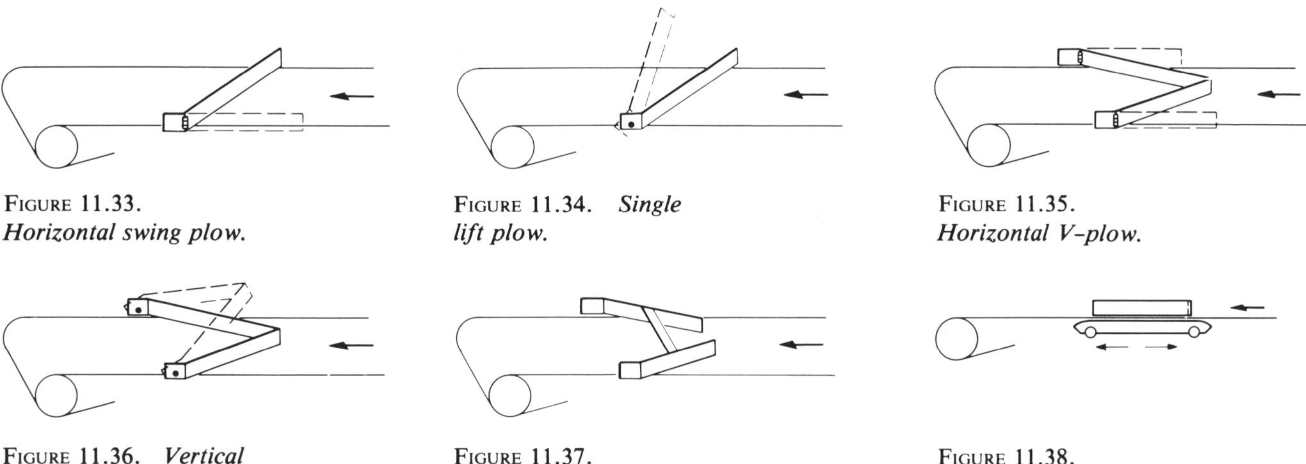

FIGURE 11.33. *Horizontal swing plow.*

FIGURE 11.34. *Single lift plow.*

FIGURE 11.35. *Horizontal V-plow.*

FIGURE 11.36. *Vertical V-plow.*

FIGURE 11.37. *Proportioning V-plow.*

FIGURE 11.38. *Traveling V-plow.*

Vertical V-Plow. A fixed V can be angularly raised or lowered to remove all material. Manual or power operation can be used. See Figure 11.36.

A variation of this plow is the open V, which allows some of the matcrial to rcmain on thc ccntral portion of thc bclt, whilc discharging thc outer portions of the material stream at both sides of the belt. Sometimes the amount of opening at the apex of the V is adjustable. Such a plow is called a proportioning V-plow. This type of plow is illustrated in Figure 11.37.

Traveling V-Plow. Moved by cable and winch in either direction, this V-plow discharges to both sides of a flat belt conveyor in any number of specific positions. See Figure 11.38.

This plow has been adapted to troughed belts by use of a flat belt support plate that lifts the belt from the troughing idlers. Such an arrangement conserves head room.

Discharge Trajectories

The path of the material discharged over the end pulley of a belt conveyor is known as its trajectory. The curvature of this path is determined by the rotational speed and radius of the end pulley, and by the force of gravity.

The proper design of transfer chutes, including the location of chute covers and wearing plates, depends upon the shape of the trajectory. Thus, the trajectory of the discharged material must be predicted as accurately as possible.

A number of authors and most catalogs prepared by belt and conveyor manufacturers have offered methods of calculating and plotting the material trajectory. However, observations and photographs of actual material trajectories from belt conveyors are not in satisfactory agreement with any of the trajectories calculated by these methods. A careful review of the literature on calculating trajectories has led to further research on this subject. The result of this research forms the basis for the method of calculating and plotting trajectories described in this manual.

The trajectories calculated by this method are in close agreement with photographs and observations of actual trajectories of material discharged from belt conveyors.

Allowance has been made for the change in the load shape of bulk material carried on a troughed belt conveyor when the belt flattens as it approaches the discharge pulley. The material tends increasingly to slump laterally toward the belt edges as the belt flattens at the pulley. Here, the load cross-section shape becomes—for all practical purposes—a segment of a circle. The cross-sectional area of this segment is equal to the average cross-sectional area of the load on the troughed portion of the conveyor.

The forces acting on the material as it reaches the pulley must be taken at the center of mass. This practically is the center of gravity of the cross-section of the load shape. A method has been developed to determine the height of this center of gravity above the belt surface, for various capacity loads, on any width of belt and 20°, 35°, and 45° three-equal-roll troughing idlers. Also, the effective radius about the center of the discharge pulley can be easily determined. The median line of the trajectory is established, thus permitting a close approximation to the upper and lower limits of the path of the discharged material stream.

If the material has an apparent density of 50 lbs or more per cu ft and is of roughly uniform particle size, the upper and lower limits of the material path will be relatively parallel to the median line for free falls of the discharged material to 7 ft below the center of the discharge pulley. If the free fall is greater than 7 ft and up to 20 ft below the center of the discharge pulley, the upper and lower limits of the material path may diverge a little.

Light, fluffy materials, very high belt speeds, and a mixture of large lumps and fines will alter the upper and lower limits of the material path. Lumps riding near the top of the material at the discharge pulley will be thrown farther from the discharge pulley. The trajectory of any such lumps can be closely approximated by calculating and plotting their individual trajectories. Air resistance will cause light, fluffy materials to spread vertically and laterally as they are being discharged over the pulley. Allowance should be made accordingly when determining the upper, lower, and lateral limits of such material trajectories.

Calculating and Plotting Normal Material Trajectories

The method of calculating and plotting the trajectories of normal materials involves careful attention to the following five considerations:

Center of Mass. The point in the mass of the material at which all forces act is the center of gravity of the load-shape cross section, at the point where the moving conveyor belt becomes tangent to the pulley.

Velocities. The material load and the belt are moving at the same linear velocity, up to the point where the belt becomes tangent to the pulley.

The tangential velocity acquired by the material taken at its cross-sectional center of gravity, as the material meets the curvature of the belt on

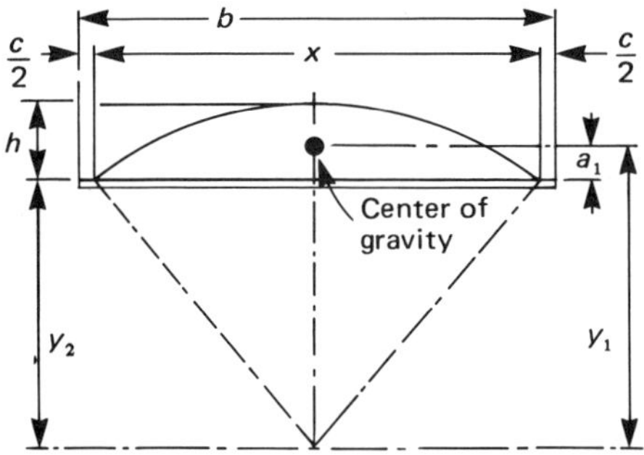

FIGURE 11.39. *Area of circular segment is equal to the cross-sectional area of the material load on the normal troughed portion of the conveyor belt.*

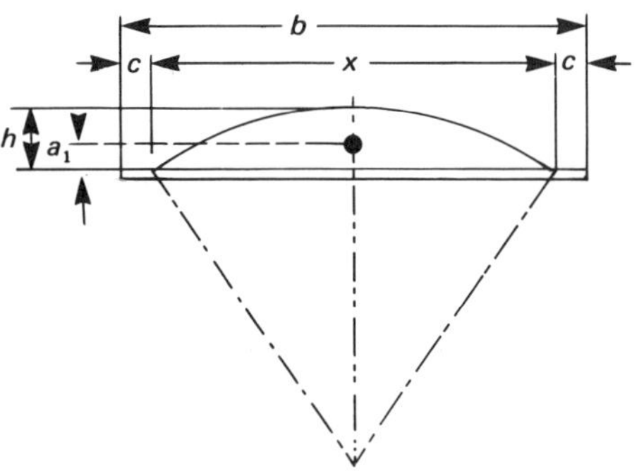

FIGURE 11.40. *To determine the center of gravity of the cross-section of the load on a flat belt, refer to Table 11-2 for values of a_1 and h.*

the pulley, is proportional to the radial distance between the cross-sectional center of gravity of the material and the center of the pulley, for any given speed of discharge-pulley rotation.

Start of Trajectory. The start of the trajectory is determined by that point on the circumference of the circle about the pulley center, formed by the cross-sectional center of gravity, where the centrifugal force at that point equals the radial component of the gravity force at that point.

Load Shape. Immediately prior to the point of tangency of the belt and pulley, the material load shape of a troughed belt conveyor can be closely approximated by a segment of a circle, as shown in Figure 11.39.

The area of the circular segment in Figure 11.39 is equal to the cross-sectional area of the material load on the normal troughed portion of the conveyor belt.

The center of gravity of the cross section of load on a flat belt can be determined as in Figure 11.40. The height of this point, a_1, and the height of h are given in Table 11-2. Note: standard edge distance applies here.

Angular Tangent Direction. The angular direction of the trajectory is determined by the forces acting on the material at its center of mass (i.e., the cross-sectional center of gravity in Figures 11.39 and 11.40).

Were it not for the effect of gravity, the median line of the trajectory would be a straight line. It is this straight line—tangent to a circle, the radius of which is the distance from the pulley center to the center of gravity of the material load-shape cross section—that determines what angular direction the trajectory will take. See Figures 11.41 through 11.47.

Fundamental Force-Velocity Relationships. Fundamentally, if the tangential velocity is V_s ft per second (fps); if g is the acceleration due to gravity (32.2 ft per sec^2); r is the radial distance, ft, from the center of the pulley to the center of mass (i.e., the cross-sectional center of gravity of the material load shape); and W is the gravity weight force of the material acting at the

TABLE 11-2. *Load Height and Center of Gravity at Discharge Pulley*

Type of idlers	Surcharge angle		*h* and a_1 Values, inches, for standard edge distance of 0.055b + 0.9 inch — Belt Width, inches: 18	24	30	36	42	48	54	60	72	84	96
20° Three-equal-length-roll troughing idlers	0°	h	1.1	1.7	2.1	2.7	3.1	3.7	4.1	4.7	5.7	6.7	7.7
		a_1	0.4	0.7	0.8	1.1	1.3	1.5	1.7	1.9	2.3	2.7	3.1
	5°	h	1.4	2.0	2.6	3.2	3.9	4.5	5.1	5.6	6.8	7.9	9.0
		a_1	0.6	0.8	1.0	1.3	1.6	1.8	2.1	2.3	2.7	3.2	3.7
	10°	h	1.6	2.4	3.1	3.7	4.5	5.2	6.0	6.7	8.1	9.6	11.1
		a_1	0.6	1.0	1.3	1.5	1.8	2.1	2.4	2.7	3.3	3.9	4.5
	20°	h	2.2	3.1	3.8	4.8	5.7	6.8	7.5	8.5	10.4	12.3	14.2
		a_1	0.9	1.3	1.6	1.9	2.3	2.7	3.0	3.4	4.2	4.9	5.6
	25°	h	2.5	3.5	4.5	5.4	6.4	7.5	8.5	9.6	11.6	13.5	15.4
		a_1	1.0	1.4	1.8	2.2	2.6	3.0	3.4	3.8	4.6	5.4	6.2
	30°	h	2.7	3.8	5.0	6.0	7.1	8.2	9.4	10.6	12.7	14.0	16.3
		a_1	1.1	1.5	2.0	2.4	2.9	3.3	3.8	4.3	5.1	6.1	7.1
35° Three- equal-length-roll troughing idlers	0°	h	1.9	2.7	3.5	4.3	5.0	6.0	6.7	7.4	9.0	10.5	12.0
		a_1	0.8	1.1	1.4	1.7	2.0	2.4	2.7	3.0	3.6	4.1	4.7
	5°	h	2.1	3.0	3.8	4.8	5.5	6.4	7.4	8.1	10.0	11.9	13.8
		a_1	0.8	1.2	1.5	1.9	2.2	2.6	3.0	3.3	4.0	4.7	5.4
	10°	h	2.3	3.2	4.2	5.2	6.1	7.2	8.2	9.0	11.0	13.0	15.0
		a_1	0.9	1.3	1.7	2.1	2.5	2.9	3.3	3.6	4.4	5.2	6.0
	20°	h	2.7	3.9	5.1	6.1	7.3	8.4	9.5	10.7	13.0	14.2	16.4
		a_1	1.1	1.6	2.1	2.5	2.9	3.4	3.8	4.3	5.2	6.1	7.0
	25°	h	3.0	4.2	5.4	6.6	7.8	9.1	10.3	11.6	13.9	15.4	17.9
		a_1	1.2	1.7	2.2	2.7	3.1	3.7	4.1	4.6	5.6	6.6	7.6
	30°	h	3.2	4.5	5.7	7.1	8.4	9.8	11.1	12.4	14.9	17.5	20.1
		a_1	1.3	1.8	2.3	2.9	3.4	4.0	4.5	5.0	6.0	7.0	8.0

Table 11-2 continues on facing page.

center of mass, then the centrifugal force acting at the center of mass of the material is as follows:

$$\text{Centrifugal force} = \left(\frac{W}{g}\right)\left(\frac{V_s^2}{r}\right) = \frac{WV_s^2}{gr}$$

When this centrifugal force equals the radial component of the material weight force, the material will no longer be supported by the belt and will commence its trajectory. At just what angular position around the pulley this will occur is governed by the slope of the conveyor at the discharge pulley, outlined in the following three cases.

Horizontal Belt Conveyor Trajectories

If the belt conveyor is horizontal to the discharge pulley, there are two conditions to consider:

TABLE 11-2 *continued.* ***Load Height and Center of Gravity at Discharge Pulley***

Type of idlers	*Surcharge angle*		*h and a_1 Values, inches, for standard edge distance of 0.055b + 0.9 inch* — *Belt Width, inches* 18	24	30	36	42	48	54	60	72	84	96
45° Three-equal-length-roll troughing idlers	0°	*h*	2.2	3.1	4.0	5.0	6.0	6.8	7.7	8.6	10.5	12.3	14.1
		a_1	0.9	1.3	1.8	2.0	2.4	2.7	3.1	3.5	4.2	5.0	5.8
	5°	*h*	2.4	3.4	4.4	5.4	6.4	7.4	8.4	9.5	11.4	13.4	15.4
		a_1	1.0	1.4	1.8	2.2	2.6	3.0	3.4	3.8	4.6	5.4	6.2
	10°	*h*	2.6	3.7	4.7	5.8	7.0	7.9	9.0	10.2	12.2	14.4	16.6
		a_1	1.0	1.5	1.9	2.3	2.8	3.2	3.6	4.1	4.9	6.8	7.7
	20°	*h*	2.9	4.1	5.4	6.6	7.8	9.1	10.3	11.5	14.0	16.5	19.0
		a_1	1.2	1.7	2.2	2.7	3.1	3.7	4.2	4.6	5.6	6.6	7.6
	25°	*h*	3.1	4.4	5.7	7.0	8.3	9.7	10.9	12.3	14.8	16.4	20.0
		a_1	1.2	1.8	2.3	2.8	3.3	3.9	4.4	4.9	5.9	6.9	7.8
	30°	*h*	3.3	4.7	6.0	7.4	8.8	10.2	11.5	13.0	15.6	17.3	20.4
		a_1	1.3	1.9	2.4	3.0	3.5	4.1	4.6	5.2	6.3	7.5	8.7
Flat belt idlers	5°	*h*	0.35	0.42	0.50	0.66	0.80	0.90	1.04	1.15	1.40	1.65	1.95
		a_1	0.14	0.17	0.20	0.27	0.32	0.36	0.42	0.46	0.57	0.69	0.71
	10°	*h*	0.65	0.85	0.95	1.35	1.60	1.80	2.10	2.30	2.80	3.30	3.80
		a_1	0.26	0.35	0.38	0.54	0.65	0.73	0.85	0.93	1.13	1.35	1.70
	15°	*h*	1.00	1.30	1.43	2.00	2.40	2.67	3.16	3.45	4.20	5.10	6.00
		a_1	0.40	0.52	0.58	0.81	0.97	1.08	1.28	1.39	1.70	2.09	2.48
	20°	*h*	1.30	1.70	1.90	2.70	3.20	3.56	4.20	4.60	5.55	6.45	7.35
		a_1	0.52	0.69	0.77	1.09	1.29	1.44	1.70	1.86	2.26	2.66	3.07
	25°	*h*	1.60	2.15	2.37	3.35	4.00	4.43	5.30	5.74	6.93	7.99	8.05
		a_1	0.65	0.87	0.96	1.35	1.61	1.79	2.14	2.32	2.80	3.29	3.78
	30°	*h*	1.94	2.57	2.83	4.03	4.80	5.30	6.30	6.86	8.30	9.78	11.26
		a_1	0.78	1.04	1.14	1.63	1.94	2.14	2.54	2.77	3.35	3.95	4.55
	35°	*h*	2.25	3.00	3.30	4.70	5.60	6.20	7.40	8.00	9.70	11.50	13.30
		a_1	0.91	1.21	1.33	1.90	2.26	2.50	2.98	3.23	3.91	4.65	5.76

(1) If the tangential speed is sufficiently high (that is, when the centrifugal force is equal to or greater than W), the material will leave the belt at the initial point of tangency of the belt with the pulley, as shown in Figure 11.41, or:

$$\frac{V_s^2}{gr} \geqq 1$$

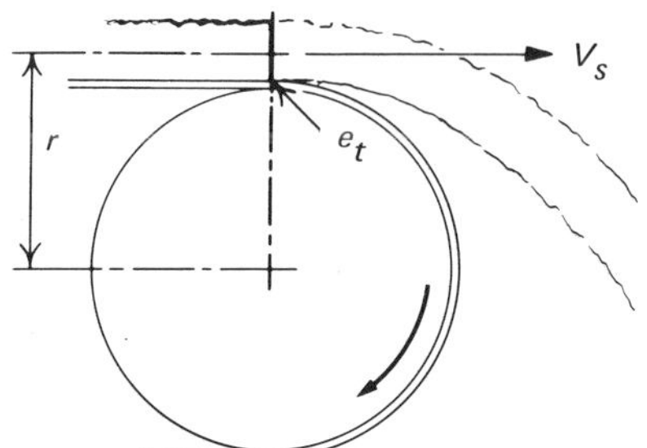

FIGURE 11.41. *When the belt speed is sufficiently high, the material leaves the belt at the point of tangency of the belt with the pulley.*

where: e_t = point where the material leaves the belt
V_s = tangential velocity, fps, of the cross-sectional center of gravity of load shape
g = acceleration of gravity, ft per second per second, or ft per sec^2
r = radius, ft, from the center of pulley to the cross-sectional center of gravity of the load shape.

(2) If the tangential speed is not high enough for the material to leave the belt at the initial point of tangency (that is, when V_s^2/gr is less than 1), then the material will follow part way around the pulley for an angular distance, γ:

$$\frac{V_s^2}{gr} = \cos \gamma$$

where: e_t = point where the material leaves the belt
γ = angle, in degrees, between the vertical centerline, through the pulley, to the point, e_t, where the material starts its trajectory.

This is illustrated clearly in Figure 11.42.

Inclined Belt Conveyor Trajectories

For a belt inclined to the discharge pulley, there are four conditions to consider:

(1) If the tangential speed is sufficiently high, or when $V_s^2/gr > 1$, the material leaves the belt at the initial point of tangency of the belt and pulley, as shown in Figure 11.43,

where: e_t = point where the material leaves the belt
ϕ = angle, in degrees, of incline of the belt conveyor to the horizontal.

(2) If the combination of the belt incline, the pulley diameter, the load depth, and the tangential belt speed is such that $V_s^2/gr > \cos \phi$ but is still less

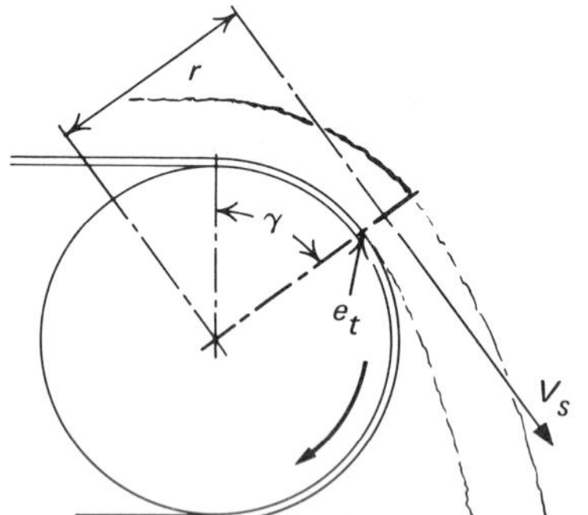

FIGURE 11.42. *When the belt speed is not high enough, the material will follow part way around the conveyor.*

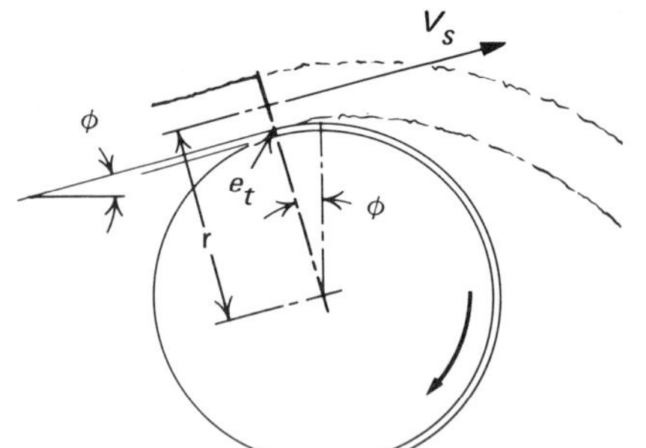

FIGURE 11.43. *When, in an inclined conveyor, the tangential speed is high (see text), the material will leave the belt at the point of tangency of the belt and pulley.*

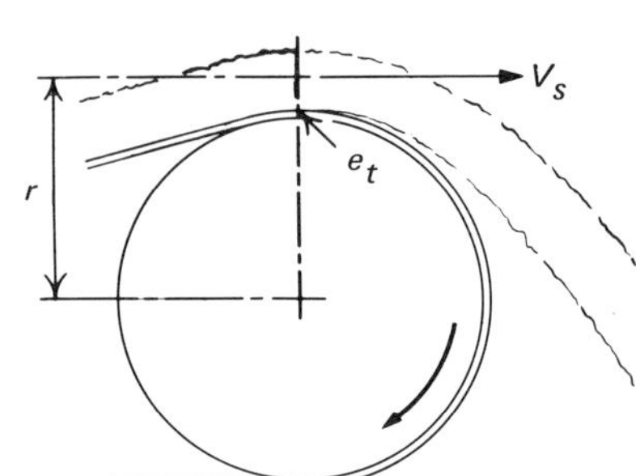

FIGURE 11.44. *When, in an inclined conveyor, the tangential velocity is equal to a specific value (see text), the material will leave the belt at the top of the end pulley.*

than 1, the material may leave the belt at the initial point of tangency of the belt and the pulley. However, the curved surface of the belt on the pulley may interfere with the theoretical trajectory path of the material. The material may then re-engage the belt and be carried farther around the pulley before it assumes its final trajectory. See condition (4), below.

(3) If the tangential speed is such that $V_s^2/gr = 1$, the material will leave the belt at the vertical centerline through the pulley, as shown in Figure 11.44,

where: e_t = point where the material leaves the belt

(4) If the tangential speed is sufficiently low, or when $V_s{}^2/gr < \cos \phi$, the material will travel partially around the pulley an angular distance, γ, beyond its top center to the point where $V_s^2/gr = \cos \gamma$. This is shown in Figure 11.45,

where: e_t = point where the material leaves the belt
ϕ = angle, in degrees, of incline of the belt conveyor to the horizontal
γ = angle, in degrees, between the vertical centerline, through the pulley to the point, e_t, where the material starts its trajectory.

Declined Belt Conveyor Trajectories

If the belt conveyor is declined toward the discharge pulley, there are two conditions to consider:

(1) If the tangential speed is sufficiently high, or when $V_s^2/gr \geqq \cos \theta$, the material will leave the belt at the initial point of tangency of the belt and pulley, as shown in Figure 11.46,

where: e_t = point where the material leaves the belt
θ = angle, in degrees, of decline of the belt conveyor.

(2) If the tangential speed is insufficient to make the material leave the belt at the initial point of tangency of the belt and pulley, the material will

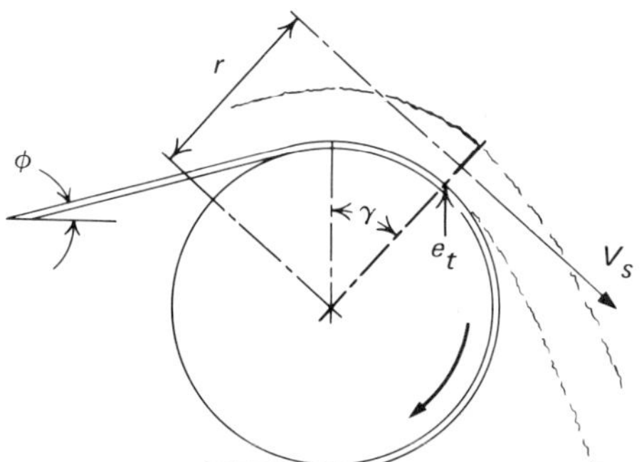

FIGURE 11.45. *When, in an inclined conveyor, the tangential velocity is low (see text), the material will follow part way around the end pulley.*

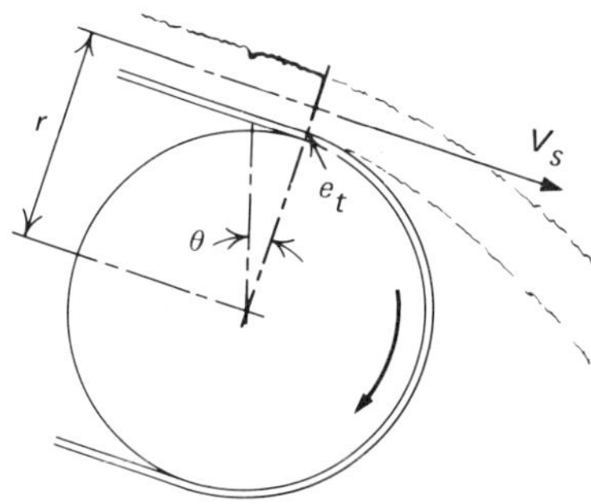

FIGURE 11.46. *When, in a declined conveyor, the tangential velocity is high (see text), the material will leave the belt at the point of tangency of the belt and pulley.*

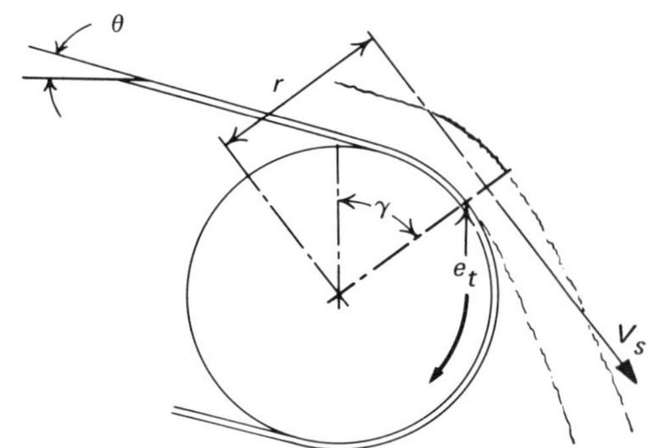

FIGURE 11.47. *When, in a declined conveyor, the tangential velocity is low (see text), the material will follow part way around the end pulley.*

follow partly around the pulley until $V_s^2/gr = \cos\gamma$, as shown in Figure 11.47,

where: e_t = point where the material leaves the belt
γ = angle, in degrees, between the vertical centerline through the pulley to point e_t.

Plotting the Trajectory

Before the trajectory of the discharged material can be plotted, it is necessary to calculate the values of V_s and r in order to solve the expression:

$$\frac{V_s^2}{gr}$$

It is also necessary to find the height of the flattened load of material on the belt, so that the upper limit of the material path can be plotted.

Referring to Figures 11.39 and 11.40,

if: a_1 = height, inches, above the belt surface of the center of gravity of the cross-section shape of the load, at the point where the pulley is tangent to the belt
h = height, inches, above the belt surface of the top of the load, at the point where the belt is tangent to the pulley
r = radius, ft, from the center of the pulley to the center of gravity of the circular segment load cross section

then:
$$r = \frac{(a_1) + (\text{belt thickness, inches}) + (\text{pulley radius, inches})}{12}$$

The values of a_1 and h have been tabulated for the various belt widths, idler end-roll angles, and surcharge angles, for troughed conveyor belts loaded to the standard edge distance ($0.055b + 0.9$ inch), as listed in Table 11-2.

The tangential velocity, V_s should be calculated from the relation:

$$V_s = \frac{(2)(\pi)(r)(\text{rpm of end pulley})}{60}$$

V_s should *never* be calculated from the nominal speed of the belt.

Table 11.3. *Fall Distance for Time Intervals*

Time interval (fractions of seconds)	Fall distance (inches)	Time interval (fractions of seconds)	Fall distance (inches)
1/20 th	1/2	12/20 ths	69 1/2
2/20 ths	1 15/16	13/20 ths	81 1/2
3/20 ths	4 3/8	14/20 ths	94 9/16
4/20 ths	7 3/4	15/20 ths	108 1/2
5/20 ths	12 1/16	16/20 ths	123 1/2
6/20 ths	17 3/8	17/20 ths	139 7/16
7/20 ths	23 5/8	18/20 ths	156 7/16
8/20 ths	30 7/8	19/20 ths	172 1/2
9/20 ths	39 1/16	20/20 (one second)	193 1/4
10/20 ths	48 1/4		
11/20 ths	58 3/8		

Determination of Angular Position of Tangent Line. In order to determine the angular position of the straight line tangent to the circle of radius, r, it is necessary to solve the expression:

$$\frac{V_s^2}{gr} = \cos \phi, \text{ or } \cos \theta, \text{ or } \cos \gamma$$

Then, from the cosine, determine the angles ϕ, θ, or γ in degrees.

Of course, if $\frac{V_s^2}{gr} = 1$, or more, the angle is zero.

See "Horizontal belt conveyor trajectories," page 280, "Inclined belt conveyor trajectories," page 282, and "Declined belt conveyor trajectories," page 283, to apply the angular values in degrees and to find the point, e_t, where the material leaves the belt and starts its trajectory.

To set up the graphical diagram, draw, to some convenient scale, the pulley rim; the belt thickness; the path of the belt, horizontal, inclined, or declined; and a circle, with radius, r, from the center of the pulley. At the determined point, e_t, draw a straight line tangent to the circle of radius r (i.e., perpendicular to the line through the pulley center fixed by the angle ϕ, θ, or γ).

Distance of Material Fall From Tangent Line. The action of gravity will prevent the material from following the tangent line, so the trajectory will lie below it. To measure and plot the amount of fall, it will be convenient to divide the time of fall of the material, resulting from the effect of gravity, into 1/20th seconds. To find the actual distance of fall from the tangent line at successive 1/20–second intervals, consult Table 11-3.

Measurement of the Time Interval. The determination of the interval of time along the tangent line depends upon the calculated tangential velocity, V_s (at radius r).

It will be helpful, in making the layout of the trajectory diagram, to recognize that the distance increments for each 1/20th of a second of time correspond to 0.6 inches for each foot per second of the tangential velocity, V_s.

For example, if the calculated tangential velocity, V_s, is 1 fps, lay out the time intervals on the tangent line from point e_t at 0.6 inch; if the tangential velocity is 2 fps, lay out the intervals at 1.2 inches; if 3 fps, lay out the intervals at 1.8 inches; etc. If the tangential velocity is some fraction of a foot per second, multiply this fraction by 0.6 inch and lay out the intervals accordingly.

1. Start the layout of time intervals on the tangent line from point e_t, the start of the line tangent to the circle of radius r. Number each interval consecutively, 0 for the point of tangency (point e_t), 1 for the first 1/20th second, 2 for the next, and so on.
2. Draw a series of parallel vertical lines downward a suitable distance from each numbered time interval and directly on the tangent line (except the zero number).
3. Lay out on these vertical lines the corresponding distance of fall from the tangent line. To do this, measure vertically downward from each numbered point on the tangent line.
4. Draw a smooth curve through the fall points. This is the *median* line of the trajectory of the material.

Limits of the Trajectory Path. Having established the median line of the trajectory, lay out a top-of-the-trajectory line with the distance $(h - a_1)$, using Table 11-2. Distance $(h - a_1)$ is the radius of partial circles drawn above and around each fall point. The top limit of the trajectory of normal materials will be a smooth curve drawn tangent to these partial circles. The value of h must be to the same scale as the diagram.

Similarly, the underside limit of the material path should be a smooth curve, tangent to partial circles drawn below and around each fall point. The circles will have a radius equal to the value of a_1, which also must be to the same scale as the diagram.

See "Discharge trajectories," page 277, with respect to long falls below the discharge pulley; light, fluffy materials; or large lumps mixed with fines. Allowance for aberrations in the trajectory limits of such materials must be made when designing discharge chutes. For the individual trajectories of single large lumps, use r as the distance from the center of the lump to the center of the pulley. Calculate the tangential velocity, V_s, of the lump as follows:

$$V_s = \frac{(2)(\pi)(r)(\text{rpm of end pulley})}{60}$$

The lateral dimension, or width, of the trajectory path of the material will be very close to the length of the circular segment chord x in Figures 11.39 and 11.40. This is approximately [$b - 0.055b - 0.9$ inch] for troughed

FIGURE 11.48. *Trajectory formed as material is discharged over the end pulley of an horizontal troughed belt conveyor. Here, shuttle belt conveyors project from openings in the face of the dock to load iron ore pellets into the ship's hatches.*

belts, or [$b - 2(0.055b + 0.9$ inch)] for flat belts, where b is the belt width in inches.

The lateral dimension, or width, of the material path is affected by the height of fall and the characteristics of the material; again refer to "Discharge trajectories," page 277.

Examples of Trajectories

The following seven examples of plotted trajectories and trajectory limits of uniform materials discharged over the ends of troughed belt conveyors—Figures 11.52 through 11.58—may be helpful. All these examples are based on 30-inch troughed conveyor belts operating horizontally, inclined, or declined, at various speeds (see Figures 11.52 through 11.58), on 20° three-equal-roll troughing idlers, and with 24-inch-diameter discharge pulleys.

Also, refer to Figures 11.48, 11.49, 11.50, and 11.51. These are actual photographs, taken in the field, of conveyor trajectories as materials are discharged over the end pulley of horizontal and inclined troughed belt conveyors.

FIGURE 11.49. *Sized aggregate pours over end pulley of this inclined conveyor.*

FIGURE 11.50. *Iron-ore pellets discharging from boom belt of Great Lakes self-unloader ship at 10,000 tph.*

FIGURE 11.51. *Closeup of trajectory from head pulley of conveyor, speed 760 fpm, handling iron-ore pellets.*

FIGURE 11.52. *Example of discharge trajectory: Drawing No. 1.*

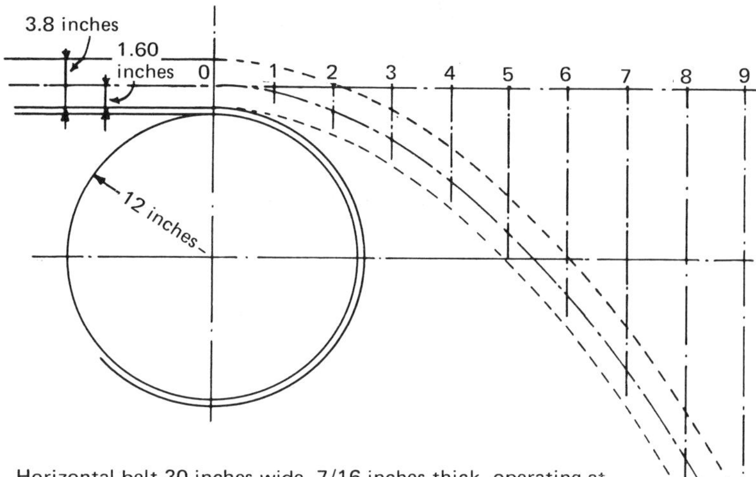

Horizontal belt 30 inches wide, 7/16 inches thick, operating at 400 fpm on three-equal-roll 20° standard troughing idlers, loaded to 20° surcharge

Pulley at 61.4 rpm, 1.023 rps

From Table 11-2:

$h = 3.8$ inches

$a_1 = 1.60$ inches

$r = 14.038$ inches, or 1.1697 ft

Tangential velocity, $V = (1.023)(6.28)(1.1697 = 7.52$ fps)

$$\frac{V^2}{gr} = 1.502 \text{ which is greater than 1}$$

See Figure 11.41, Discharge at top of pulley
Velocity spaces on tangential line
(7.52)(0.6 inches) = 4.52 inches
Fall distances from Table 11-3.

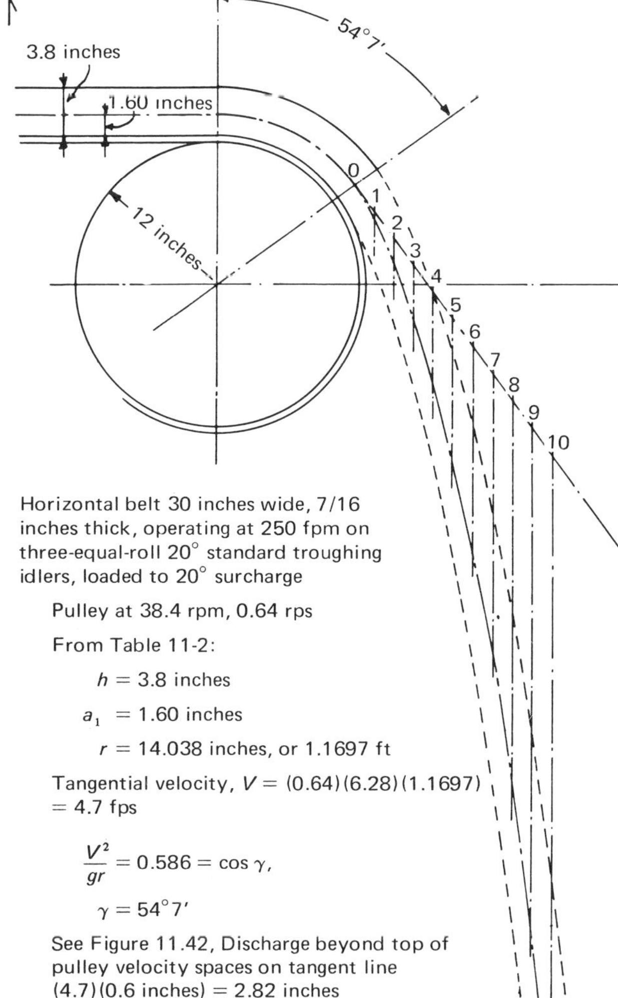

Horizontal belt 30 inches wide, 7/16 inches thick, operating at 250 fpm on three-equal-roll 20° standard troughing idlers, loaded to 20° surcharge

Pulley at 38.4 rpm, 0.64 rps

From Table 11-2:

$h = 3.8$ inches

$a_1 = 1.60$ inches

$r = 14.038$ inches, or 1.1697 ft

Tangential velocity, $V = (0.64)(6.28)(1.1697) = 4.7$ fps

$$\frac{V^2}{gr} = 0.586 = \cos \gamma,$$

$$\gamma = 54°7'$$

See Figure 11.42, Discharge beyond top of pulley velocity spaces on tangent line
(4.7)(0.6 inches) = 2.82 inches
Fall distances from Table 11-3.

FIGURE 11.53. *Example of discharge trajectory: Drawing No. 2.*

FIGURE 11.54. *Example of discharge trajectory: Drawing No. 3.*

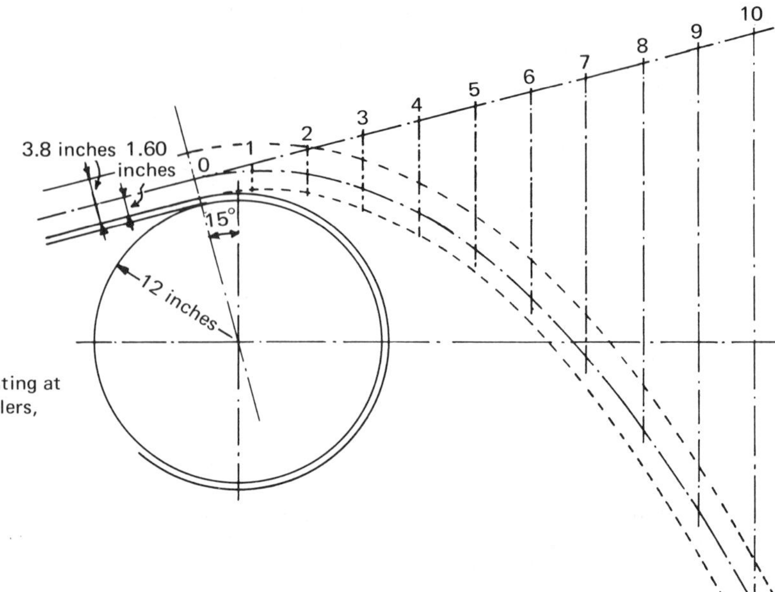

Inclined belt, 30 inches wide, 7/16 inches thick, operating at 400 fpm on three-equal-roll 20° standard troughing idlers, loaded to 20° surcharge

Pulley at 61.4 rpm, 1.023 rps

From Table 11-2:

$h = 3.8$ inches

$a_1 = 1.60$ inches

$r = 14.038$ inches, or 1.1697 ft

Tangential velocity, $V = (1.023)(6.28)(1.1697) = 7.52$ fps

$$\frac{V^2}{gr} = 1.502, \text{ which is greater than 1}$$

See Figure 11.43, Discharge at tangency of belt with pulley
Velocity spaces on tangent line
(7.52)(0.6 inches) = 4.52 inches
Fall distances from Table 11-3.

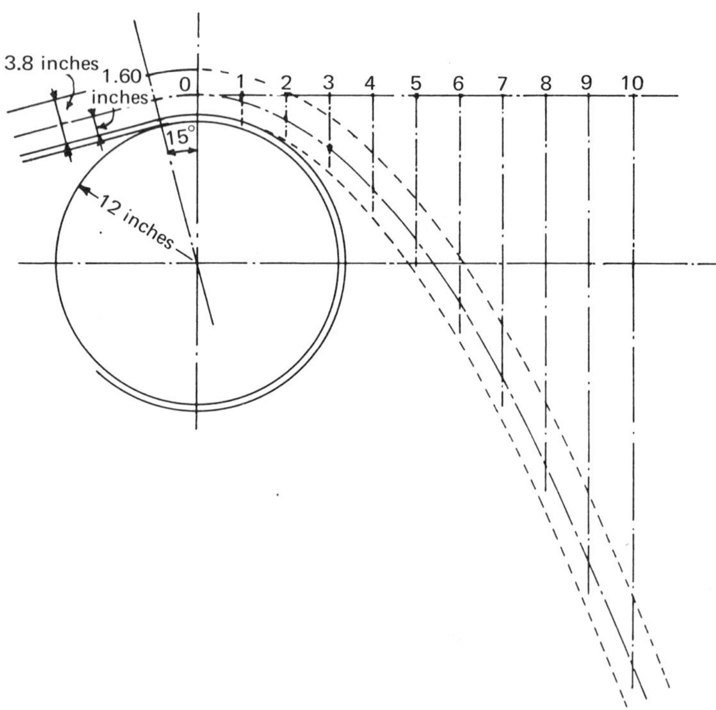

Inclined belt, 30 inches wide, 7/16 inches thick, operating at 327 fpm on three-equal-roll, 20° standard troughing idlers, loaded to 20° surcharge

Pulley at 50.2 fpm, 0.837 rps

From Table 11-2:

$h = 3.8$ inches

$a_1 = 1.60$ inches

$r = 14.038$ inches, or 1.1697 ft

Tangential velocity, $V = (0.837)(6.28)(1.1697) = 6.15$

$$\frac{V^2}{gr} = 1$$

See Figure 11.44, Discharge at top of pulley
Velocity spaces on tangent line
(6.15)(0.6 inches) = 3.690 inches
Fall distances from Table 11-3.

FIGURE 11.55. *Example of discharge trajectory: Drawing No. 4.*

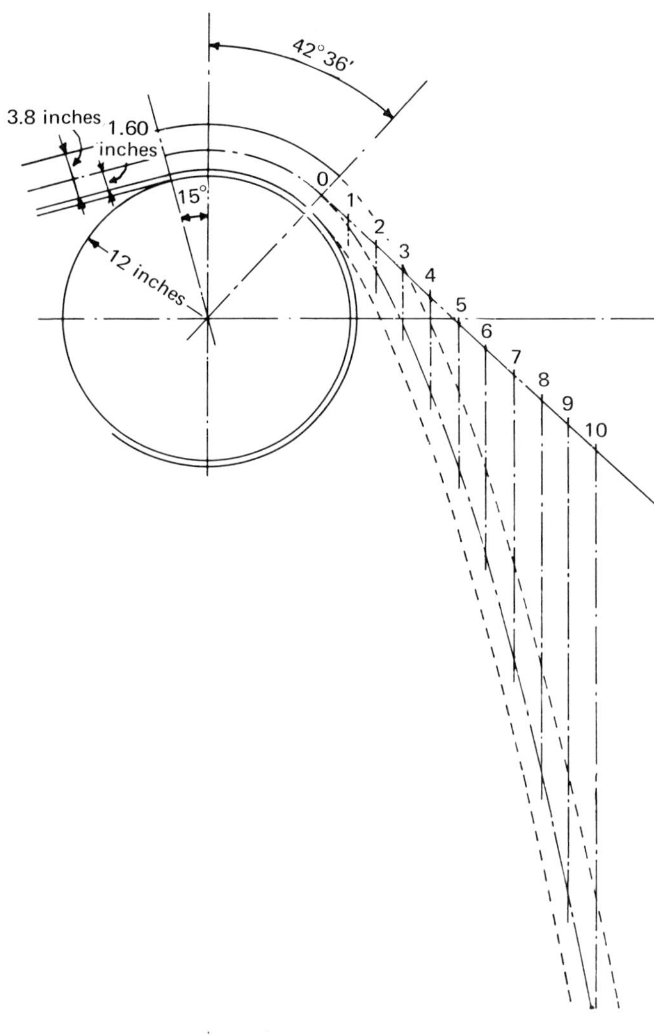

FIGURE 11.56. *Example of discharge trajectory: Drawing No. 5.*

Inclined belt, 30 inches wide, 7/16 inches thick, operating at 280 fpm on three-equal-roll 20° standard troughing idlers, loaded to 20° surcharge

Pulley at 43 fpm, 0.717 rps

From Table 11-3:

$h = 3.8$ inches

$a_1 = 1.60$ inches

$r = 14.038$ inches, or 1.1697 ft

Tangential velocity, $V = (0.717)(6.28)(1.1697) = 5.27$ fps

$$\frac{V^2}{gr} = 0.736 = \cos\gamma, \ \gamma = 42°36'$$

See Figure 11.45, Discharge beyond top of pulley
Velocity spaces on tangent line
(5.27)(0.6 inches) = 3.162 inches
Fall distance from Table 11-3.

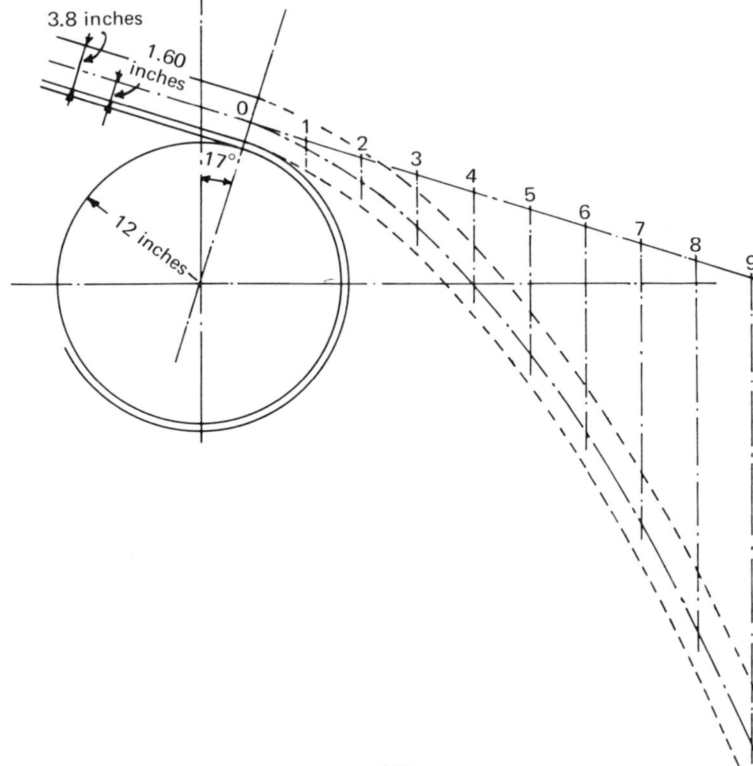

Declined belt, 30 inches wide, 7/16 inches thick, operating at 400 fpm on three-equal-roll 20° standard troughing idlers, loaded to 20° surcharge

Pulley at 61.4 rpm, 1.023 rps

From Table 11-2:

$h = 3.8$ inches

$a_1 = 1.60$ inches

$r = 14.038$ inches, or 1.1697 ft

Tangential velocity, $V = (1.023)(6.28)(1.1697) = 7.52$ fps

$$\frac{V^2}{gr} = 1.502, \text{ which is greater than 1}$$

See Figure 11.46, Discharge at tangency of belt with pulley
Velocity spaces on tangent line
(7.52)(0.6 inches) = 4.52 inches
Fall distances from Table 11-3.

FIGURE 11.57. *Example of discharge trajectory: Drawing No. 6.*

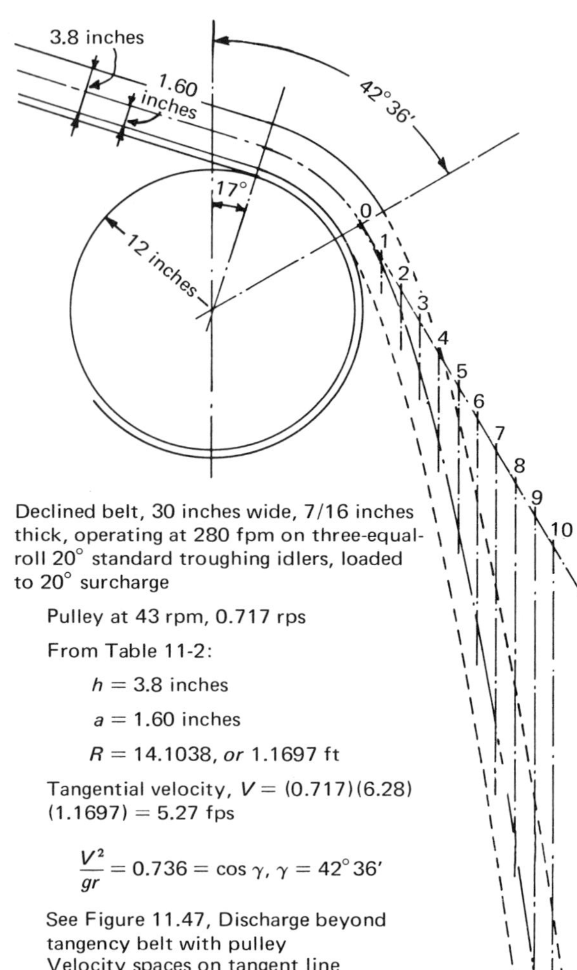

FIGURE 11.58. *Example of discharge trajectory: Drawing No. 7.*

Chapter 12

Motors and Controls

Contents

Motor Selection

Having determined the basic horsepower requirements, the selection of motors for belt conveyor drives then depends on many factors. These include starting characteristics; type and voltage of power supply; ambient and atmospheric conditions; single or multi-speed requirements; special service conditions; whether the conveyor is inclined, declined, or particularly if it has one or more vertical curves.

Motor Ratings

The motor selected should have a continuous nameplate rating, at least equal to the horsepower required by the conveyor, divided by the efficiency of all drive components.

Open drip-proof motors rated 200 hp or less *may have* a service factor of 1.15, which will allow them to carry 15% overload continuously at safe temperatures. Totally enclosed fan-cooled and explosion-proof motors normally carry a National Electrical Manufacturers Association (NEMA) service factor of 1.00. Some manufacturers supply certain totally enclosed motors with special insulation and a 1.15 service factor. But these motors are not NEMA standard and should be checked with the motor manufacturer.

Where a service factor of 1.15 is available, it is not recommended that the additional capacity be utilized to supply any portion of the normal load and *especially should not be used for additional locked-rotor or acceleration torque.* If the calculated horsepower at the motor shaft is slightly in excess of a standard NEMA horsepower rating and a complete analysis indicates that starting the belt conveyor is absolutely possible under all operating conditions, the conveyor purchaser may approve use of a portion of the service factor when carrying the maximum load, rather than insisting on the next larger standard motor rating.

Torque Characteristics

A belt conveyor drive must provide sufficient torque at standstill to overcome the static forces and to accelerate the loaded conveyor to running speed within the time limitation imposed by the motor manufacturer. Likewise the accelerating torque must not impart tensions to the conveyor belt which exceed the manufacturer's allowable rating for acceleration. The amount of torque imparted to the conveyor belt will vary depending on the relative WK^2 values of the drive and the moving parts of the conveyor system as well as the profile of the conveyor. Also since motor torque varies as a square of the impressed voltage, the stiffness of the power supply system should be considered in any analysis of accelerating torque.

Squirrel-cage alternating current induction motors represent the simplest and most economical means of driving belt conveyors. Unfortunately, NEMA standard motor designs do not exactly meet ideal speed torque conditions required by a belt conveyor. Other means have to be employed to control the torque. These are usually reduced-voltage control or auxiliary devices which are discussed later in this chapter. An ideal speed torque curve for a belt conveyor is shown in Figure 12.1.

NEMA classifies polyphase squirrel-cage motors into four designations with respect to speed-torque characteristics. Of these, two designs, B and C, satisfy all but a few unusual applications. A comparison of the torque characteristics of these two designs is shown in Tables 12-1 through 12-5, with typical speed-torque curves shown in Figure 12.1.

It should be noted that Tables 12-1 through 12-4 provide a listing of *minimum* torque values. Normally, the actual torque values provided by the motor manufacturers are greater than those listed in order to provide design margin. Since the conveyor designer is concerned with *limiting* as well as supplying sufficient torque, specific needs and limitations should be discussed with the manufacturer.

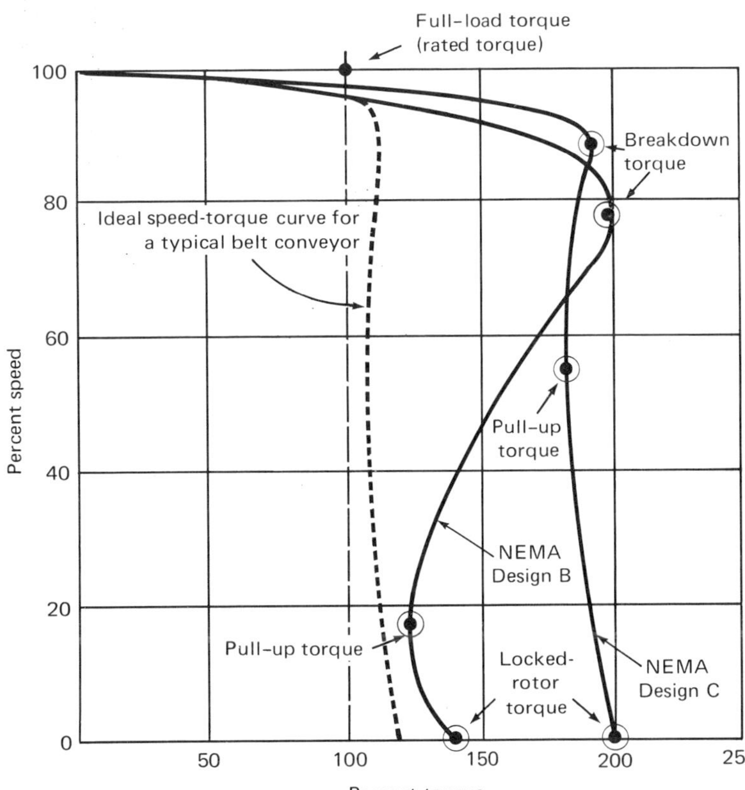

FIGURE 12.1. *Typical speed torque curves for 50-hp, 1,800-rpm NEMA Design B and C motors; ideal speed torque curve for a typical belt conveyor drive.*

TABLE 12-1. ***Designs A & B.***
Minimum Locked-Rotor Torque. % of Full-Load Torque

Hp	*Synchronous speed, rpm*							
	60 hertz	3,600	1,800	1,200	900	720	600	514
	50 hertz	3,000	1,500	1,000	750	....	....	....
½		....	....	....	140	140	115	110
¾		....	....	175	135	135	115	110
1		....	275	170	135	135	115	110
1½		175	250	165	130	130	115	110
2		170	235	160	130	125	115	110
3		160	215	155	130	125	115	110
5		150	185	150	130	125	115	110
7½		140	175	150	125	120	115	110
10		135	165	150	125	120	115	110
15		130	160	140	125	120	115	110
20		130	150	135	125	120	115	110
25		130	150	135	125	120	115	110
30		130	150	135	125	120	115	110
40		125	140	135	125	120	115	110
50		120	140	135	125	120	115	110
60		120	140	135	125	120	115	110
75		105	140	135	125	120	115	110
100		105	125	125	125	120	115	110
125		100	110	125	120	115	115	110
150		100	110	120	120	115	115	....
200		100	100	120	120	115	....	....
250		70	80	100	100	....	....	....
300		70	80	100	....	....	....	....
350		70	80	100	....	....	....	....
400		70	80	....	....	....	....	....
450		70	80	....	....	....	....	....
500		70	80	....	....	....	....	....

TABLE 12-2. ***Design C.***
Minimum Locked-Rotor Torque, % of Full-Load Torque

Hp	*Synchronous speed, rpm*			
	60 hertz	1,800	1,200	900
	50 hertz	1,500	1,000	750
3		....	250	225
5		250	250	225
7.5		250	225	200
10		250	225	200
15		225	200	200
20-200, inclusive		200	200	200

Table 12–3. ***Designs A & B.***
Minimum Breakdown Torque, % of Full-Load Torque

Hp	Synchronous speed, rpm							
	60 hertz	3,600	1,800	1,200	900	720	600	514
	50 hertz	3,000	1,500	1,000	750	....	....	....
½		....	....	....	225	200	200	200
¾		....	....	275	220	200	200	200
1		....	300	265	215	200	200	200
1½		250	280	250	210	200	200	200
2		240	270	240	210	200	200	200
3		230	250	230	205	200	200	200
5		215	225	215	205	200	200	200
7½		200	215	205	200	200	200	200
10-125, inclusive		200	200	200	200	200	200	200
150		200	200	200	200	200	200	....
200		200	200	200	200	200	....	....
250		175	175	175	175	....	....	....
300-350		175	175	175	....	....	....	....
400-500, inclusive		175	175	....	....	....	....	....

Table 12–4. ***Design C.***
Minimum Breakdown Torque, % of Full-Load Torque

Hp	Synchronous speed, rpm			
	60 hertz	1,800	1,200	900
	50 hertz	1,500	1,000	750
3		....	225	200
5		200	200	200
7½-200, inclusive		190	190	190

Table 12–5. ***Pull-up Torque***

The pull-up torque of Designs A and B, single-speed, polyphase squirrel-cage motors, with rated voltage and frequency applied, is not less than the following:

Locked-rotor torque from Table 12-1	*Minimum pull-up torque, %*
110% or less	90% of Column 1
Greater than 110%, but less than 145%	100% of full-load torque
145% or more	70% of Column 1

The pull-up torque of Design C motors, with rated voltage and frequency applied, is not less than 70% of the locked-rotor torque from Table 12-2.

Table 12-1 shows that the minimum locked rotor torque for a Design B motors rated 200 hp or less varies from 100% to 275% of the full-load torque, depending on horsepower and speed. For conveyor drives requiring higher locked-rotor torques, NEMA Design C motors can be used. The speed torque curve of Design C motors approximates the ideal straight-line speed torque curve (see Figure 12.2). However, because NEMA C motor accelerating torque remains reasonably constant at values between 190% and 250% of full-load torque, with the possible exception of "pull-up" torque, this output makes it possible to overstress a conveyor belt during the starting period. The excess torque disadvantage can be overcome by reduced voltage starting. The load torque for a conveyor with a steady load is normally a constant torque from zero to full speed. Therefore, it is important to consider the *minimum motor torque* between "locked-rotor" torque and "breakdown" torque. This is called "pull-up" torque and should never be less than the conveyor full-load torque. Table 12-2 lists the minimum NEMA motor pull-up torque values.

Motors rated in excess of 200 hp for Design C and 500 hp for Design B are not covered by NEMA standards. Also, between 200 and 500 hp, not all motors at all speeds are part of NEMA standards. The standard Design B motor above 100 hp may have a starting torque of 100% or less; therefore, the motor manufacturer should be consulted before such motors are selected for conveyor drives.

The following illustrates a typical application involving motor torque considerations. Refer to Chapter 6, Problem 1, "Inclined belt conveyor," Figure 6.20. The first torque consideration to be satisfied is that the motor locked-rotor torque should exceed twice the conveyor friction torque requirements plus the lift requirement. The tension, T_e, for this conveyor is 15,853 lbs. This includes the lift, which represents 7,995 lbs. The friction requirements are therefore:

$$15{,}853 - 7{,}995 = 7{,}858 \text{ lbs.}$$

The horsepower required for starting the conveyor is:

$$\text{Belt horsepower at start} = 2\,\frac{(7{,}858)(500)}{33{,}000} + \frac{(7{,}995)(500)}{33{,}000} = 359$$

Converting the starting belt horsepower to starting locked-rotor motor torque, assuming a motor full-load speed of 1,750 rpm and adding for drive pulley and speed reduction losses:

$$\frac{(359 + 1.52 + 12.09)(5{,}250)}{1{,}750} = 1{,}118 \text{ lb-ft}$$

A dual-motor drive is indicated as the most economical drive for the conveyor in Problem 1. The total calculated horsepower at the motor shafts is 253.8. It is assumed that a 150 hp primary motor and a 125 hp secondary motor will be used. The rated torque of the combined motors, at 1,750 rpm, will be:

$$\frac{(275)(5{,}250)}{1{,}750} = 825 \text{ lb-ft}$$

Therefore, the motors selected should have a minimum combined locked-rotor torque of 1,118/825 times, or 135% of, full-load torque.

The above example is based on constant voltage at nameplate rating, to be supplied to the motor during starting. If this is not the case, then corrective factors should be included on the basis that the motor torque varies directly as the square of the voltage.

Another important consideration is the characteristic of the motor speed torque curve. *The curve should not droop below a line drawn from the locked-rotor torque requirement to the running torque requirement at full speed.* Assuming that 1,800 rpm motors will be used in this application, reference to Table 12-1 shows that a NEMA B motor of 150 hp rating at rated voltage will have a locked-rotor torque of 110% of full-load torque. Likewise, a 125 hp motor will have a locked-rotor torque of 110% of full-load torque. Obviously, these motors will *not* meet the starting torque requirement for the conveyor in question. There are a number of possible solutions:

1. Use NEMA B squirrel-cage motors and fluid, electromagnetic, or other type of couplings which will allow the motors to accelerate at zero load and engage the connected load at the proper point on the speed-torque curve.
2. Use NEMA C squirrel-cage motors with or without reduced voltage starting, depending on limitations placed on maximum allowable starting tension by the belting manufacturer.
3. Use a 150 hp wound-rotor motor and a 125 hp squirrel-cage motor.
4. Use two wound-rotor motors.

All of the above solutions would require careful analysis to ensure sufficient accelerating torque without exceeding the maximum starting tensions imposed by the belting manufacturer.

Reduced-Voltage Starting

In any of the methods of reduced-voltage starting, the torque is reduced by the square of the voltage. In other words, the motor torque = (full-voltage torque) $\times$ (applied volts)^2/(rated volts)^2. This also reduces the in-rush current which accompanies full-voltage starting. Reduced-voltage starting of squirrel-cage motors can be accomplished by the use of primary resistance, auto-transformer, or reactor starters.

Reduced-voltage starting is illustrated by the speed-torque curve in Figure 12.2. With voltage at the motor terminal reduced to 80% by either primary resistance or auto-transformer-type reduced-voltage starters, the torque is reduced to 64% of its value, as shown in Figure 12.2. The timer used to transfer to full voltage should be set to allow the conveyor to accelerate to close to full speed in order to minimize the torque bump when transferring to full speed. Also, closed-circuit transition, available on autotransformer starters, helps reduce this torque bump.

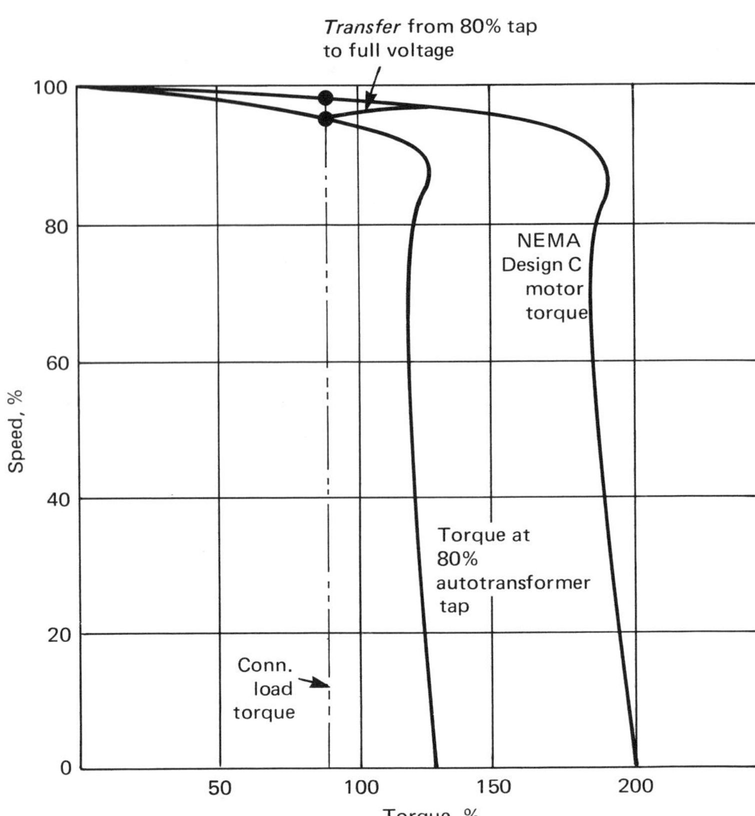

Figure 12.2. *Reduced-voltage starting characteristics using autotransformer or primary resistor.*

A smooth start without the transition shock can be achieved by a "solid-state starter" in which silicon-controlled rectifiers (SCRs) are used with regulators to control the voltage applied to the motor. The two types of regulators most commonly used provide either "current-limit starting" or "linear-timed acceleration." Typical characterisitics for the two types of regulators on solid-state starters used with NEMA Design C motors are shown in Figures 12.3 and 12.4.

As implied in the discussion of Table 12–1, care must be taken to make certain that there is adequate torque available to start the belt *under the most severe loading conditions.* This is especially true if a reduced-voltage starter of any type is used, or if there is a possibility of a voltage dip in the power supply to the motor when starting. For example, if the selected motor must deliver 100% rated motor torque to run a loaded belt, a slightly higher torque may be necessary to overcome static friction in order to start the loaded belt, and some additional torque is necessary to accelerate the total drive inertia. If a 100 hp 1,800 rpm NEMA Design B motor is being considered, and there is a possibility of 10% voltage dip at the motor terminals when full voltage is applied, the guaranteed starting torque is only $(.9)^2 \times 110 = 89.1\%$, which is not enough to ensure that the loaded belt will start. If the selected motor is stepped up to 125 hp, the loaded belt requires only 80% of rated motor torque, but the guaranteed locked-rotor torque of the 125 hp motor is only 110%, resulting in 89.1% if there is a 10% voltage dip with a full-voltage starter. The conveyor manufacturer must determine whether or not that is enough to ensure starting the loaded belt.

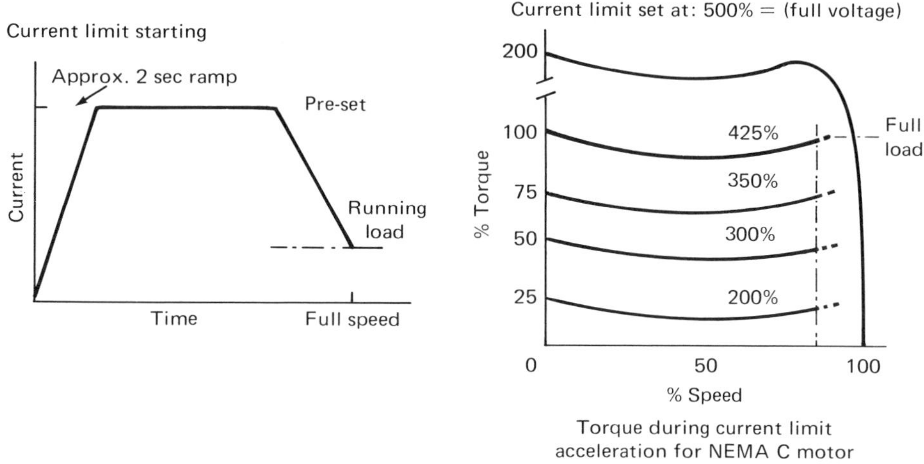

FIGURE 12.3. *Reduced-voltage solid-state starter.*

In comparison, note that a 125 hp NEMA Design C motor will provide a locked-rotor torque of 200%. If a solid-state starter with a current-limit regulator adjustable up to 425% current is being considered and the locked-rotor current at full voltage is 600%, the guaranteed starting torque is $200 \times (425/600)^2 = 100\%$. Again, a loaded conveyor requiring 100% motor torque for steady-state running would probably not start. However, if a linear-timed regulator is used for the same installation, the motor voltage will increase to the value required for breakaway. If necessary, the SCRs will phase full on and provide full voltage, although in the above example it is quite likely that the belt will start before the voltage builds up to 80%, where the locked-rotor torque will be $200 \times (.8)^2 = 128\%$. As soon as the drive tachometer associated with the linear-timed regulator indicates that the

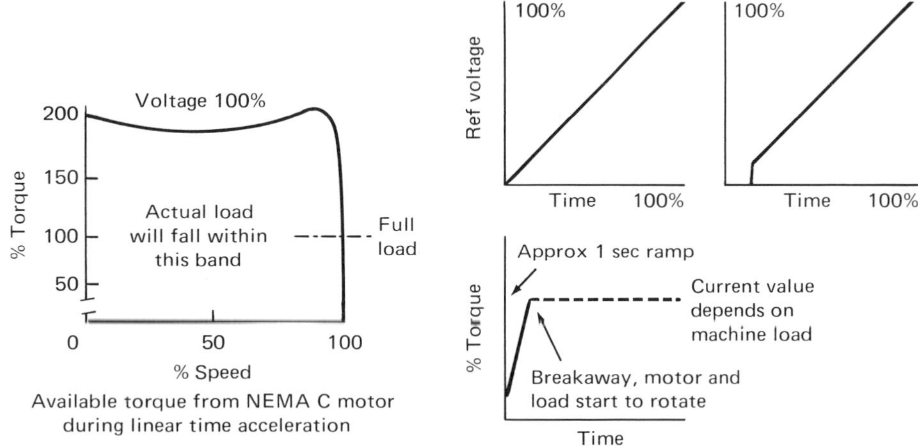

FIGURE 12.4. *Linear timed acceleration.*

motor is rotating, the voltage to the motor is reduced immediately to the value necessary to accelerate the drive at the selected rate. An empty belt will accelerate at the same rate as a fully loaded belt, which, of course, means that the motor torque and the forces in the conveyor drive will be at lower values than those required for the loaded belt. The conveyor manufacturer must determine what the setting of the linear-timed regulator should be to remain within the desired maximum motor-torque limitation. When acceleration is complete, the SCRs are locked "on" in full conduction, providing full line voltage to the motor.

Accelerating Time and Starting Frequency

Conveyors with large mass often require a long period of time to come up to normal running speed. When applying squirrel-cage motors to such conveyors, it is necessary to check their *thermal capacity.* A general rule of thumb requires NEMA Design B motors to be accelerated to full speed in 15 seconds and NEMA Design C motors in 10 seconds, although it is not unusual for a motor and solid-state starter to be suitable for 30-second acceleration. The motor manufacturer should be consulted where acceleration times are close to or exceed these values. Locked-rotor time should not exceed 6 seconds.

If frequent starting is likely to be required, the suitability of the motor and control being used should be checked with the manufacturer. Starting duty imposes severe mechanical stresses within a squirrel-cage motor, as well as a rapid increase in winding temperature. NEMA allows only *2 starts* in succession for motors of 250 to 500 hp, starting with the motor at ambient temperature. Smaller motors can be started more often; however, consideration should be given to a wound-rotor motor if the conveyor is to be started frequently. AC motors can be provided with built-in heat-responsive detectors, which can be used to sound an alarm and/or shut down the power supply.

Regenerative Braking

Standard squirrel-cage motors, when driven beyond their synchronous speed by external means, will become generators and therefore exert a braking torque. The amount of torque is identical to the accelerating torque, but in reverse. For this reason, the squirrel-cage motor is used on a declined conveyor, if the belt and its load are such that the motor is driven beyond its synchronous speed.

The faster the motor is forced to run beyond its synchronous speed, the greater its restraining torque. This is apparent from the speed-torque curve, as it will be noted that a NEMA Design B motor has its greatest torque at the breakdown point.

There are three factors to check in drives of this nature; first, the motor must have sufficient continuous torque to restrain the load. If it does not, then an overload condition exists and the motor will be disconnected from the power supply through the action of its overload device. A brake is necessary as back-up protection. Also, centrifugal switches, set for a critical speed, are commonly used to shut down the motor and apply the brake.

Second, the power developed by the motor acting as a generator must be absorbed by other drives or devices capable of using electric power. The power distribution system must accommodate this situation.

The third factor is the thermal adequacy of the brake(s) to stop a loaded conveyor in the event of a power failure or other emergency. Brake torque will decrease when the wheel is hot. To make certain that at least 90% of rated brake torque is available, the hp-secs. generated in the brake wheel must not exceed 50% of the value that results in a wheel temperature of 125 °C. If the brake wheel becomes so hot that the torque drops below the value required to stop the loaded belt, a "runaway condition" can result.

Wound-Rotor Motors

For long, high-capacity conveyors or for difficult starting conditions where squirrel-cage motors are not adequate, wound-rotor motors can be used. The wound-rotor motor permits torque control from a few to 20 steps of acceleration by the addition of external resistance to the secondary winding. Magnetic contactors or motorized drum switches are used to short out secondary resistance as the driving motor and conveyor come up to speed. The magnetic contactors can be actuated automatically by timing, frequency, or current relays.

Proper selection of resistance values and time or current values permits development of a specific pattern of accelerating torque to suit a particular conveyor. The speed-torque-curve in Figure 12.5 illustrates how a wound-round motor with 11 steps can be used to provide an average torque of 160% during acceleration, with very small plus or minus variation. Use of additional steps would decrease the variance even further.

Special wound-rotor motor drives can be used to refine further the acceleration of a conveyor. These involve both primary and secondary resistance, and the selective removal of each during starting to provide virtually constant torque. Also, a reactor and resistance network in the secondary of a wound-rotor motor will produce a highly desirable set of speed-torque characteristics.

Variable-Speed Drives

The majority of conveyors require only single-speed operation. There are cases, however, where two or more speeds are necessary. Typical examples

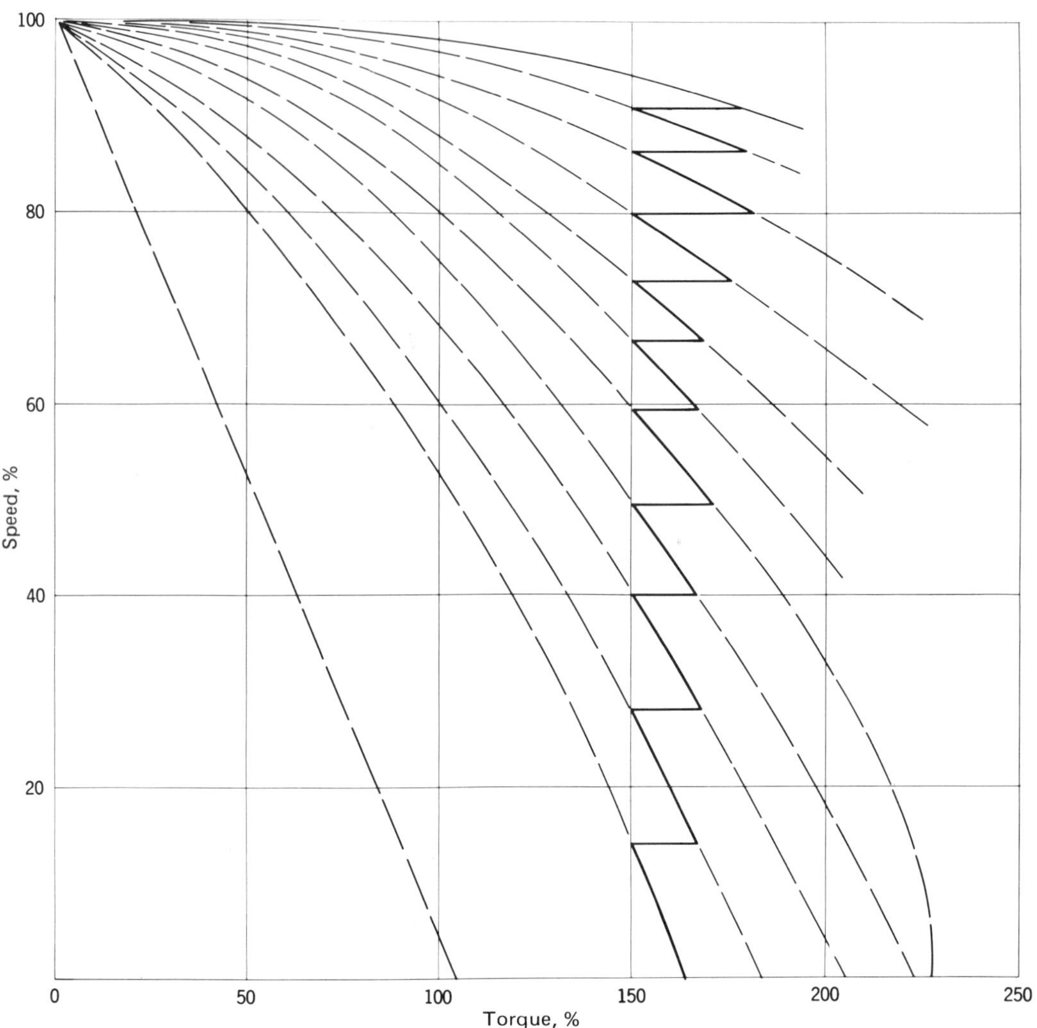

FIGURE 12.5. *Speed torque curves, 11-point starting, wound-rotor motors.*

are belt conveyor feeders or conveyor belts handling several materials of greatly differing weights. Two- and four-speed squirrel-cage motors are available in either constant-torque or constant-horsepower design. Constant-speed squirrel-cage motors connected to mechanical variable speed transmissions offer a simple method of providing variable speed to a conveyor. These can be arranged for manual, remotely operated, or automatic speed change.

Direct-current motors operating with an adjustable voltage system also can be used for adjustable speed, with either a motor-generator set or a static power supply. These adjustable voltage drives have been used on several high-capacity installations, as well as on many feeders, where accurate control of the flow of material is essential for efficient operation of the facility. Adjustable-frequency operation of squirrel-cage motors can be considered, especially if several motors start and stop simultaneously.

Another method of controlling the speed of conveyor drives is to employ constant-speed squirrel-cage motors and adjustable electromagnetic or fluid-type couplings. These are discussed in Chapter 6, "Devices for acceleration, deceleration, and torque control," page 169.

Ambient Conditions and Temperature Rise

Ambient and atmospheric conditions affect the selection of a motor. Open and totally enclosed fan-cooled motors are rated at a specific temperature rise at 40 °C ambient, when carrying their rated horsepower. Measured by resistance, a Class A insulated motor is rated at 60 °C rise, Class B at 80 °C, Class F at 105 °C, and Class H at 125 °C. When required for high-ambient temperature or high altitude, it may be possible to specify motors having Class F insulation rated at Class B rise, etc. Where the temperature rise of a motor is important, the motor manufacturer's catalog should be consulted for current manufacturing standards, insulations, ratings, and temperature rises.

Ambient Conditions

Where ambient temperatures are in excess of 40 °C, motors with larger frames may be necessary to keep the motor insulation at a safe temperature and so rctain "normal" motor life. An increase in ambient temperature of 10 °C may shorten insulation life by 50%.

If temperature variations are present over 24-hour periods and motors do not operate continuously, it may be necessary to install heaters within the frame to reduce the electrical hazard of condensation on the windings. The heaters are energized when the motor is not operating. Also, condensate drain plugs are necessary to permit a complete removal of condensed moisture.

Altitude

Because the thinner air at high altitudes reduces the cooling capacity of motors, larger frames may be necessary to dissipate the heat. However, motors built for use at sea level will operate satisfactorily at altitudes up to 3,330 ft.

Motor Enclosures

The type of motor enclosure to be selected depends on the material being conveyed and the amount of dust in the atmosphere. While open, drip-proof motors are used to some extent, the majority of conveyor applications use totally enclosed, fan-cooled motors.

If the material conveyed is explosive or dust accumulation can create a hazardous condition, totally enclosed, explosion-proof motors are used. These carry Underwriter's Laboratories, Inc. labels. The two common labels are Class II, Group F (for use in areas containing carbon black, coal, and coke dusts), and Class II, Group G (for use in areas containing grain dusts). Epoxy-encapsulated windings may make a drip-proof motor suitable for use in both wet and dusty areas, provided the dust is not abrasive. However, these motors do not carry Underwriter's Laboratories, Inc. approval for use in explosive-material areas. Where dust is extremely abrasive, it is advisable to incorporate grease seals or flingers at the motor bearings to protect them from possible damage. In extremely corrosive atmospheres, special noncorrodible fittings are available for use on motors.

Electrical Interlocking for Conveyor Systems

Modern high-speed, high-capacity belt conveyor systems make electrical interlocking between individual units an absolute necessity. Electrical interlocking is that provision in an electrical control system whereby failure or stoppage of any conveyor automatically will stop all conveyors feeding material to it, with the further provision that before any conveyor is started, all subsequent conveyors will be running.

In a line of conveyors, the stoppage progresses back in sequence from the point of failure to the initial source of feed. Sequence interlocking is combined with sequence starting so that it is necessary to start the last unit first to receive the load and to progress back, starting unit by unit, to the source of the feed.

It is desirable and sometimes necessary to make sure that each unit has attained full rated speed before the next unit can be started. This is especially true where the system is made up of conveyors of varying mass or inertia, requiring different accelerating times. Such a system, started under load, could flood transfer points, unless each belt attained full speed prior to the start of the unit feeding it.

In its simplest form, electrical interlocking is accomplished by connecting an interlock contact of a motor starter contactor in series with the start pushbutton of the next starter in the sequence. This method has two disadvantages: (1) It does not provide for a unit to reach full speed before the next unit is started. (2) It cannot distinguish between "motor running" and "belt running" (i.e., it cannot discern a drive failure). The first disadvantage can be overcome by introducing a time-delay relay between successive conveyor starters. However, this device will not overcome the second disadvantage.

A better form of interlock is the use of a centrifugal switch, driven by an idler pulley at the last point on the belt to start moving. These switches can be used to prevent starting of the next belt, until such time as belt speed has closed the switch contacts. Conversely, they can be used to stop the next belt if speed has decreased below a given value. Many of these switches can be obtained with variable settings for speed increase and decrease.

Belt systems, where one or more units can coast greater distances than others, require special stopping features to prevent flooding of transfer points. Brakes can be used if applicable. In some installations flywheels have been used to equalize coasting periods.

In addition to the interlocking between individual conveyors, several items require consideration in any system interlock.

Gates

Diverting gates must be correctly positioned to insure proper material flow. This can be accomplished through the use of limit switches. For a two-position gate, two switches are desirable to ensure full activation in either direction.

Tramp Iron Magnets

Where used, the proper method of interlocking a tramp iron magnet is through the use of a direct-current relay contact, in series with the pushbutton of the conveyor starter with which it is associated.

Sampling Equipment

Sampling equipment should be so arranged that it can be taken out of interlock if desired. When in interlock, it should be energized simultaneously with or prior to the conveyor feeding the primary sampler.

Safety Devices

Conveyor systems should incorporate electrical safety devices to facilitate protection of operating personnel, as well as to prevent damage to the mechanical portions of the conveyors. The most common devices normally used are the following:

Chute-Level Switches. Some transfer points and certain materials dictate the use of chute-level switches. These are intended to operate when the chute becomes nearly plugged, and are arranged to shut down the conveyor discharging to the chute. Similar switches are used in hoppers, bins, and below the discharge points of stackers.

Side-Slip Switches. On long conveyors or where belt training can be a problem, special limit switches are used to detect belt misalignment. These switches can be arranged to shut down the belt or to sound an alarm.

Emergency Stop Switches. Pull-cord switches are located along the walkway side (or sides) of conveyors, and are intended for emergency use. Maintained-contact type switches are preferred to prevent accidental restart

FIGURE 12.6. *The 8½-mile conveyor system is monitored by closed-circuit television at each of the 17 transfer points. Indicator panel in camera view at each point signals sources of trouble by flashing light.*

of the conveyor. These switches require manual reset to make the motor control circuit operable.

Travel Limit Switches. Conveying equipment moving during normal operation requires end-travel and over-travel limit switches to maintain such movement within safe limits. Examples of this type of equipment include trippers, shuttle conveyors, and stackers.

Warning Horns. Audible devices are normally used to warn operating personnel that the conveyor system is being placed in operation or that equipment is in the travel mode.

Centrifugal Switches. These devices are discussed above in regard to electrical interlocking.

Closed-Circuit Television. Figure 12.6 illustrates a modern and efficient control center employing television to monitor a complex and extended belt conveyor system.

Control Centers. Figure 12.7 shows an example of a well-developed master control center and console which provide for efficient operation of belt conveyor systems.

Computer Control and Multiplexing of Belt Conveyor Systems

Because of recent developments in electrical and electronic control, master control systems can now be applied to manage, monitor, optimize, and record the performance of large and complex conveyor systems.

The growth of this recent technology can be traced to the period when relay-based control was replaced by transistors and integrated circuitry. Pro-

FIGURE 12.7. *Master control center and console.*

grammable controllers and minicomputers can be directed by supervisory and management-level computerization to optimize the operation of a system based on continuous data processing. Simultaneously, the computer can perform diagnostic, production, and maintenance monitoring.

Programmable Controllers

The "Programmable Controller" (PC) is a system consisting of solid-state components that can be easily programmed to control repetitive processes and functions. Figure 12.8 illustrates the application of a PC to a belt conveyor system. The central processing and memory units contain the

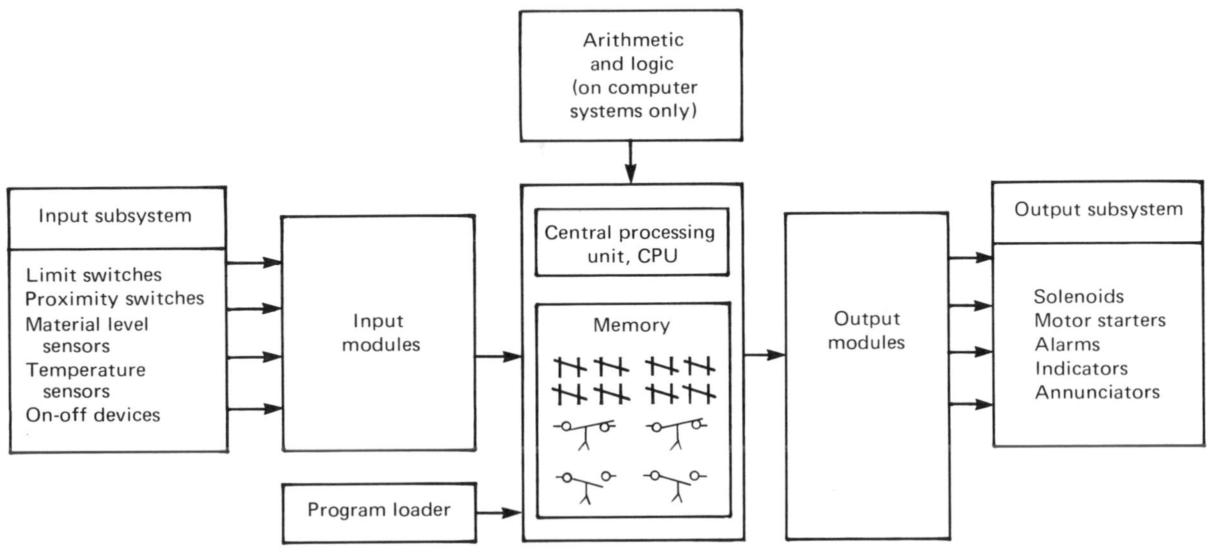

FIGURE 12.8. *Programmable controller (PC).*

operating program that controls the PC internal components and the conveyor system operational program. Although Figure 12.8 illustrates the memory as groups of contacts and limit switches, the logic is generally stored in semiconductor memory. The input and output voltage levels can be the same as with relay logic.

Many PC systems can be programmed using ladder or elementary diagrams. With very little training, plant electricians or semiskilled maintenance personnel can program PCs and even higher-level devices.

Typical functions of a PC include:

- Simple relaying;
- Timing and counting;
- Shift registering;
- Adding, subtracting, multiplying, dividing;
- Comparing;
- Producing message printouts—production reports;
- Communicating with a computer.

A battery backup is available for units utilizing semiconductor memory so that the program content will not be lost when electric power is turned off.

Some of the benefits of using PCs are:

- Lower design and contruction costs when compared to hard-wired or relay logic systems;
- Ease of expanding and checking out new programs to meet changes in application requirements;
- Reduced costs and time required for field start-up;
- Fast operating speed, especially in regard to controlling simultaneous events.

It is important to note that field changes to the logic can generally be made without wiring changes. The program is generally placed into the PC using a keyboard unit. It plugs into the PC and is disconnected after the complete program is entered. Most units allow simulation of the program to check the logic and the correctness of program entry. Portions of the logic can be changed during or after program entry (see Figure 12.9).

The cost of a PC system can vary greatly according to the number and type of functions, the amount of memory, and the relative speed of execution. Therefore, selection should be based on a careful evaluation of the system requirements.

Computerized Belt Conveyor Systems

A control system involving a large number of relays and logic functions must be flexible. Where there is considerable computation required, a computer may be the best application. The computer stores instructions and data in a memory, directs and controls the conveyor system operation, and can record the results in the memory if desired.

For many material-handling systems, a small computer—or minicomputer—can control the system and operate in real time, particularly

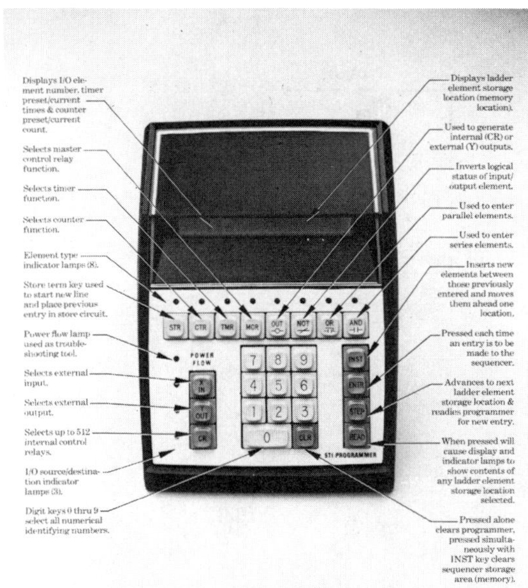

FIGURE 12.9. *Programmer panel explanation.*

where a change in input may require some logic or arithmetic function to be performed immediately. The input data typically comes from limit switches, pushbuttons, sensors, or other devices which tell the computer the state of the system at any point in time. Analog signals, such as material level height,motor speed, or scale information, must be converted to digital form for use by the computer. If the desired equipment control signal must be analog then the signal from the computer is converted to analog.

Computers can control peripheral devices such as teletypes, printers, magnetic tape cassettes, PCs, or other computers and controllers.

Remote Control Through Multiplexing

A multiplex (MUX) system receives many signals and transmits them over a coaxial cable or twisted pair of wires to a distant or remote location, where it then redistributes the signals. Figure 12.10 illustrates the system's limited wiring requirements. The economics of a MUX system are very favorable, especially when the cost is compared with that of a conventional wire and conduit system over appreciable distances. Most MUX units incorporate trouble-shooting aids which permit fast diagnosis of a failure. Figure 12.10 illustrates a simplex system used for transmission in one direction only. Hardware is also available for a duplex operation with nonsimultaneous transmission in either direction, or a full duplex system which allows simultaneous transmission in both directions.

Several factors must be considered in choosing a multiplex system appropriate for use in a particular conveyor system. Some of the most important considerations are noted below.

"Stand-Alone," Send-and-Receive Units versus a Central MUX Unit. For small systems, the "stand-alone" unit is generally lower in cost, whereas the

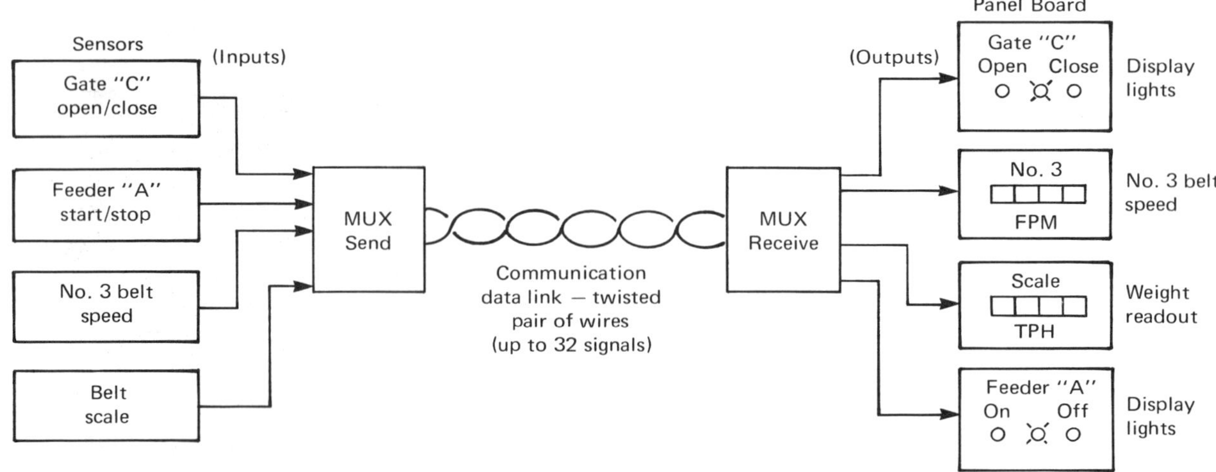

FIGURE 12.10. *Multiplex wiring system.*

central MUX unit approach is more economical for use in large systems. See Figure 12.11.

Choice of Code or Method of Transmission to Ensure Message Reliability. The system must not only reject incorrect transmissions but also ensure that data will arrive at its destination. One difficulty is intense lightning storms.

Rate and Distance of Transmission and Type of Wire. With some relatively low-speed transmission systems, which send only a few thousand bits of information per second, messages can be sent over long distances, with the use of standard wire, and in relatively noisy electrical environments. On the other hand, high-speed transmission systems, handling several hundred thousand bits of information per second, require a special wire and are limited to shorter transmission distances. At additional cost, a repeater can be added to increase the distance.

System Wire Configuration. The MUX unit can consist of individual wires between send/receive stations, a series wire configuration, or a daisy-chain wire configuration. Most system hardware is designed to operate with one particular type of wire configuration. Also, with some system configurations, transmission can continue after a wire is broken.

It should be noted that the inputs to the MUX units are digital, "on" or "off." Analog signals require conversion to digital (A/D conversion) prior to transmission. Similarly, if a multiplexed signal is to operate analog devices, a digital to analog (D/A) conversion is required.

Computer control and multiplexing offer the control engineer a means for energy management by which various elements of a conveyor

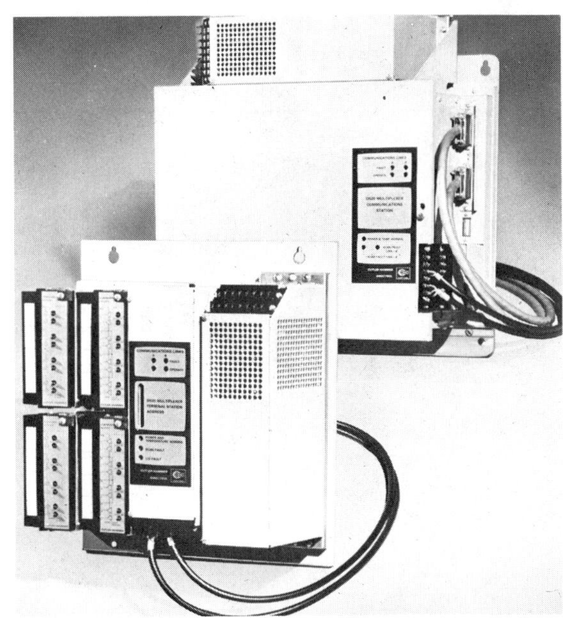

FIGURE 12.11. *Stand-alone signal multiplexing system*

system can be automatically turned off by the computer when not required for operation. With perhaps little or no additional equipment, important savings can be realized in terms of energy costs. Increasingly, belt conveyor material handling systems will use computers for real-time control, data processing, and management information.

Chapter 13

Operation, Maintenance, and Safety

Contents

The preceding chapters of this book have provided engineering information that can form the basis for the proper application and design of high-quality belt conveyors and conveyor systems. To acquire optimum performance, dependability, and cost economy, equal importance and consideration must be given to the installation, safe operation, and proper maintenance of the system and all related equipment.

Generally, each conveyor system is installed in a different location and facility and will have unique performance requirements, design features, and operating environment. All of these special considerations must be studied and evaluated when developing the best operating procedures and the safest work place.

Because of the number and range of these variables, this chapter can provide only a brief overview of some of the most important and generally applied practices. Conveyor manufacturers are often able to offer assistance in establishing the best operating maintenance and safety programs.

Operation

Characteristically, the operation of truck and rail haulage systems for bulk materials requires a relatively large number of trained personnel. Operators are needed for scheduling and master control of the system, as well as to dispatch individual units and drive large vehicles. Employees are also required for such additional operations as loading and discharge and weighing of the material.

Many belt conveyors, when appropriately designed, installed, and operated, will perform continuously and dependably with as few as one or two operators. One basic requisite is that the material being handled by the conveyor has the originally specified physical properties and is fed uniformly and at the design rate.

Performance of a system can be continuously monitored by a combination of modern electrical controls, built-in safety sensors and devices, closed-circuit TV, and other signal systems. Figure 12.7 illustrates an operator at a central control panel equipped so that he can monitor performance even at remote areas.

Depending on the length and complexity of a system, one or perhaps two trained mechanics should patrol the system at regular intervals to detect any conditions or components that need attention. On a 5½-mile overland system, two mechanics and a supervisor can easily perform such inspections.

At the time of installation and during trial runs of a large and complex conveyor system, it is advantageous to offer a program of instruction and discussion for all personnel who will be involved in operation, safety, and maintenance of the system. Such programs should be repeated at sufficiently frequent intervals so that trained personnel can update their knowledge and new employees can become indoctrinated.

Maintenance

It is important that belt conveyor maintenance be performed only by well-trained and competent personnel who are provided with proper test equipment and good tools. They should receive training in the value and conduct of a comprehensive preventive maintenance program.

Although one stuck idler roll may not appear important, maintenance personnel should realize that under a high-speed belt handling abrasive material, its shell could soon wear through, presenting a knife edge which could severely damage an expensive belt. Well-trained personnel would be able to detect impending failure in such a case, and correct the malfunction before any damage could occur.

The conveyor belt often represents a high proportion of the total conveyor cost. Since the composition and construction of the belting makes it vulnerable to accidental damage and/or accelerated wear, belt operation and maintenance deserve special attention in a good training program to minimize replacement and repair cost.

Premature wear or accidental damage may result from loading improper grades, sizes, or volumes of conveyed material onto the belt. Also, foreign materials, such as tramp iron, spikes, timbers, and the like, when entrained in the flow, can cause expensive shutdowns and costly repairs.

Table 13-1 lists the causes and cures of a wide range of operational problems that relate to the belt. It can serve as a good reference both in a training course and as a maintenance instruction.

It is highly recommended that before initial operation of a conveyor on production runs there should be careful and detailed inspection of the conveyor and all of its components. Following such a close inspection, test runs should be made to allow careful observation of actual operation before conveying material. During these inspections and test runs, the alignment of all the mechanical components and the operating alignment of the belt on the carrying and return idlers should be checked. Refer to Chapter 5 for information on belt and idler alignment.

Inspection should also establish that there are no construction materials, tools, or projecting members that can rub, tear, or cut the belt when it starts up. Chute skirtboards should have been installed so that they do not touch the belt. The rubber edging strips on the skirts should be adjusted so that they rest only lightly on the belt surface. Belt scrapers should be observed and a final adjustment made, if necessary.

Modern electrical control systems can incorporate computers and other automatic means for measuring performance and controlling such functions as weighing, mixing, blending, and material flow paths. Sensors and other devices for indicating maintenance requirements and unsafe conditions may be an integral part of an electric control system. The electrical control system should be given a thorough checkout and dry run, or ''bug out,'' during the initial test runs.

TABLE 13-1. ***Belt Conveyor Troubleshooting—Causes and Cures***

COMPLAINT	CAUSE *In Order of Probable Occurrence*					
Belt runs off at tail pulley	7	15	14	17	21	—
Entire belt runs off at all points of the line	26	17	15	21	4	16
One belt section runs off at all points of the line	2	11	1	—	—	—
Belt runs off at head pulley	15	22	21	16	—	—
Belt runs to one side throughout entire length at specific idlers	15	16	21	—	—	—
Belt slip	19	7	21	14	22	—
Belt slip on starting	19	7	22	10	—	—
Excessive belt stretch	13	10	21	6	9	8
Belt breaks at or behind fasteners; fasteners tear loose	2	23	13	22	20	10
Vulcanized splice separation	13	23	10	20	2	9

COMPLAINT	CAUSE *In Order of Probable Occurrence*					
Excessive wear, including rips, gouges, ruptures and tears	12	25	17	21	8	5
Excessive bottom cover wear	21	14	5	19	20	22
Excessive edge wear, broken edges	26	4	17	8	1	21
Cover swells in spots or streaks	8	—	—	—	—	—
Belt hardens or cracks	8	23	22	18	—	—
Covers become checked or brittle	8	18	—	—	—	—
Longitudinal grooving or cracking of top cover	27	14	21	12	—	—
Longitudinal grooving or cracking of bottom cover	14	21	22	—	—	—
Fabric decay, carcass cracks, ruptures, gouges (soft spots in belt)	12	20	5	10	8	24
Ply separation	13	23	11	8	3	—

1 **Belt bowed**—Avoid telescoping belt rolls or storing them in damp locations.* A new belt should straighten out when "broken in" or it must be replaced.

2 **Belt improperly spliced or wrong fasteners**—Use correct fasteners. Retighten after running for a short while. If improperly spliced, remove belt splice and make new splice.* Set up regular inspection schedule.

3 **Belt speed too fast**—Reduce belt speed.

4 **Belt strained on one side**—Allow time for new belt to "break in." If belt does not break in properly or is not new, remove strained section and splice in a new piece.*

5 **Breaker strip missing or inadequate**—When service is lost, install belt with proper breaker strip.

6 **Counterweight too heavy**—Recalculate weight required and adjust counterweight accordingly.* Reduce takeup tension to point of slip, then tighten slightly.

7 **Counterweight too light**—Recalculate weight required and adjust counterweight or screw takeup accordingly.

8 **Damage by abrasives, acid, chemicals, heat, mildew, oil**—Use belt designed for specific condition. For abrasive materials working into cuts and between plies, make spot repairs with cold patch or with permanent repair patch. Seal metal fasteners or replace with vulcanized step splice. Enclose belt line for protection against rain, snow, or sun. Don't over-lubricate idlers.

9 **Differential speed wrong on dual pulleys**—Make necessary adjustment.*

10 **Drive underbelted**—Recalculate maximum belt tensions and select correct belt. If line is over-extended, consider using two-flight system with transfer point. If carcass is not rigid enough for load, install belt with proper flexibility when service is lost.

11 **Edge worn or broken**—Repair belt edge. Remove badly worn or out-of-square section and splice in a new piece.

12 **Excessive impact of material on belt or fasteners**—Use correctly designed chutes and baffles. Make vulcanized splices. Install impact idlers. Where possible, load fines first. Where material is trapped under skirts, adjust skirtboards to minimum clearance or install cushioning idlers to hold belt against skirts.*

13 **Excessive tension**—Recalculate and adjust tension. Use vulcanized splice within recommended limits.

14 **Frozen idlers**—Free idlers. Lubricate. Improve maintenance. (Don't over-lubricate.)

15 **Idlers or pulleys out-of-square with center line of conveyor**—Realign. Install limit switches for greater safety.

16 **Idlers improperly placed**—Relocate idlers or insert additional idlers spaced to support belt.*

17 **Improper loading, spillage**—Feed should be in direction of belt travel and at belt speed, centered on the belt. Control flow with feeders, chutes and skirtboards.

18 **Improper storage or handling**—Refer to the manufacturer for storage and handling tips.

19 **Insufficient traction between belt and pulley**—Increase wrap with snub pulleys. Lag drive pulley. In wet conditions, use grooved lagging. Install correct cleaning devices for safety. See item 7, above.

20 **Material between belt and pulley**—Use skirtboards properly. Remove accumulation. Improve maintenance.

21 **Material build-up**—Remove accumulation. Install cleaning devices, scrapers, and inverted "V" decking.* Improve housekeeping.

22 **Pulley lagging worn**—Replace worn pulley lagging. Use Grooved Lagging for wet conditions. Tighten loose and protruding bolts.

23 **Pulleys too small**—Use larger-diameter pulleys.

24 **Radius of convex vertical curve too small**—Increase radius by vertical realignment of idlers to prevent excessive edge tension.

25 **Relative loading velocity too high or too low**—Adjust chutes or correct belt speed. Consider use of impact idlers.

26 **Side loading**—Load in direction of belt travel, in center of conveyor.

27 **Skirts improperly placed**—Install skirtboards so that they do not rub against belt.

* Consult supplier for additional recommendations and procedures.

Good housekeeping is essential for dependable operation and low-cost maintenance. A build-up of material on the deck can brake and eventually stop idler rolls, resulting in increased belt tension and possible damage to the belt. Spillage onto the return belt can also injure it as lumps are squeezed between the belt and pulleys. Scrapers on the return belt at the point where it enters the foot pulley may be desirable in some cases.

A build-up of sticky or frozen material on pulleys or idlers can cause belt misalignment and other malfunctions that can damage a belt. Pulley scrapers and/or the application of soft rubber as a pulley lagging may help to alleviate the condition. ***Caution:* No one should be allowed to attempt to remove the adhering material manually unless the conveyor has been stopped and the master electrical control locked off.**

As with all good machinery, a well-developed and mandatory lubrication program is essential for low maintenance cost and dependable operation. Because of the relatively large number of bearings in the idler rolls, and their influence on belt tensions and horsepower requirements, their lubrication is very important. Life expectancy will be enhanced by following the idler manufacturer's recommendations as to the type of lubricant, the amount and frequency of application, and the type of greasing equipment to be used.

To optimize the dependability and productivity of conveyors and to minimize maintenance costs, it will be advantageous to stock certain types of repair parts. A well developed maintenance program will provide for emergency repair of both mechanical and electrical types of equipment, and include provision of parts, parts lists, and appropriate drawings.

Climatic conditions may require some additional considerations. Special lubricants are sometimes necessary for subzero operation so as to avoid overload of the drives and undesirably large increases in belt tensions. On applications where the belt may periodically be covered with moisture or frost, it may be advisable to operate the belt empty for a brief period at start-up. Application of a belt scraper on the drive pulley side of the belt, just ahead of the point where the belt engages the drive pulley, may be a great help.

It has been shown that modern belt conveyor electrical control systems employ relatively sophisticated units. Chapter 12 explains and illustrates some of the modern automation that has been employed, including programmable controllers, computers, multiplexing, and solid-state control. Although these types of equipment are highly durable, the proper maintenance and servicing of large and complex systems can only be accomplished by specialists in this field.

CEMA member companies can be of assistance in solving special problems.

Safety

The many years of experience of leading conveyor engineers and manufacturers have shown that the development and maintenance of a safe work place requires the combined effort and cooperation of the several organizations that may become involved in a belt conveyor installation and operation.

Belt conveyor safety generally begins with sound design that, as far as is practical, avoids foreseeable dangers and hazards. Diligence in safety considerations must be applied during the course of manufacture, installation, and establishment of operating and maintenance policies and procedures.

Generally, the accidents that cause personal injury in connection with a belt conveyor are not the result of faulty design or component failure. It has been found that most accidents are caused by human carelessness, negligence, or lack of training in operations and awareness of possible hazards. In as many instances, the conveyor equipment will have a perfect safety record for some years before an improperly trained or careless worker will cause, and likely become involved in, an accident.

After the design and installation are developed and supervised by qualified engineers familiar with recognized safety features and requirements, the next priority should be personnel training. Operating and maintenance personnel and their supervisors should be initially, and then subsequently, instructed in safe operating procedures, recognizable hazards, precautions, and the maintenance of a safe work place. They should also be provided with the proper tools and equipment to operate and maintain the conveyor in an adequately safe condition. Employees who do not have proper training should be made aware of, and forbidden to enter, hazardous areas.

Since World War II, the Conveyor Equipment Manufacturers Association has taken an active role in the development of the most widely recognized and accepted safety standard for conveyors, entitled "Safety Standard for Conveyors and Related Equipment." The seventh edition was approved through the American National Standards Institute for issuance as ASME Standard B20.1-1990. B20.1 suggests that ANSI Standard B15.1, "Safety Standard for Mechanical Power Transmission Apparatus," should be used in conjunction with it.

The stated purpose of Standard B20.1 is to present certain guidelines for the design, construction, installation, operation, and maintenance of conveyors and related equipment. These guidelines and recommended safety practices will be of assistance in establishing an appropriately safe work place. It is important to realize that the best design and safety features can be useless in conjunction with faulty maintenance and operating practices.

The broad scope and fine detail of ASME Standard B20.1 preclude its inclusion in this book. However, it is highly recommended that those responsible for assuring safety in the engineering, manufacture, installation, operation, and/or maintenance of belt conveyor systems and equipment acquire and use Standard B20.1 as a reference and guide.

Guidelines for Safe Operation and Maintenance

The following general safety guidelines are not extracted directly from, and cannot take the place of, the more complete and detailed information available in the ANSI Standards B20.1 and B15.1. A brief listing is presented here simply to illustrate the type of safety considerations generally applicable to belt conveyor installations and equipment.

1. At a time close to completion of installation, all personnel and supervisors should be given a complete indoctrination in the use of the system and all of its equipment. Field inspection and classroom techniques are two valuable types of training.
2. A formal safety training program for operations, maintenance, and supervisory personnel will go a long way toward establishing and-maintaining the highest standards of safety in the work place.
3. Concurrent with completion of the installation and the trial runs of all belt conveyors and associated equipment, a ''Safety Checkup'' is recommended. The checkup should include all mechanical and electrical operating equipment, plus the structures, walkways, ladders, stairs, headrooms, and access ways. It is at this time that a detailed physical inspection of the facility and the installed conveyor equipment will often reveal the need for additional guarding, safety devices, and warning signs.
4. At no time should the conveyors be used to handle material other than that originally specified. Capacity and belt speed design ratings should not be exceeded.
5. Only trained personnel should be allowed to operate the conveyor system. They should have complete knowledge of conveyor operation, electrical controls, safety and warning devices, and the capacity and performance limitations of the system.
6. The location and operation of all emergency control and safety devices should be made known to all personnel. Surrounding areas should be kept free of obstructions or materials that could impede ready access and a clear view of such safety equipment at all times.
7. A program should be established to provide frequent inspections of all equipment. Guards, safety devices, and warning signs should be maintained in their proper positions and in good working order. Only competent and properly trained and authorized persons should adjust or work on safety devices.
8. A ''walking inspection'' of a belt conveyor system is a good means by which well-trained maintenance personnel can often detect potential problems from any unusual sounds made by such components as idlers, pulleys, shafts, bearings, drives, belts, and belt splices.
9. Hands and feet should never come in contact with any conveyor component, and no one should be allowed to ride on a moving or operable conveyor. Poking at or prodding material on the belt or any component of a moving conveyor should be prohibited. **Contact with, or work on, a conveyor must occur only while the equipment is stopped, with the electrical control locked off.**
10. No person should be allowed to ride on, step on, or cross over a moving conveyor, nor to walk or climb on conveyor structures, without using the walkways, stairs, ladders, and crossovers provided.

11. Good housekeeping is a prerequisite for safe conditions. All areas around a conveyor, and particularly those surrounding drives, walkways, safety devices, and control stations, should be kept free of debris and obstacles, including inactive or unused equipment, components, wiring, and obsolete or nonapplicable warning signs or posted instructions.
12. Any conveyor found to be in an unsafe condition for operation, or one that does not have all guards and safety devices in excellent condition, should not be used unless adequate supplementary safety devices are installed.
13. All persons should be barred by appropriate means from entering an area where falling material may present a hazard. Warning signs and barricades can be used.
14. First-class maintenance is a prerequisite for the safest operation of conveyors. Maintenance, including lubrications, should be performed with the conveyor stopped and locked out. Special lubricating equipment, lube extensions, pipes, and the like can be installed so as to permit lubrication of an operating conveyor without any foreseeable hazards.
15. Good lighting contributes to a safe working environment.
16. During the life of a belt conveyor system, its operational conditions and environment may require changes. There should be a continuing effort to detect and treat promptly any new possible safety hazards associated with these changes. If such a hazard cannot be readily eliminated, warning signs, barricades, or posted instructions should be installed.

With the increasing use of belt conveyors in the transportation of bulk materials, the number and severity of accidents have been reduced. When conveyors are used as a means of transport in the place of vehicular units, such as railcars and trucks, the problem of traffic-related accidents is minimized. Also, environmentally related health problems can be easily limited by the elimination of dust hazards. Indeed, belt conveyors have substantially reduced the hazards present in practically any other method of bulk materials handling. The further reduction in number and severity of accidents will be a direct result of applying and enforcing the safe practices of design, installation, operation, and maintenance such as have been described here and in ANSI Standard B20.1.

Appendix A

Guide for Use of SI (Metric) Units

Table A-1 is reproduced with permission of The American Society of Mechanical Engineers from the ASME Guide SI-1, *Orientation and Guide for Use of SI (Metric) Units,* ninth edition, March 24, 1982.

Portions of the text and Figure A.1 are extracted from the American National Standard ANSI/IEEE 268-1982, *Metric Practice,* and are reprinted with permission.

In the preparation of *Belt Conveyors for Bulk Materials,* the Bulk Conveyor Engineering Committee of the Conveyor Equipment Manufacturers Association carefully studied the practicality of several possible schemes of incorporating metric practice into the book.

It was found practically impossible to integrate a soft conversion of English and SI units throughout the text including the extensive engineering data, formulae, and examples. Appendix A is therefore presented to provide assistance to the reader who wishes to develop certain basic belt conveyor calculations in SI terms.

Conversion Factors

Table A–1 offers a list of commonly used conversion factors extracted from ASME guide SI–1, *Orientation and Guide for Use of SI (Metric) Units.* Conversion factors are presented for ready adaptation to computer readout and electronic data transmission. Each factor is written as a number greater than one and less than ten, with six or fewer decimal places. This number is followed by the letter E (for exponent), a plus or minus symbol, and two digits indicating to power of 10 by which the number must be multiplied in order to obtain the correct value.

Thus: 3.523 907 E – 02 becomes $3.523\ 907 \times 10^{-2}$

or 0.035 239 07

Similarly:

3.386 389 E + 03 becomes $3.386\ 389 \times 10^{3}$

or 3 386.389

An asterisk (*) after the last significant digit indicates that the conversion factor is exact and that all subsequent digits are zero. All other conversion factors have been rounded. Where less than six decimal places are shown, greater precision is not warranted.

Example

To convert from:	*to:*	*Multiply by:*	*Result:*
pound–force per square foot	Pa	4.788 026 E + 01	1 lbf/ft^2 = 47.880 26 Pa
inch	m	2.540 000*E – 02	1 inch = 0.0254 m (exactly)

The conversion factors for other compound units can easily be generated from numbers given in the alphabetical list by the substitution of converted units.

Example To find conversion factor of lb·ft/s to kg·m/s:

first convert 1 lb to 0.453 592 4 kg
and 1 ft to 0.3048 m
then substitute: (0.453 592 4 kg)(0.3048 m)/s
= 0.138 255 kg·m/s
thus the factor is 1.382 55 E−01

Example To find conversion factor of oz·in^2 to kg·m^2:

first convert 1 oz to 0.028 349 52 kg
and 1 in^2 to 0.000 645 16 m^2
then substitute: (0.028 349 52 kg)(0.000 645 16 m^2)
= 0.000 018 289 98 kg·m^2
thus the factor is 1.828 998 E−05

TABLE **A-1.** ***Commonly Used Conversion Factors***

Quantity	*Conversion*	*Factor*
Plane angle	degree *to* rad	1.745 329 E−02
Length	in *to* m	2.54* E−02
	ft *to* m	3.048* E−01
	mile *to* m	1.609 344*E+03
Area	in^2 *to* m^2	6.451 600*E−04
	ft^2 *to* m^2	9.290 304*E−02
Volume	ft^3 *to* m^3	2.831 685 E−02
	US gallon *to* m^3	3.785 412 E−03
	in^3 *to* m^3	1.638 706 E−05
	oz (fluid, US) *to* m^3	2.957 353 E−05
	liter *to* m^3	1.000 000 E−03
Velocity	ft/min *to* m/s	5.08* E−03
	ft/sec *to* m/s	3.048* E−01
	km/h *to* m/s	2.777 778 E−01
	mile/h *to* m/s	4.470 4* E−01
	mile/h *to* km/h	1.609 344*E+00
Mass	oz (avoir) *to* kg	2.834 952 E−02
	lb (avoir) *to* kg	4.535 924 E−01
	slug *to* kg	1.459 390 E+01
Acceleration	ft/s^2 *to* m/s^2	3.048* E−01
	std. grav. m/s^2	9.806 65* E+00
Force	kgf *to* N	9.806 65* E+00
	lbf *to* N	4.448 222 E+00
	poundal *to* N	1.382 550 E−01

The factors are written as a number greater than one and less than ten with six or fewer decimal places. The number is followed by the letter E (for exponent), a plus or minus symbol, and two digits which indicate to power of 10 by which the number must be multiplied to obtain the correct value.

For example

1.745 329 E−02 is 1.745 329$\times 10^{-2}$ or 0.017 453 29

*Relationships that are exact in terms of the base units.

TABLE A-1. *continued.* *Commonly Used Conversion Factors*

Quantity	*Conversion*	*Factor*
Bending, torque	kgf-m *to* N·m	9.806 65* E+00
	lbf-in *to* N·m	1.129 848 E−01
	lbf-ft *to* N·m	1.355 818 E+00
Pressure, stress	kgf/m^2 *to* Pa	9.806 65* E+00
	$poundal/ft^2$ *to* Pa	1.488 164 E+00
	lbf/ft^2 *to* Pa	4.788 026 E+01
	lbf/in^2 *to* Pa	6.894 757 E+03
Energy, work	Btu (IT) *to* J	1.055 056 E+03
	Calorie (IT) *to* J	4.186 8* E+00
	ft lbf *to* J	1.355 818 E+00
Power	*hp* (550 ft lbf/s) *to* *W*	7.456 999 E+02
Temperature**	°C *to* *K*	$t_K = t_{°C} + 273.15$
	°F *to* *K*	$t_K = (t_{°F} + 459.67)/1.8$
	°F *to* °*C*	$t_{°C} = (t_{°F} - 32)/1.8$
Temperature interval	°C *to* *K*	1.0** E+00
	°F *to* *K* *or* °*C*	5.555 556 E−01

The factors are written as a number greater than one and less than ten with six or fewer decimal places. The number is followed by the letter E (for exponent), a plus or minus symbol, and two digits which indicate to power of 10 by which the number must be multiplied to obtain the correct value.

For example
1.745 329 E−02 is $1.745\ 329 \times 10^{-2}$ or 0.017 453 29

*Relationships that are exact in terms of the base units.

** °F should be converted to °C. Absolute temperatures should be converted to K.

Metric Practice and Units for Mass, Force, and Weight

The following information is very pertinent to the application of metric practice and units to the basic calculations of belt tensions and horsepower requirements as presented in Chapter 6. It is extracted from the American National Standard ANSI/IEEE 268-1982, *Metric Practice.*

Mass, Force, and Weight

The principle departure of SI from the gravimetric system of metric engineering units is the use of explicitly distinct units for mass and force. In SI, the term *kilogram,* kg, is restricted to the unit of mass, and *kilogram-force* (from which the suffix *force* was, in practice, often erroneously dropped) is not used. In its place, the SI unit of force, the Newton, N, is used (see Figure A.1). Likewise, the Newton, rather than kilogram-force, is used to form derived units which include force.

Example	Pressure or stress	$N/m^2 = Pa$ (pascal)
	Energy	$N \cdot m = J$ (joule)
	Power	$N \cdot m/s = W$ (watt)

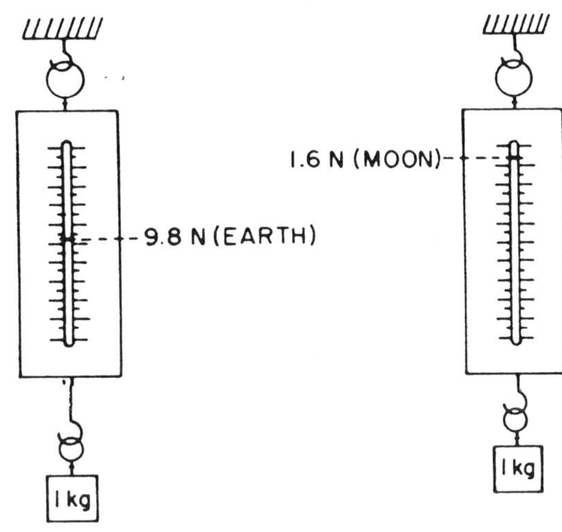

FIGURE A.1
Illustration of Difference Between Mass (Unit: kilogram) and Force (Unit: newton)

Considerable confusion exists in the use of the term *weight* as a quantity to mean either force or mass. In commercial and everyday use, *weight* nearly always means *mass;* thus, when one speaks of a person's *weight,* the quantity referred to is actually *mass.* The nontechnical use of the term *weight* in everyday life will probably persist. In science and technology, however, the term *weight of a body* has usually meant the force which, if applied to the body, would give it an acceleration equal to the local acceleration of free fall. The adjective *local* in the expression *local acceleration of free fall* has usually meant a location on the surface of the earth. In this context the *local acceleration of free fall* has the symbol g (sometimes referred to as *acceleration due to gravity*) with observed values of g differing by over 0.5% at various points on the earth's surface. The use of *force of gravity* (mass × acceleration due to gravity) instead of *weight* as a quantity, this expression should be avoided in technical practice except under circumstances in which its meaning is completely clear. When the term is used, it is important to know whether mass or force is intended, and to use SI units properly by using kilograms for mass and Newtons for force.

Similarly the use of the same name for units of force and mass causes confusion. When non-SI units are used, a distinction should be made between force and mass. For example, *lbf* to denote force in gravimetric engineering units, and *lbs* for mass.

SI (Metric Units) for Belt Conveyor, Belt Tension, and Horsepower Calculations

The following information and a sample problem will be of assistance in applying SI metric units to the calculation of belt tensions and horsepower requirements of belt conveyors.

Metric Use of CEMA Belt Tension Formulae

The basic measurement of force in the United States is pounds force, lbf. This comes from the expression:

$$F = Ma$$

Where: $M = \dfrac{W}{g}$

And: $a = \text{ft/sec}^2$

$W = \text{lbs weight}$

The units are:

$$F = \frac{W}{g}\,a = \left(\frac{\text{lbs}}{\text{ft/sec}^2}\right)\left(\text{ft/sec}^2\right) = \text{pounds force, lbf}$$

In the SI system the corresponding units are:

Newtons, N for force
kilograms, kg for mass
meters/sec², m/sec² for acceleration

By definition, a Newton is the force required to accelerate a mass of one kilogram by one meter per second per second:

$$N = \text{kg} \times \text{m/sec}^2$$

Any expression for force in Newtons must have the units kg × m/sec². Page 92 provides the summary of the components of the effective belt tension, T_e.

T_x	= idler friction	$= L \times K_x \times K_t$
T_{yc}	= belt flexure, carrying idlers	$= L \times K_y \times W_b \times K_t$
T_{yr}	= belt flexure, return idlers	$= L \times 0.015\ W_b \times K_t$
T_{ym}	= material flexure	$= L \times K_y \times W_m$
T_m	= material lift (+) or lower (−)	$= \pm H \times W_m$
T_p	= pulley resistance	= See Chapter 6
T_{am}	= accelerate material	$= 2.8755 \times 10^{-4} \times Q \times (V - V_0)$
T_{ac}	= accessories	= See Chapter 6

$$T_e = T_x + T_{yc} + T_{yr} + T_{ym} \pm T_m + T_p + T_{am} + T_{ac}$$

$$T_e = LK_t(K_x + K_yW_b + 0.015\ W_b) + W_m(LK_y \pm H) + T_p + T_{am} + T_{ac}$$

To calculate T_e in SI units, it is necessary to express all of the above values of tension in Newtons, N, and weight in kilograms, kg, and linear measure in meters, m.

W_b = Weight of belt, kg/m
W_m = Weight of material, kg/m
L = Length, m
H = Lift or lower, m, (plus for lift, minus for lower)
K_t = Factor, same as in other units
K_y = Factor, same as in other units
A_i = Force to rotate idlers, N
S_i = Idler spacing, m

$$K_x = .00068\ (W_b + W_m)\ 9.807 + \frac{A_i}{S_i},\ \text{N/m}$$

$$= .006669\ (W_b + W_m) + \frac{A_i}{S_i},\ \text{N/m}$$

(To convert A_i values in Chapter 6 to Newtons, N, multiply by 4.45.)

V = Belt speed, m/sec
V_0 = Initial velocity of material, m/sec
Q = Capacity, kg/sec
1 metric tonne = 1,000 kg

$$W_m = \frac{.2778 \text{ metric tonnes per hour}}{V} \quad or$$

$$W_m = \frac{\text{kg per second}}{V},\ \text{kg/m}$$

$$\text{k}W = \frac{T_e \times V}{1000},\ \text{kilowatts}$$

T_e = Effective tension, N

$$\text{Metric Horsepower} = \frac{\text{k}W}{.735}$$

$$\text{US Horsepower} = \frac{\text{k}W}{.7457}$$

The preceding tension formulae should be expressed in Newtons, *N*.

$$T_x = L \times K_x \times K_t$$
$$T_{yc} = 9.807\ L \times K_y \times W_b \times K_t$$
$$T_{yr} = 9.807\ L \times 0.015 \times W_b \times K_t$$
$$= 0.1471\ L \times W_b \times K_t$$
$$T_{ym} = 9.807\ L \times K_y \times W_m$$
$$T_m = \pm 9.807\ H \times W_m$$
$$T_{am} = Q \times (V - V_0)$$

T_p, T_{ac}—Calculate in pounds of tension as directed in Chapter 6 and convert to Newtons, N, of tension. Pound-force, lbf × 4.45 = Newtons, N.

$$T_e = LK_t[K_x + 9.807\ W_b(K_y + .015)] + 9.807\ W_m(LK_y \pm H) + T_p + T_{am} + T_{ac}$$

A sample problem follows for the purpose of showing the procedure and the comparison of values. In working such a problem, either the T_e formula shown above or the individual formulae can be used. For the purpose of comparison the individual formulae were used in this sample problem. A list of frequently used conversion formulae is given on page 331 along with some comparison values of belt tension and belt velocities.

Problem Comparing Application of US and SI Units

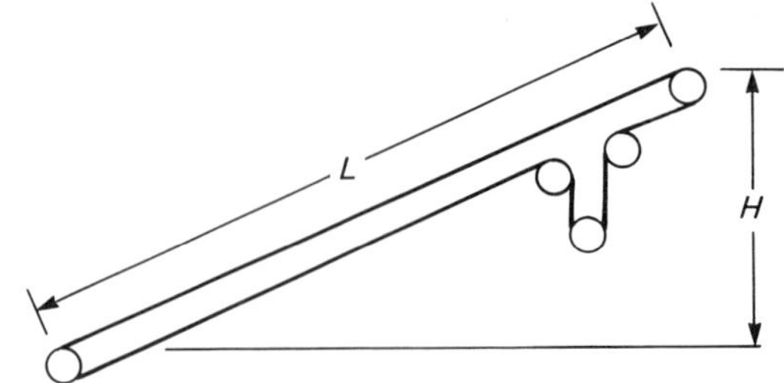

English or US		*Metric or SI*
5,760 tph	CAPACITY, Q	1,451.5 kg/sec
1,000 ft	LENGTH, L	304.80 m
240 ft	LIFT, H	73.15 m
30 lbs/ft	BELT, W_b	44.64 kg/m
320 lbs/ft	MATERIAL, W_m	476.21 kg/m
1.5 lbf	A_i	6.67 N
5 ft	S_i	1.52 m
1.2	K_t	1.2
0.018	K_y	0.018
600 fpm	VELOCITY, V	3.048 m/sec
60 in	BELT WIDTH, b	1.52 m

$$K_x = .00068(30 + 320) + \frac{1.5}{5} \qquad K_x = .006669(44.64 + 476.21) + \frac{6.67}{1.52}$$

$$K_x = 0.538 \text{ lbs/ft} \qquad K_x = 7.862 \text{ N/m}$$

$$W_m = \frac{33.33 \times 5{,}760}{600} = 320 \text{ lbs/ft} \qquad W_m = \frac{1451.5}{3.048} = 476.21 \text{ kg/m}$$

TENSIONS IN POUNDS, lbf

$$\begin{aligned}
T_x &= 1{,}000 \times 0.538 \times 1.2 &&= 646\\
T_{yc} &= 1{,}000 \times .018 \times 30 \times 1.2 &&= 648\\
T_{yr} &= 1{,}000 \times .015 \times 30 \times 1.2 &&= 540\\
T_{ym} &= 1{,}000 \times 0.018 \times 320 &&= 5{,}760\\
T_m &= 240 \times 320 &&= 76{,}800\\
T_p &= 2 \times 200 + 3 \times 150 &&= 850\\
T_{am} &= 2.8755 \times 10^{-4} \times 5{,}760 \times 600 &&= 994\\
T_{ac} &= 3 \times 60 &&= 180\\
&& T_e &= 86{,}418 \text{ lbs}
\end{aligned}$$

TENSIONS IN NEWTONS, N

$$\begin{aligned}
T_x &= 304.8 \times 7.862 \times 1.2 &&= 2{,}876\\
T_{yc} &= 304.8 \times .018 \times 44.64 \times 1.2 \times 9.807 &&= 2{,}882\\
T_{yr} &= 304.8 \times .1471 \times 44.64 \times 1.2 &&= 2{,}402\\
T_{ym} &= 304.8 \times .018 \times 476.21 \times 9.807 &&= 25{,}623\\
T_m &= 73.15 \times 476.21 \times 9.807 &&= 341{,}625\\
T_p &= (2 \times 200 + 3 \times 150) \times 4.45 &&= 3{,}783\\
T_{am} &= 1{,}451 \times 3.048 &&= 4{,}423\\
T_{ac} &= 3 \times 60 \times 4.45 &&= 801\\
&& T_e &= 384{,}415 \text{ N}
\end{aligned}$$

$$\text{US hp} = \frac{86{,}418 \times 600}{33{,}000} = 1{,}571 \qquad \text{kW} = \frac{384{,}415 \times 3.048}{1{,}000} = 1{,}172$$

$$\text{Metric hp} = \frac{1{,}172}{.735} = 1{,}594$$

Frequently Used Conversion Formulae

Pound-force, lbf	×	4.4482	=	N
Mass, lbs	×	.4536	=	kg
Length, ft	×	.3048	=	m
Velocity, fpm	×	.0051	=	m/sec
Mass per length, lbs/ft	×	1.4882	=	kg/m
Acceleration, ft/sec^2	×	.3048	=	m/sec^2
Area, ft^2	×	.0929	=	m^2
Volume, ft^3	×	.0283	=	m^3
Horsepower, hp (US)	×	745.7	=	W (Watts)

Comparison of Belt Tensions

N	lbf	N	lbf	N	lbf	N	lbf
1,000	225	10,000	2,248	50,000	11,240	200,000	44,962
2,000	450	15,000	3,372	60,000	13,489	250,000	56,202
3,000	674	20,000	4,496	70,000	15,737	300,000	67,443
4,000	899	25,000	5,620	80,000	17,985	350,000	78,683
5,000	1,124	30,000	6,744	90,000	20,233	400,000	89,924
6,000	1,349	35,000	7,868	100,000	22,481	450,000	101,164
8,000	1,798	40,000	8,992	150,000	33,721	500,000	112,404

Comparison of Belt Velocities

m/sec	fpm	m/sec	fpm	m/sec	fpm	m/sec	fpm
1.00	197	2.00	394	3.00	591	4.50	886
1.25	246	2.25	443	3.25	640	5.00	984
1.50	295	2.50	492	3.50	689	5.50	1,083
1.75	344	2.75	541	4.00	787	6.00	1,181

Appendix B

Nomenclature

The symbols and meanings listed here are those employed throughout the text for the presentation of data, equations, calculations, and definitions.

For reasons of simplification and clarity, letter symbols used only as linear dimensions in graphic illustrations as well as symbols used only for shafting calculations in Chapter 8 are omitted.

A	Area, ft^2, unless otherwise specified
AL	Adjusted idler load, lbs
A_i	Force required to rotate idler, lbs
A_t	Cross sectional area of material on a troughed belt, ft^2
a	Acceleration, ft/sec^2
aa	Skirtboard height above center of belt, in
a_1	Vertical distance, surface of belt to center of gravity of load at discharge pulley, in
B_m	Belt modulus of elasticity, lbs/in of width per ply
b	Width of conveyor belt, in
C_s	Skirtboard friction factor
C_w	Wrap factor on drive pulley or pulleys
C_{wp}	Wrap factor for the primary drive pulley
C_{ws}	Wrap factor for the secondary drive pulley
c	Edge distance, edge of material to edge of belt, in
d_m	Apparent density of material, lbs/ft^3
e	Base of Naperian logarithms, 2.718
e_t	Point on the curved belt around a discharge pulley, at which the material starts its trajectory
F	Force, lbs
F_a	Accelerating or decelerating force, lbs
F_d	Braking force at beltline, lbs
F_f	Frequency factor, minutes for belt to traverse one cycle
F_r	Resultant force on idlers at convex vertical curve, lbs
f	Coefficient of friction between pulley and belt surfaces
fpm	Abbreviation: feet per minute
fps	Abbreviation: feet per second
ft	Abbreviation: feet
ft^2	Abbreviation: square feet, area
ft^3	Abbreviation: cubic feet, volume
ft/sec^2	Abbreviation: feet per second per second, acceleration or deceleration
g	Acceleration due to gravity, 32.2 ft/sec^2

H	Vertical distance that material is lifted or lowered, ft
H_c	Vertical distance from tail pulley to start of concave vertical curve, ft
h	Vertical distance, surface of belt to top of load at discharge pulley, in
h_s	Depth of material touching skirtboard, in
hp	Abbreviation: horsepower
Hd	Lift to drive pulley (Ref: see pages: 113, 114, and 115)
IL	Actual idler load, lbs
in	Abbreviation: inch
K_t	Ambient temperature correction factor
K_x	Frictional resistance of idlers and sliding resistance between belt and idler rolls, lbs/ft
K_y	Factor for resistance of belt and load to flexure while moving over idlers
K_1	Lump adjustment factor, idlers
K_2	Environmental and maintenance factor, idlers
K_3	Service factor, idlers
K_4	Belt speed correction factor, idlers
L	Length of belt conveyor, centers of terminal pulleys, ft
L_b	Length of one skirtboard, ft
L_c	Length of belt conveyor from tail pulley to start of concave vertical curve, ft
lbs	Abbreviation: pounds
lbs/ft	Abbreviation: pounds per foot of length
lbs/ft^2	Abbreviation: pounds per square foot, pressure
lbs/ft^3	Abbreviation: pounds per cubic foot, density
M	Mass, slugs
M_e	Mass equivalent, slugs
NEMA	Acronym: National Electric Manufacturers Association
N_{ri}	Number of return idlers
n	Number of spaces between idlers at convex vertical curve
P	Total force against one skirtboard, lbs
p	Number of plies in belt
P_f	Pulley face width, in
Q	Quantity of material conveyed, tph (short tons, 2,000 lbs)
R	Resultant radial load, lbs (pulleys)
R_1	Mechanical advantage ratio, takeups
RMA	Acronym: Rubber Manufacterers Association
r	Radial distance center of dischare pulley to center of gravity of load, in
r_1	Minimum radius of concave vertical curve, ft
r_2	Minimum radius of convex vertical curve, ft
rpm	Abbreviation: revolutions per minute
rpm$_b$	Abbreviation: revolutions per minute of brake shaft

rpm_p	Abbreviation: revolutions per minute of drive pulley shaft
rps	Abbreviation: revolutions per second
S_i	Troughing idler spacing, ft
S_{ic}	Maximum troughing idler spacing on convex vertical curve, ft
sec	Abbreviation: seconds
T	Tension to overcome skirtboard friction, lbs
T_a	Tension induced in belt by accelerating forces, lbs
T_{ac}	Total of the tensions from friction of conveyor accessories, lbs
T_{am}	Tension required to accelerate material, lbs
T_b	Tension required to lift or lower the belt, lbs
T_{bc}	Tension required for belt cleaning devices, lbs
T_{1_b}	Tension in the tight side of the belt, under the head-end drive pulley, during braking, lbs
T_{2_b}	Tension in the slack side of the belt, above the head-end drive pulley, during braking, lbs
T_c	Tension in belt at beginning of a vertical curve, lbs
T_{cx}	Tension in belt at point x on carrying run, lbs
T_e	Effective belt tension at drive, lbs
T_{eb}	Equivalent braking force, lbs
T_{ep}	Effective tension at primary pulley of a two-pulley drive, lbs
T_{es}	Effective tension at secondary pulley of a two-pulley drive, lbs
T_{fcx}	Tension in belt at point x on carrying run, resulting from friction, lbs
T_{frx}	Tension in belt at point x on return run, resulting from friction, lbs
T_{hp}	Tension at head or discharge pulley, lbs
T_m	Tension needed to lift or lower material, lbs
T_{max}	Maximum tension in belt, lbs
T_{min}	Minimum tension in belt, lbs
T_0	Minimum allowable sag tension in belt for a definite idler spacing, lbs
T_p	Tension from belt flexure around pulleys plus pulley bearing friction, lbs
T_{pl}	Tension from friction of plows, lbs
T_r	Rated belt tension, lbs
T_{rx}	Tension in belt at point x on return run, lbs
T_{sb}	Tension from skirtboard friction, lbs
T_t	Tension in belt at tail pulley, lbs
T_{tr}	Tension from friction of pulleys and flexure of belt on trippers and stackers, lbs
T_{wcx}	Tension in belt at point x on carrying run, resulting from the weight of the belt plus the material carried, lbs
T_{wrx}	Tension in the belt at point x on the return run, resulting from the weight of the empty return belt, lbs
T_x	Tension from friction of carrying and return idlers, lbs
T_{yb}	Tension from belt flexure as belt rides on carrying plus return idlers, lbs
T_{yc}	Tension from belt flexure as belt rides over the carrying idlers, lbs
T_{ym}	Tension from material flexure as material rides on the belt over the carrying idlers, lbs
T_{yr}	Tension from belt flexure as belt rides over the return idlers, lbs
T_1	Tension in the belt at the tight side of the driving pulley, lbs
T_2	Tension in the belt at the slack side of the driving pulley, lbs

T_3	Tension in belt between primary and secondary drive pulleys of a dual pulley drive, lbs
TU	Abbreviation: takeup
t	Time, seconds
t_d	Actual stopping time of a belt conveyor braked or coasting to a stop, seconds
t_m	Maximum permissible stopping time, seconds
tph	Abbreviation: tons per hour (ton = 2000 lbs)
V	Design belt speed, fpm
V_0	Initial velocity of material fed onto belt, fpm
V_s	Tangential velocity of center of gravity of material discharged over head pulley, fps
W	Weight, lbs
W_b	Weight of belt, lbs/ft of length
W_c	Total weight to be accelerated by the partially loaded belt, at the beginning of a concave vertical curve, lbs
W_e	Equivalent weight of load and moving parts of a conveyor that are accelerated or decelerated by the belts, lbs
W_f	Force to overcome friction of an automatic takeup, carriage, ropes, sheaves, and other frictional resistances, lbs
W_g	Required takeup force, lbs
W_m	Weight of material conveyed, lbs/ft of belt length
W_p	Vertical component of the weight force of an automatic takeup, carriage, wheels, pulley, shaft, and shaft bearings, etc., lbs
WK^2	Moment of intertia of rotating parts, lb-in^2
W_t	Total equivalent weight of all moving parts plus the weight of the full conveyed load that must be accelerated, excluding the drive and drive pulley, lbs
Z_b	Torque rating or setting of brake, lb-ft

Belt Tension to Rotate Pulleys

This alternate method of determining belt tension to rotate pulleys is offered to those desiring a more precise value than Table 6-5 (pg. 87). This alternate method would be especially useful for calculation of smaller, lighter duty conveyors where T_p could be a large percent of conveyor total effective tension (T_e).

Following equations are very similar to those in ISO Standard 5048 First Edition 1979-09-01.

$T_{p(1)}$ = Resistance of belt to flex over pulley, lbs:

$$T_{p(1)} = 9b\left(0.8 + 0.01\frac{T}{b}\right)\frac{B_t}{D_p} \text{ (for fabric carcass belts)}$$

$$T_{p(1)} = 12b\left(1.142 + 0.01\frac{T}{b}\right)\frac{B_t}{D_p} \text{ (for steel cable belts)}$$

$T_{p(2)}$ = Resistance of pulley bearings, lubricant and seals, lbs:

(Not to be calculated for drive pulleys.)

$$T_{p(2)} = 0.01\frac{d_s}{D_p}R$$

b = width of conveyor belt, in
B_t = conveyor belt thickness, in
D_p = pulley diameter, in
d_s = pulley shaft diameter, in
R = resultant radial load (vector sum of belt tensions and pulley weight), lbs
T = belt tension at pulley, lbs

Tables C-1 and C-2 provide graphical method for quick estimate of T_p value. This method should be suitable for most conveyors and can also be used as comparison to Table 6-5.

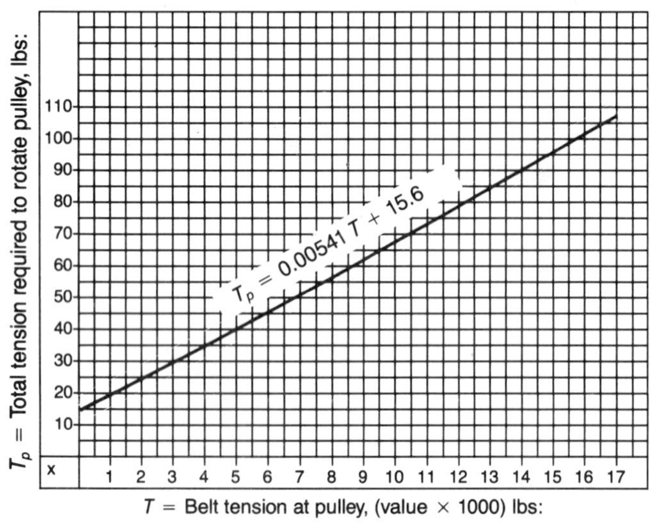

Table C-1. ***T_p Determination–Graphical Method for Fabric Carcass Belts***

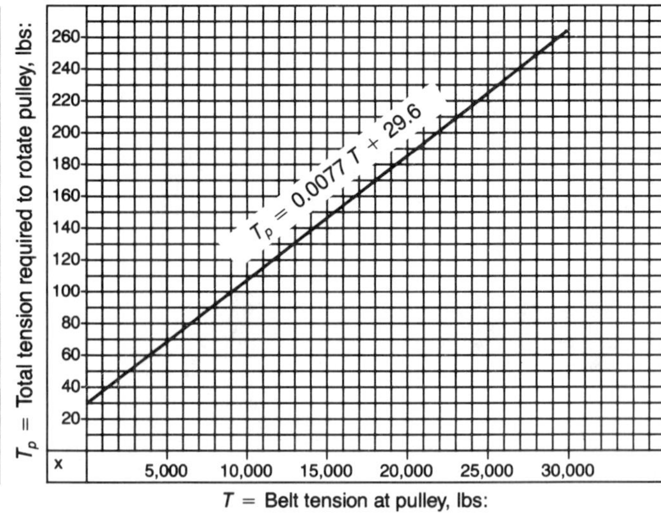

Table C-2. ***T_p Determination–Graphical Method for Steel Cable Belts***

Note: Tables C-1 and C-2 have been derived using belt thickness (B_t) = 1.00 in. If another belt thickness is used, multiply tables T_p value by actual belt thickness.

Example: T = 10,000 lbs with B_t = 0.75 in. (Ref. Table C-2 for Steel Cable Belt)

$$T_p = 107 \text{ lbs} \times \frac{0.75 \text{ in.}}{1.00 \text{ in.}} = 80 \text{ lbs}$$

Appendix D

Conveyor Installation Standards

For Belt Conveyors Handling Bulk Materials

Contents

A trouble free belt conveyor operation is the product of three (3) properly executed stages of development—

1. Design
2. Manufacturing
3. Installation

followed by an effective maintenance program. Less than satisfactory performance in any of these developmental stages will negatively impact all others, resulting in unanticipated operating problems.

CEMA has already addressed many of the design and maintenance considerations critical to proper operation. It is not CEMA's intent to specify minimum levels of manufacturing quality. Indeed, it is the responsibility of each manufacturer to produce a product of which he and the user can be proud.

This document will specify minimum standards for installation of belt conveyors—i.e., acceptable tolerances for structural and mechanical erection in addition to dynamic operating tolerances. In addition, it will provide helpful suggestions which can be utilized to meet or exceed these standards. Each item will be addressed in the sequence in which it is encountered in the field.

NOTES

1. It is important that ANSI lockout procedures be followed when making adjustments to bring conveyor machinery into tolerance (ref: ANSI B15.1, B20.1, and 244.1).

2. All mechanical tolerances should be documented by millwrights.

Conveyor Stringer Alignment

Trusses and channel frame conveyor stringers must be installed parallel, straight, square, and level to allow proper belt training. During installation, dimensional checks shall be made to insure that the following tolerances in the idler carrying chords are not exceeded:*

Parallel

A maximum tolerance of ± 1/8″ shall be allowed for the "back to back" dimension in channel frame or angle stringers. Similarly, ± 1/8″ shall be allowed between webs of I-beams, wide flange beams, or tees when used as truss chords (Figure D.1).

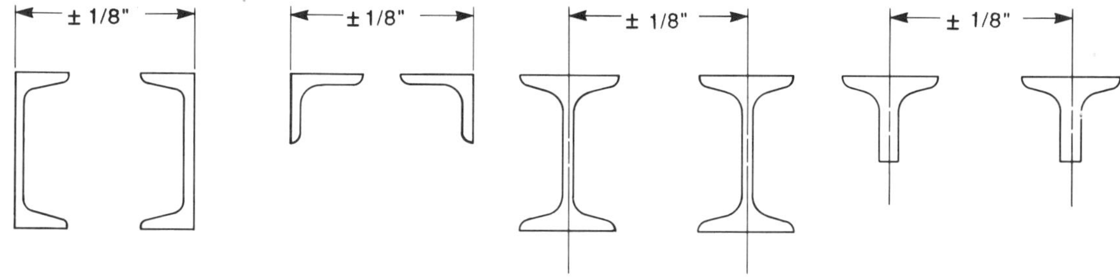

FIGURE D.1

Straightness

The maximum allowable lateral offset in conveyor stringers shall be 1/8″ in 40′ of length (Figure D.2).

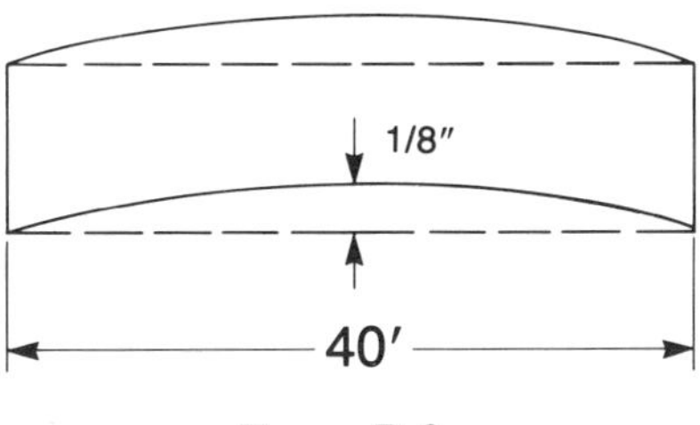

FIGURE D.2

***NOTE:** These tolerances are guidelines for design/manufacture to facilitate proper idler and belt alignment in accordance with the Idler Alignment section of this standard. The overriding issue is idler and belt alignment as opposed to structural alignment.

Squareness

A check on squareness can be made by comparing diagonal measurements between idlers, as shown in Figure D.3.

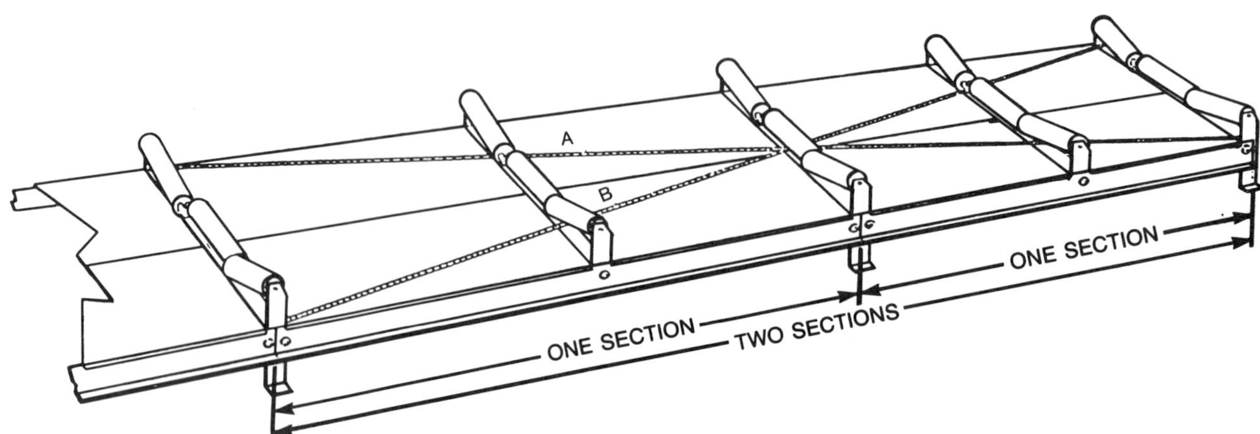

FIGURE D.3
Conveyor Frame and Idler Alignment

Lines A and B must be within 1/8″ to assure conveyor frame squareness.

Return idlers should, likewise, be installed level and parallel to the head and tail pulleys.

Level

The two idler support members shall be level within 1/8″ regardless of belt width. In addition, the elevation of stringer above support structure shall be held within ± 1/4″ (Figure D.4).

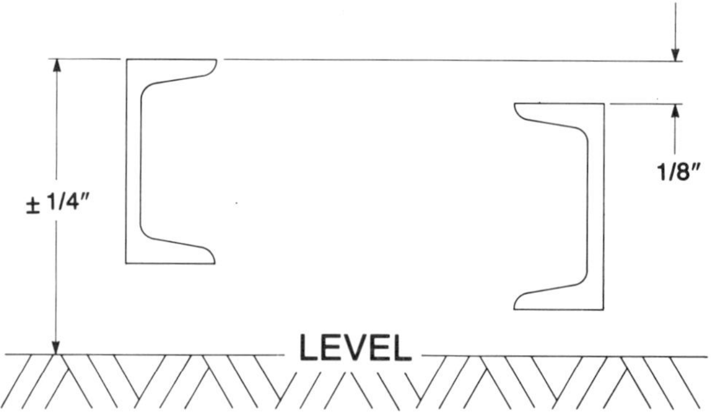

FIGURE D.4

NOTE: Designers and fabricators must compensate for mill tolerance of members to satisfy this level requirement.

Pulley and Shaft Alignment

Except for special purpose applications (such as turnover exit), all belt conveyor pulleys should be set level, and shaft centerline should be perpendicular to the centerline of the belt. Significant departure in alignment will result in unnecessary thrust load on bearings, accelerated and uneven pulley lagging wear, and belt training problems.

It is most convenient to field align pulleys after supporting steel is secured and before belting is installed. After alignment, it is good practice to matchmark bearing housing and supporting steel to enable realignment if disturbed for any reason. Intentional misalignment should *not* be used as a means to counteract other misalignment forces for training the belt.

Because of pulley manufacturing tolerances, alignment measurements should be taken on the shaft and not on pulley elements. By use of adjustable bearing stops (commonly supplied by the conveyor manufacturer) and full bearing surface shim packs (commonly supplied by the erector), the following alignment tolerances can be reasonably achieved:

Using a level and checking both sides of the pulley, shaft elevations at the bearings should be set within 1/32 of an inch (Figure D.5).

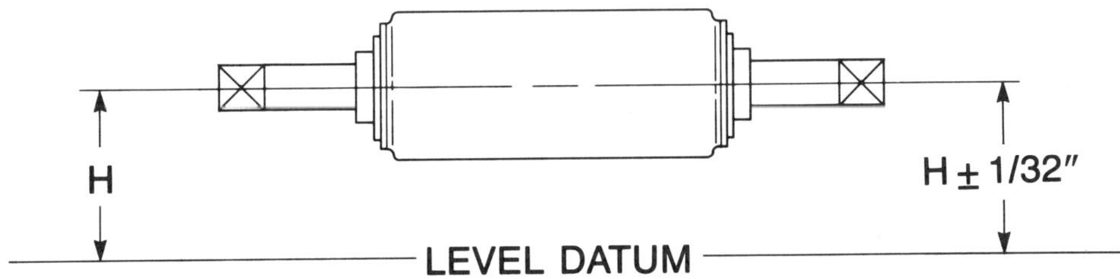

FIGURE D.5

Measuring from a line constructed perpendicular to the conveyor centerline, the shaft centerline should not deviate greater than ± 1/32 of an inch at the bearings (Figure D.6). Because of pulley locations and access thereto, it is common practice to utilize offset lines and plumb bobs to make these measurements.

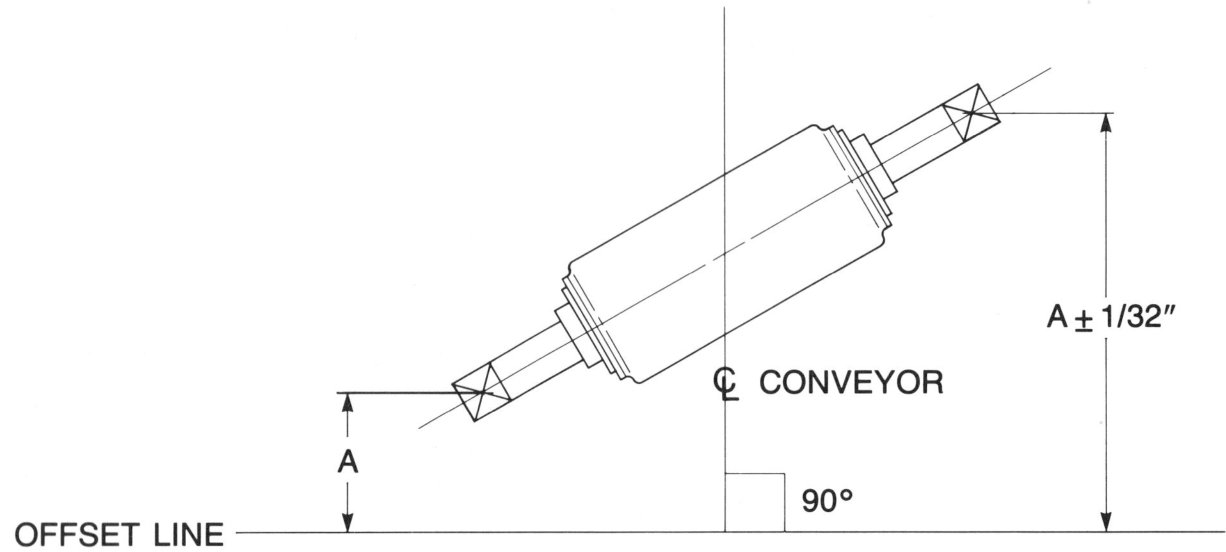

FIGURE D.6

NOTE: See Figure D.3 for squareness check.

When pulleys are mounted on shafts by the equipment supplier, they shall be secured to the shaft in accordance with the recommended practice of the locking element manufacturer. When supplied separately, they should be located on the shaft ± 1/8 of an inch of the position shown on the drawings.

If pulleys and shafts are field assembled or repositioned, pulley manufacturer recommendations must be followed to avoid dishing of end discs and residual end disc prestress.

With some locking elements, it may be necessary to check and tighten clamping bolts several times to insure security of hub and shaft assembly. Recommendations of the manufacturer should be followed with regard to checks and bolt torques.

Pulleys, shafts, bearings, and supporting steel may be shipped in various degrees of preassembly. Due to manufacturing and erection tolerances, final bearing setting should be done in the field after supporting steel has been secured.

Procedures and tolerances for bearing settings should be obtained from the bearing manufacturer's catalogs. Improper setting of both fixed and expansion type bearings may result in thrust pre-load, causing premature bearing failure.

Reducer/Motor Base Installation Tolerances

Characteristic of all operating mechanical apparatus, correct manufacturing and installation alignment tolerances are necessary to prolong operating life and prevent unnecessary breakdowns. Also, since many reducer manufacturers do not factory oil their reducers, it is imperative that oil levels be checked and maintained by field personnel.

Reducer/motor base installation guidelines generally fall into two categories—concrete or structural steel. Either should be sufficiently rigid to minimize vibration and to maintain alignment between the motor and the driven equipment within the manufacturer's recommended tolerances. Base application information may apply to either motors or reducers, or to both in combination.

Fabricated Structural Bases

Structural bases require a surface which, if not flat, can be shimmed to proper position in the structure, using .003″ to .005″ shim packs. After base plate is secured, flatness should be verified and shims added, if necessary.

Concrete Supports

Concrete foundation-type supports are normally the most rigid and secure. However, the installation must be correct within certain limitations to provide acceptable tolerances in the final arrangement with a minimum of shimming.

"Sole plates" can provide a means of adjusting and aligning the motors and reducers or common bedplate to an acceptable tolerance prior to final grouting. Reducer and motor mounting plate surface should be flat to tolerances acceptable to the manufacturer. The bottom of the sole plate needs no tolerance requirement inasmuch as the grout provides a fulling bearing surface.

If the output from the reducer is direct-coupled to a driven unit, the reducer shaft centerline should be held to a *minus* tolerance only, which would allow shimming by the erector if required. The same is true of the motor shaft in relation to the reducer input shaft.

It is mandatory that alignment checks be made after the installation is complete, but before operation. Misalignment should be no more than 50% of the misalignment tolerance allowed by the manufacturer of the coupling being aligned.

Structural Steel Supports

Although combination fabricated structural bases are normally built as a unit, it is not always necessary to provide continuous support in the structural steel below the units; however, it is necessary to provide an integrated and sufficiently rigid connection between reducer base and support steel to accommodate every bolt hole.

The supporting structure must minimize deflection and distortion under load, preventing warping of the motor/reducer base and subsequent unacceptable misalignment.

Flexible Coupling Alignment

Flexible couplings are used in conveyor drive trains to transmit torque from one rotating element to another. In doing so, they protect expensive driving and driven machinery from misalignment, shock loads, vibration, and thrust loads.

It is true that most couplings will operate under severe misalignment for a given period of time; however, the penalty is reduction in useful life. For this reason, manufacturers of couplings have established misalignment tolerances which should be strictly adhered to during field assembly.

A prerequisite to proper alignment is a rigid and level base on which the drive and driven elements are to be set. If they are to be mounted on a steel frame, a common base plate is necessary to prevent movement between independently supported steel members.

Prior to coupling installation, the installer must thoroughly familiarize himself with all of the manufacturer's published requirements of the motor, speed reducer, and couplings. Field breakdown and checking of alignment during assembly to the support structure shall be accomplished using dial indicators, machine test levels, or other means, with only commercial shim stock used for final alignment.

Three types of alignment must be checked: (1) Angular; (2) Parallel; (3) Axial.

Angular (Figure D.7)

Defined as the movement of the input and output halves of the coupling in such a manner as to permit a rocking and/or sliding action of the element which connects the coupling halves.

After the coupling halves have been mounted to the driver and driven shafts, the two units should be positioned so that the distance between the coupling faces is equal to the "normal" coupling gap.

The coupling halves are then aligned by positioning a spacer block of thickness equal to the required gap between the faces. The spacer block should check gap at a minimum of 90° intervals. Once this is accomplished, the gap should be measured with a feeler gauge.

Parallel (Figure D.8)

Defined as the movement of the input and output halves of the coupling in a way which maintains parallelism between the faces of each coupling half but allows the shafts to occupy separate center lines.

The driving and driven equipment should be aligned so that a straight edge can be placed on both coupling flanges at 90° intervals and remain parallel to the equipment shafts. Care should be taken when tightening set screws and bolts to insure the proper torque is achieved.

Axial (Figure D.9)

End float for both the driving and driven shafts is sometimes required for expansion and various other reasons. End float on most couplings can be limited to any required distance by the use of limited end float kits; however, these kits should only be used as recommended by the manufacturer.

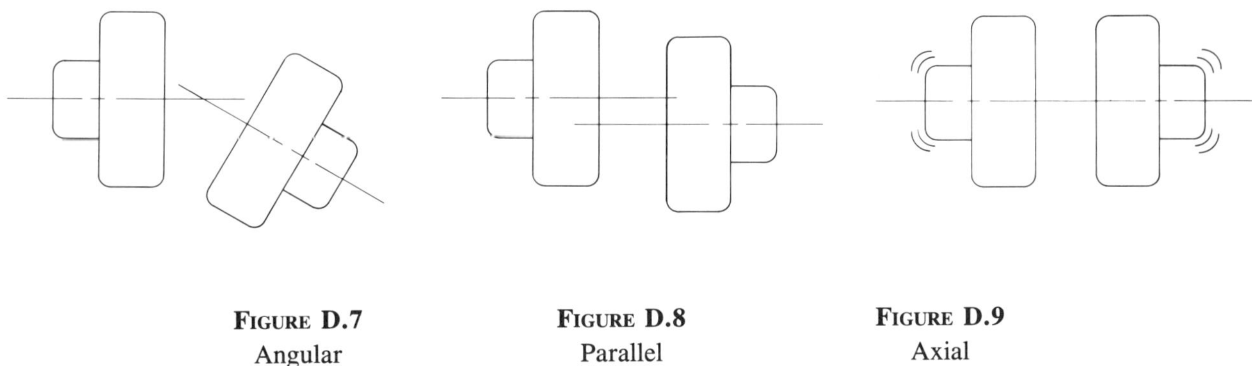

FIGURE D.7 Angular

FIGURE D.8 Parallel

FIGURE D.9 Axial

After the couplings have been aligned (cold) and the drive run in to obtain normal operating temperature, it shall be shut down and alignment re-checked by the following procedure:

Break all couplings and re-check ''hot'' alignment. These readings shall not exceed 75% of the manufacturer's maximum allowable tolerances. If not within these limits, procedure should be repeated as often as needed to achieve necessary ''hot'' readings.

Idler Alignment

The objective in aligning idlers is to achieve settings which are square and in line to the conveyor centerline and parallel to each other (Figure D.3).

Idlers shall be set from a previously squared and leveled terminal pulley (preferably the discharge pulley, although this is not essential). A suggested method is to use a tight wire on the conveyor centerline or offset to it, stretched to form a true centerline reference. This line should be at least 100′ long and referenced to the "squared" starting pulley.

Idlers shall be placed at design spacing and square to the tight wire. After a span of about 50′ has been filled with idlers, the 100′ line shall be relocated so that there is 50′ overlap on the first position. Repositioning of the tight wire should continue until the entire conveyor length is filled with idlers.

Belt Alignment

Belt training is usually minimized if:

a. Conveyor has been installed straight and level with tolerances stated herein.
b. All pulleys and idlers are square with the conveyor centerline.
c. Belt splices are correct and square.
d. There are no defects in the conveyor belting.
e. All idler rolls turn freely.

However, all these conditions are rarely present concurrently, and some training of the belt is usually required.

NOTE

Belt training should be supervised by one person.

Training the belt is a process of adjusting idlers and the method of loading in a manner which will correct any tendency of the belt to run off to one side or the other.

When the belt continues to run off consistently over a fixed length of conveyor, the cause is probably in the alignment or leveling of the conveyor structures, idlers, or pulleys in that area.

If one or more segments of the belt run off at all points along the conveyor, the cause is more likely the belt itself, a splice, or the method of loading. When a belt is loaded off center, the lightly loaded edge will rise on the inclined idler roll that it contacts.

All pulleys should be level and at 90 degrees to the centerline of the conveyor. They should be maintained this way and never shifted to train the belt.

Training the belt with the troughing idlers is accomplished by shifting the idler axis with respect to the path of the belt, commonly known as "knocking idlers."

This method is effective when the entire belt runs to one side along a fixed length of the conveyor. The belt can be centered by "knocking" ahead (in the direction of belt travel) the end of the idler to which the belt runs (see Figure D.10).

Shifting idlers in this manner should be spread over a length of conveyor preceding the region of trouble. In no event shall any idler be shifted more than 1/4″ in any direction from its squared position.

NOTE

Compensation by idler "knocking" may have adverse effects on reversible belts; therefore, avoid "knocking" on reversible belts. Instead, use extreme care in initial alignment.

A belt might be made to run straight with half the idlers "knocked" one way and half the other, but this would increase rolling friction between the belt and idlers. For this reason, all idlers must be initially squared with the path of the belt in accordance with the Idler Alignment section of this standard, and minimum shifting should be used for training. If the belt is over-corrected by shifting idlers, it should be restored by moving the "knocked" idlers, not by shifting additional idlers in the other direction.

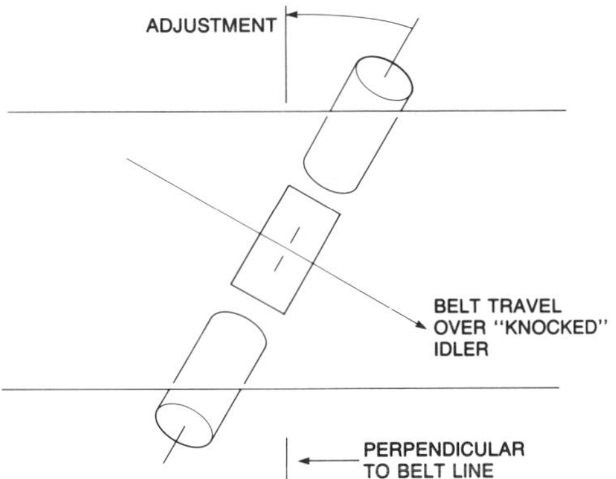

FIGURE D.10

Empty Run-in

The conveyor belt should first be aligned (trained) without material load. Most alignment problems, with the exception of off-center loading, will be detected during this empty run-in period.

After assuring that all elements have been aligned true, as stipulated in the Idler Alignment section of this standard, the belt should be slowly ''inched'' along to provide indication of where corrections of the types described may be required. The first corrections must be those at points where the belt is in danger of being damaged. Once the belt is clear of all danger points, a sequence of training operations can be followed.

The best procedure to use in the training sequence is to start with the return run and work toward the tail pulley. This assures early centering of the belt on the tail pulley so that it can be centrally loaded. Normally the belt can be trained properly onto the tail pulley by manipulation of the return idlers.

With the empty belt trained satisfactorily, good operation with load is usually assured. Disturbances which appear with load are then usually due to off-center loading or to accumulation of material on snub pulleys and return idlers.

If all corrective measures have been applied, and a particular section of belt continues to run to one side along the entire conveyor length, it can be concluded that the belt may be cambered at that point or may have a splice that is not properly squared. The only corrective measure for a ''non-squared'' splice is to re-splice. On the other hand, slight camber in a belt may remove itself after an adequate run-in period under load.

Full Load Run-in

A belt may be considered properly aligned if after 8 hours of continuous operation under full design load, the belt edges remain within the width of the pulley face. Also, when extra wide face pulleys are used, a further criteria that the belt runs within the confines of the normal troughing, or return, idler should be followed.

NOTE: Disc return idlers shall be exempt from the above as the standard roll faces supplied are approximately equal to the belt width. In this case, the belt shall be made to track within 1-1/2″ of the conveyor centerline. Idler brackets shall not restrict this movement.

To obtain the above alignment under load, it will be necessary to assure that the load is centered and that all belt scrapers, plows, and skirt seals exert uniform pressure on the belt. Minor adjustments can be made to these devices to improve tracking.

Carrying and return training idlers should be considered in correcting belt alignment problems before rejecting the belting or splice.

HELPFUL HINTS

a. One person should be responsible for training belts. This person should supervise all adjustments.

b. Periodically check the conveyor to be sure it is level. This requirement is apparent if the conveyor becomes misaligned with no apparent change in loading.

c. After the belt is run in, an electrician should take readings on voltage, amperes, or wattage. This information can be used for future comparison and a quick trouble check.

   Higher readings in the future may indicate excessive drag due to belt misalignment or frozen belt idler rolls.

d. Do not overtrain the belt. Overtraining will result in increased roll shell wear, increased belt cover wear, and increased power consumption.

Skirtboard Adjustment

Skirtboards are an important element of a well designed belt conveyor. Their primary function is to avoid spillage over belt edges; however, when utilized at loading points, they center the material on the belt and retain dust.

It is recommended that the steel side plates be designed to allow vertical adjustment from 1/2″ to 1-1/2″ above the conveyor belting. They must not make contact with the belt, yet be set low enough to protect the skirtboard rubber seal from excessive pressure and material wear. A vertical distance of 1″ is normally acceptable.

The skirtboard rubber seal, when properly adjusted, should retain a light contact with the conveyor belt, thus minimizing friction while retaining all material within its boundaries. Hand pressure under the rubber should be adequate to flex the rubber. Excessive pressure will burn grooves in the belt rubber cover and, therefore, must be avoided.

A slight gap between skirt rubber and belting can be expected to occur as the belt is loaded. This gap cannot be totally eliminated, and excessive adjustment will cause accelerated wear.

Idler Lubrication

Modern conveyor idlers have evolved through better bearings, improved lubricants, and more effective seals. Likewise, the lubrication requirements have changed to accommodate current designs. The following considerations should be reviewed when establishing a lubrication program tailored to a particular installation:

Manufacturer's Recommendation

The manufacturer's data can provide a solid basis for designing the lubrication program.

Type of Lubricant

The proper lubricant will be heavily influenced by operating conditions, operating environment, and the quality of maintenance program desired. Operating conditions such as operating speed, idler loading, type and size of material being handled, and number of operating hours per year should be considered.

The operating environment is one of the greatest contributors to determining the lubrication program. Temperature, dust, material abrasiveness, washdown techniques, and washdown frequency all affect the lubrication that flushes contaminants from the bearings. Special synthetic lubricants are available for subzero operating temperatures.

Idler Construction

The type and efficiency of the seals directly reflect on the lubrication schedule. Following are several recommendations which will contribute to establishing a successful lubrication program:

A. Remove several idlers from service every six months, disassemble them, and inspect to determine if adequate lubrication has been attained, if migratory contamination and wear have occurred, or if pressure relief plugs are fouled.

   If these idlers demonstrate that adequate lubrication has been attained, they could be put back into service and the conveyor operated as normal. If contamination was found in this sampling, all idlers on the conveyor should be lubricated. Also, fouling of pressure relief plugs in sampling should encourage cleaning of all plugs.

B. Thoroughly clean all grease fittings prior to lubricating the idlers to eliminate the possibility of contamination introduced during maintenance.

C. Establish a good maintenance recordkeeping system. Record the date on which samples were taken and the condition of the samples taken and verify trends that may be occuring within the system. Also record idler failures, including location, to establish system trends and effectiveness of the lubrication program.

The establishment of a formal lubrication program as above will increase the likelihood of many years of successful, trouble-free operation.

Index